CHEMICAL AND GEOLOGICAL

ESSAYS

BY

THOMAS STERRY HUNT, LL. D.,

Fellow of the Royal Society of London; Member of the National Academy of Sciences of the United States, the Imperial Leopoldo-Carolinian Academy, the American Philosophical Society, the American Academy of Sciences, the Geological Societies of France and Belgium and of Ireland; Officer of the Order of the Legion of Honor, etc., etc., etc.

BOSTON:
JAMES R. OSGOOD AND COMPANY,
LATE TICKNOR & FIELDS, AND FIELDS, OSGOOD, & CO.
1875.

UNIVERSITY PRESS: WELCH, BIGELOW, & CO.,
CAMBRIDGE.

TO

JAMES HALL,

IN RECOGNITION OF MANY YEARS OF FRIENDSHIP,

This Volume is Dedicated

BY

THE AUTHOR.

PREFACE.

In choosing from a large number the following papers for republication, it may be well to state the considerations which have guided the author in his selection. His researches and his conclusions as to the chemistry of the air, the waters, and the earth in past and present times, the origin of limestones, dolomites, and gypsums, of mineral waters, petroleum, and metalliferous deposits, the generation of silicated minerals, the theory of mechanical and chemical sediments, and the origin of crystalline rocks and vein-stones, including erupted rocks and volcanic products, cover nearly all the more important points in chemical geology. They have, moreover, been by him connected with the hypothesis of a cooling globe, and with certain views of geological dynamics, making together a complete scheme of chemical and physical geology, the outlines of which will be found embodied in essays I. – XIII. of the present collection. It was at one time proposed to rewrite for this volume the first seven of these, giving them a more connected form, and thereby avoiding some little repetition; but it is thought better to reproduce them in the shape in which they originally appeared, and this chiefly for the reason that they seem to the author to have a certain historic value, and serve to fix the dates of the origin and development of views, some of which, after meeting for a time with neglect or with active opposition, are now beginning to find favor in the eyes of the scientific world. That such will be the ultimate fate of others herein contained, and not yet generally received, the author is persuaded. It has been his fortune to enunciate, in very many cases, views for which his fellow-workers were not prepared, and after a lapse of years to find these views propounded by others as new discoveries or original conclusions. Naturally desirous, however, of vindicating his claims to priority in certain of these matters, he feels that the best way of attaining this result is to reprint the original essays. It should be said that two of these, namely, IV. and XII., were given as popular lectures, and are thus unlike the others in method and style.

The reproduction of the papers on the Geology of the Alps and the History of Cambrian and Silurian requires, it is conceived, no explanation, inasmuch as, apart from their general interest, they serve to throw great light upon many questions raised in the essay on the Geognosy of the Appalachians as to the origin and age of their rocky strata.

As regards the five papers which are placed at the end of the volume, the author reprints them for the reason that, incomplete and fragmentary as they are, they have a certain value in the history of chemical theory; and, moreover, contain, in his opinion, the germs of a philosophy of chemistry and mineralogy which he hopes one day to develop himself or to see developed by others.

In preparing this collection for the press, the author has been compelled by the limits assigned to the volume to omit several papers which would else have found a place here, and to abridge others. In some cases, paragraphs have been rewritten and additions made, which are distinguished by being placed in brackets. Explanatory notes are given, and introductory and historical sketches prefixed, with references both to other papers in this volume and to many which have been omitted. Read with these aids, and with the help of the table of contents and index, this volume will, it is believed, suffice to give clear and connected notions of the author's views on the various questions herein discussed.

T. S. H.

BOSTON, Mass., September, 1874.

TABLE OF CONTENTS.

I.

THEORY OF IGNEOUS ROCKS AND VOLCANOES (1858).

PAGE
The chemistry of a cooling incandescent globe 1
The primitive ocean and primitive crystalline rock 2
Origin of eruptive rocks; views of Bunsen, Phillips, and Durocher . . 3
Softening of crystalline stratified rocks 4
Poulett Scrope and Scheerer on aqueo-igneous fusion 5
Daubrée and the author on the origin of mineral silicates . . . 6
Views as to the condition of the earth's interior 7
Existence of a solid anhydrous nucleus maintained 7
Intervention of sedimentary rocks in volcanic phenomena . . . 8
Origin of the volatile products of volcanoes 8
Sir J. F. W. Herschel on the cause of volcanic action 9
Its relation to recent sedimentary deposits 9
Note on the decomposition of crystalline rocks 10
Note on the deposition of clays 10

II.

ON SOME POINTS IN CHEMICAL GEOLOGY (1859).

Ancient and modern sea-waters compared 11
Origin and geological importance of alkaline carbonates . . . 12
Different relations of potash and soda 12
Deposits of iron-oxide as evidences of organic life 13
Deposits of alumina; emery and bauxite; their origin 13
Supposed aqueous origin of basic and acidic eruptive rocks . . 14
Babbage and Herschel on the effects of internal heat 15
Theory of volcanic and plutonic phenomena 15
Note on the views of Keferstein 16
Geological distribution of volcanoes 17

III.

THE CHEMISTRY OF METAMORPHIC ROCKS (1863).

Preface; objections to the name of metamorphic rocks 18
Probable relations between the age and constitution of crystalline rocks 19
Sub-aerial and sub-aqueous decay of feldspars 20

Chemistry of alkaline natural waters 21
Relations of the soil to potash-salts and phosphates 22
Origin of insoluble metallic sulphides 23
Deoxidation of metals, sulphur, and carbon through vegetation . . 23
Twofold origin of carbonates of lime and magnesia 23
The two types of igneous rocks; their sedimentary origin . . . 24
Rock-metamorphism defined and distinguished from pseudomorphism 24
Relation of alkaline waters to crystalline silicates 25
Local metamorphism; views of Daubrée and Naumann . . . 26
Progressive change in silico-aluminous sediments 27
Chemical relations of certain mineral silicates 28
Various series of crystalline stratified rocks 29
Laurentian, Labrador, Green Mountain, and White Mountain series . 30
The hypothetical granitic substratum; granitic veins 33
Crystalline rocks of Europe and North America compared . . . 34

IV.

THE CHEMISTRY OF THE PRIMEVAL EARTH (1867).

The spectroscope and the nebular hypothesis 35
Dissociation defined; terrestrial chemical elements 37
Probable existence of more elemental forms of matter in the stars . . 37
Chemical and physical constitution of the sun 37
Chemical history of the cooling earth 38
Probable solidification from the centre 39
Primitive atmosphere and ocean; their composition 40
Their action on the primitive crust 40
Mutual relations of carbonic acid, clay, limestone, and sea-salt . . 41
Waters of the ancient ocean 41
Carbonic acid of the ancient atmosphere 42
Its relations to life and to climate 42
Formation of gypsums and magnesian limestones 43
Secondary and aqueous origin of granites 43
Action of internal heat; volcanoes 44
Hopkins, Pratt, and Sir William Thomson on the earth's interior . 45
Controversies of the neptunists and plutonists 45

Appendix.

The earth's climate in former ages 46
Tyndall on the relation of gases and vapors to radiant heat . . . 46
Former predominance of carbonic acid in the air 47
Note on the amount of carbonic acid now fixed in limestones . . 47

V.

THE ORIGIN OF MOUNTAINS (1861).

Hall on palæozoic sediments in eastern North America 49
Eastern origin of these mechanical sediments 49

Varying thickness of palæozoic strata 50
Relation of mountains to sedimentation 50
Continental as opposed to local elevation 52
Views of Buffon, Montlosier, and Constant-Prevost 52
Views of Humboldt, Von Buch, and Elie de Beaumont 52
Lesley on the topography of mountains 52
Relations of mountains to synclinals and to erosion 52
Hall's views of the origin of mountains 54
Relations of subsidence to foldings of strata 55
Condensation consequent on the crystallizing of sediments 56
The hypothesis of a solid contracting nucleus maintained 57
Relation of this nucleus to water-impregnated sediments 57
The softening of these produces lines of weakness in the crust . . . 57
Relation of this process to corrugations 57
Relations of volcanic and plutonic phenomena to sedimentation . . 58

VI.

THE PROBABLE SEAT OF VOLCANIC ACTION (1869).

Discussion of the views of Hopkins and Scrope on volcanoes . . . 59
Views of Lemery and Breislak, of Davy and Daubeny 62
Views of Keferstein and Sir J. F. W. Herschel 62
Exposition of the author's view 63
Disintegration of the primitive crust 63
Hopkins on internal heat and its increase in descending 64
Sorby on the relations of heat and pressure to fusion and solution . . 65
Chemical differences in eruptive rocks 66

APPENDIX.

Geographical distribution of modern volcanoes 68
Distribution of ancient eruptive rocks; their geological relations . . 69

VII.

ON SOME POINTS IN DYNAMICAL GEOLOGY (1858).

LeConte on the reconstruction of geological theory 70
His views compared with those of the author 71
Hall's theory of mountains; the criticisms of Dana, Whitney, and LeConte 73
Views of Hall and the author misunderstood 73
LeConte's theory of mountains considered 74
Continental elevation and erosion; Montlosier and Jukes 74
Hall on some North American mountains 75
Origin and structure of the Appalachians 75
Their crystalline strata not palæozoic but eozic 75
Evidences of an eastern pre-palæozoic continent 75
Dry climate of eastern North America in palæozoic times 76
Oscillations of continents; their cause 76

Source of heat in plutonic phenomena 77
The notion of its chemical origin untenable 77
Henry Wurtz on a mechanical source of heat 78
Experiments and conclusions of Mallet 78
His views on the origin of volcanic products 79

VIII.

ON LIMESTONES, DOLOMITES, AND GYPSUMS (1858-1866).

Introductory note; letter to Elie de Beaumont 80
Cordier's views of the origin of limestones and dolomites 81
Their identity with those of the author 82
Chemistry of evaporating lakes and sea-basins 83
Alkaline waters of rivers and springs 84
Separation of lime-salts from sea-water; gypsum and rock-salt . . 85
Origin of sulphuretted hydrogen and native sulphur 87
Origin of deposits of magnesian limestones 88
Their deposition necessarily in isolated basins 88
Hall on the organic remains in magnesian limestones 88
Deposits of pure carbonate of lime 89
Generation of dolomite; its crystallization 89
Note on chemically deposited silica 89
Conclusions as to the chemistry of gypsum and dolomite 90
Conditions of temperature for the production of dolomite 91
Relative solubilities of gypsum and magnesian bicarbonate . . . 91
Influence of carbonic acid on the formation of gypsum 91
Geographical and climatic conditions for the production of dolomite . 91
Recent conclusions of Ramsay as to magnesian limestones 92

IX.

THE CHEMISTRY OF NATURAL WATERS.

Part I. — General Principles.

Atmospheric waters and the result of vegetable decay 94
Action of waters on the soil; researches of Way and Voelcker . . 95
Eichhorn on the replacement of protoxide bases in silicates . . . 96
Possible relations of saline waters to the soil 97
Relations of organic matters to oxides of iron and manganese . . 98
Solution and deposition of alumina 98
Origin of sulphuretted hydrogen and sulphurets 99
Decomposition of silicates; studies of Ebelmann 100
Kaolinization of feldspars and other minerals 101
Relation of soda and potash salts to sediments 101
Carbonic acid and water as agents in decomposing rocks 102
Marine salts in solution in sedimentary strata 103
Porous nature of sandstones and dolomites 103
Calculations as to the volume of waters held in rocky strata . . . 104
Solid salts and bitterns from sea-water in the rocks 105

Action of bicarbonate of soda on calcareous and magnesian salts . . 105
Origin of sulphates in natural waters 106
Indifference of gypsum solutions to dolomite 106
Decomposition of gypsum by hydrous magnesian carbonate . . 107
Results of the gradual evaporation of sea-water 107
Composition of the ancient seas 108
Separation of the lime salts from sea-water 109
Decomposition of sulphate of magnesia by bicarbonate of lime . . 109
Twofold origin of gypsum 110
Twofold origin of magnesian carbonate 110
Sulphuric and hydrochloric acid in waters 111
Carbonic acid in waters 112
Ammonia and nitrogen in rocks and waters 113
Classification of natural waters 113

PART II. — ANALYSES OF VARIOUS NATURAL WATERS.

Waters of the first class or bitter salines; analyses 116
Their resemblance to bitterns; absence of sulphates 117
Predominance of chlorides of calcium and magnesium 118
Probable constitution of the Cambrian ocean 119
Brines of ancient saliferous deposits 119
Note on analyses of saline waters 120
Silicate of magnesia; its formation and chemical relations . . . 122
Waters of the second and third classes; analyses 123
Waters of the fourth class or alkaline waters; analyses 125
Waters of the Ottawa River; analysis 126
Variations in the composition of mineral springs 127
Comparative analyses of the Caledonia waters 129
Sulphuric-acid springs of New York and Ontario 130
Neutral sulphated waters; their sources 132
Sulphate of magnesia in waters 134

PART III. — CHEMICAL AND GEOLOGICAL CONSIDERATIONS.

Salts of the alkaline metals in natural waters 135
Salts of calcium and magnesium; relations of chlorides and carbonates . 137
Results of evaporation; deposition of carbonates of lime and magnesia 138
Solubility of carbonate and bicarbonate of lime 139
Supersaturated solutions of carbonates of lime and magnesia . . 140
Salts of barium and strontium in waters 141
Iron, manganese, alumina, and phosphates in waters 142
Bromides and iodides in waters 142
Relations of chlorides and iodides to earthy minerals 143
Sulphates in natural waters; their frequent absence 144
Soluble sulphides in natural waters 145
Borates; waters of a borax-lake 146
Carbonates; studies of the Caledonia waters 147
Waters with a deficiency of carbonic acid 149
Silica; its amount in various waters 150
Silicates of lime and magnesia deposited from waters 151

Organic matters in water; their nature and origin 152
Geological relations of mineral waters 153
Palæozoic formations of the St. Lawrence basin 154
Relations of mineral waters to the various formations 156
Contiguity of dissimilar mineral springs 157
Temperatures of the mineral waters of Canada 157
Results of the evaporation of these waters 158

SUPPLEMENT.

Waters with a predominance of chloride of calcium 158
Waters with soluble sulphides; mode of analysis 159

APPENDIX.

On the porosity of rocks and its significance 164
Mode of determining the density and porosity of rocks 165
Table of the density and porosity of various rocks 166

X.

ON PETROLEUM, ASPHALT, PYROSCHISTS, AND COAL.

Geological relations of petroleum 168
Origin and source of petroleum 170
The oil-bearing limestone of Chicago; its analysis 172
Large amount of petroleum contained in the limestone 173
Bitumens; their analyses and chemical composition. 175
Wall on the bitumens of Trinidad and Venezuela 176
Conversion of organic matters into coals and bitumen 177
Pyroschists or bituminous shales; their nature defined 177
Their geological and chemical relations 178
Chemical similarity of animal and vegetable tissues 179
Note on the constitution and artificial production of albuminoids . . 180
Dawson on the origin of coal 181
Comparative analyses of epidermal tissues 181
On the gaseous hydrocarbons found in nature 182

XI.

ON GRANITES AND GRANITIC VEIN-STONES (1871-1872).

Granite and its varieties defined 184
The relations of granite to gneiss 185
Stratiform structure in various erupted rocks 186
Feldspar-porphyries; their characters and distribution 187
Granitoid gneisses of New England; true granites 188
Granitic vein-stones; theories as to their origin 189
Views of Scheerer, Scrope, and Elie de Beaumont 189
The concretionary origin of granitic veins 192
Granitic vein-stones of the White Mountain rocks described . . . 194
Their banded structure; disturbance of the strata by veins . . . 196

Evidences of the progressive formation of such veins 198
Rare minerals in the granitic vein-stones of New England . . . 200
Geodes in granites in New Brunswick and Italy 201
Granitic veins distinguished from dikes 202
Volger and Fournet on the filling of granite veins 202
Recent age of some concretionary veins 203
Note on the salt-wells of Goderich in Ontario 204
On the conditions of the crystallization of quartz 205
On the emerald-bearing veins of New Grenada 205
Recent production of crystalline zeolites 205
The Laurentian series; its lithological characters 206
Vein-stones in the Laurentian rocks 208
These vein-stones compared with those of Scandinavia 209
Minerals of the Laurentian vein-stones 210
Note on the occurrence of leucite 210
The concretionary character of these vein-stones shown 211
Incrustation and skeleton-crystals described 211
Crystals with rounded angles; their significance 212
Feldspathic veins of the Laurentian rocks 214
Complex nature of the Laurentian vein-stones 215
Vein-stones with apatite and with graphite 216
Paragenesis of their mineral species 216
Concretionary copper-bearing veins of the Blue Ridge 217
Supposed eruptive origin of crystalline limestones 218

XII.

THE ORIGIN OF METALLIFEROUS DEPOSITS.

Preliminary statement of the theory of ore-deposits 220
Distribution and diffusion of the chemical elements 221
Separation and concentration of certain elements 222
Note on the solvent powers of water 223
The terrestrial circulation compared with that of animals 224
History of the diffusion and concentration of phosphates 225
Potash and iodine; their elimination from sea-water 226
Intervention of organic life in all these processes 226
History of the diffusion and the concentration of iron 227
Relation of iron-oxides to ancient vegetation 229
Formation of iron-pyrites and other sulphides 230
Diffusion of copper, silver, and lead in the ocean 231
Reduction of copper from its solutions 232
Ore-deposits in beds and in fissures 233
The process of deposition in veins 234
Uniformity of operations in nature 235

Appendix.

Sonstadt on the iodine in sea-water 237
On gold in the ocean; Sonstadt and Henry Wurtz 237

XIII.

THE GEOGNOSY OF THE APPALACHIANS AND THE ORIGIN OF CRYSTALLINE ROCKS.

The relations of geology to the sciences 240

PART I.—THE GEOGNOSY OF THE APPALACHIAN SYSTEM.

History of the Appalachian mountain system 241
Eaton on the classification of rock-formations 241
Jackson and Rogers on the rocks of New England 242
The Adirondack or Laurentide series; Laurentian 243
The Green Mountain series; Huronian 243
The White Mountain series; Montalban 244
Rogers on the crystalline rocks of Pennsylvania 245
His hypozoic and azoic series probably identical 247
Crystalline rocks of New York and New England 248
Crystalline rocks of Virginia, the Carolinas, and Tennessee . . . 249
Emmons on the crystalline rocks of western New England . . . 250
Note on the decay of these rocks to the southwest 250
The Taconic rocks of Emmons distinguished from the primary . . 251
The Taconic system described and defined 252
Views of Mather and Rogers on the Taconic rocks 254
Rogers and Safford on the primal rocks of Virginia and Tennessee . . 255
Relations of the Taconic to the Champlain division 256
The organic remains of the Taconic rocks 257
The rocks of the so-called Quebec group 259
The Red sand-rock of western Vermont 260
Lower palæozoic rocks of Labrador and Newfoundland 261
Lower palæozoic rocks in the Champlain and Mississippi valleys . . 261
Note on the palæozoic formations in the Rocky Mountains . . . 262
Stratigraphical breaks in the lower palæozoic series 263
Continuation of the Taconic controversy 264
The Upper and Lower Potsdam of Billings 266
Lower palæozoic rocks of Europe 266
Identity of Taconic with Lower and Middle Cambrian 268
The Huronian or Urschiefer distinct from Cambrian 269
Crystalline schists of Anglesea and the Rhine 270
Crystalline rocks of the Scotch Highlands 271
Comparative studies of crystalline formations 272
Crystalline schists of Lakes Huron and Superior 274
The crystalline schists of the Appalachians, pre-Cambrian 276
Credner on the Eozoic formations of North America 277
History of the Norian or Labrador rocks 279
Relations of the various crystalline formations 281
Hitchcock on the geology of the White Mountains 282

PART II.—THE ORIGIN OF CRYSTALLINE ROCKS.

Mineralogy of the two classes of crystalline rocks 283
Theories of the source of eruptive rocks 284

Mechanical disintegration and recomposition of rocks 285
Crystalline silicated rocks of stratified formations 285
Two hypotheses to explain their origin 286
Alleged pseudomorphous change of plutonic rocks 286
The doctrine of pseudomorphism by alteration 287
Symmetrical and asymmetrical envelopment of minerals 288
Difficulties of the doctrine of pseudomorphous alteration 290
Scheerer's doctrine of polymeric isomorphism 291
Delesse and Naumann on pseudomorphism 292
Supposed plutonic origin of crystalline stratified rocks 294
Views of Naumann and of Macfarlane on primary rocks 294
Hypothesis of the aqueous origin of crystalline rocks 296
Generation of silicates in cases of local alteration 297
Aqueous deposition of feldspathic minerals 298
Alleged palæozoic age of many crystalline schists 299
The author's view of their origin defined 300
Early views of the aqueous origin of crystalline rocks 301
Evidences of life at the time of their deposition 301
Evidences of life afforded by meteoric stones 302
Discovery of the Eozöon Canadense 302
Silicates injecting this and various other organisms 303
Observations of Gümbel, Hoffmann, and Dawson 303
Credner and Gümbel on the origin of crystalline schists 304
Gümbel on diagenesis and epigenesis 305
Note containing the views of Gümbel on this question 305
Note on the crystalline aggregation of finely divided matter 305
Conditions of early times favoring diagenesis 306
The origin and formation of dolomites 307
Influence of carbonic acid on the production of dolomite 308
Supposed generation of dolomite by Von Morlot and Marignac . . 308
Two classes of dolomites; their origin 310
Relations of one class of these to gypsum and rock-salt 310
Former climate of eastern North America 310
Magnesian silicates of Syracuse, New York 310
Supposed organic origin of limestones 311
Their true relation to organic life 311
Relations of phosphates to organisms 311
Phosphatic nature of Lingula, Orbicula, Conularia, etc. 312
Relations of silica to organic life 312

Appendix.

Reply to Mr. Dana's criticisms 313
The question of the transmutation of species stated 313
Warrington Smyth's opinion of epigenesis 313
The views of Delesse on pseudomorphism defined 314
Delesse on the eruptive origin of serpentine 316
He subsequently adopts the view of its aqueous origin 317
A revolution in the theory of crystalline rocks 317
Scheerer's views explained and defended 318

Dana's teachings as to pseudomorphism 319
He affirms the doctrine of epigenic metamorphism 320
The old doctrine of diagenesis explained and defended 321
The views of Naumann examined 322
Various illustrations of the doctrine of transmutation 324
King and Rowney on the supposed transformations of serpentine . 325
Genth on the supposed alterations of corundum 326
Dana and Emmons on the Taconic rocks 326
On the relations of the pre-Cambrian schists 327

XIV.

THE GEOLOGY OF THE ALPS.

The researches of Alphonse Favre 328
The crystalline rocks of Mont Blanc 329
The uncrystalline rocks around it 331
Association of carboniferous and liassic fossils 332
Difficulties presented by folded and inverted strata 334
Sismonda on the anthracitic system of the Alps 334
Section presented by the Mont Cenis Tunnel 335
Age of the crystalline schists with anhydrites 336
Examples of inverted strata in the Alps 337
On the supposed recent age of the crystalline schists 338
The recomposed crystalline rocks of the Alps 339
The true crystalline schists of great antiquity 341
Little or no evidence of metamorphism in the Alps 342
The fan-like structure of the Alps explained 343
Grand section across Chamonix and Mont Blanc 343
Geological history of Mont Blanc 344

Appendix.

Antiquity of the crystalline schists of Mont Cenis 347
Favre on the origin of crystalline schists 347
De Beaumont and Pillet on the rocks of Mont Cenis Tunnel 348

XV.

HISTORY OF THE NAMES CAMBRIAN AND SILURIAN IN GEOLOGY.

Part I. — Silurian and Upper Cambrian in Great Britain.

The Graywacke formation of the older geologists 350
Early studies of Sedgwick in North Wales 350
Early researches of Murchison in Wales 351
Cambrian as first defined by Sedgwick 352
Silurian as first defined by Murchison 353
Examination of the Berwyns by Murchison and Sedgwick 354
Identity of Cambrian and Lower Silurian fossils 355

Publication of Murchison's Silurian System 355
Difficulty of distinguishing between Cambrian and Silurian 355
Sedgwick's views and position misrepresented 357
Errors of Murchison's sections exposed 358
His Silurian system based upon a series of mistakes 362
Sedgwick's proposed compromise in nomenclature 363
Unauthorized alteration of Sedgwick's geological map 364
Further history of Sedgwick's wrongs 364

PART II. — MIDDLE AND LOWER CAMBRIAN.

Ancient fossiliferous rocks of Scandinavia 365
The early studies of Hisinger ; curious errors 366
Section of the rocks of Kinnekulle 367
Angelin on the crustacea of Scandinavia 367
Barrande on the fossiliferous rocks of Bohemia 368
The so-called primordial Silurian 369
The fossils of the Lingula flags of Wales 370
Fossiliferous rocks of the Malvern Hills 370
Subdivision of the Lingula flags ; the Menevian beds 371
Fossils of Lower Cambrian or Harlech rocks 372
True boundary between Cambrian and Silurian 374
Breaks in the succession of the lower rocks 375
Note on the Tremadoc rocks 375
Ramsay on stratigraphical breaks 376
General considerations on breaks in geological series 377
Note on the thickness of British Cambrian and Silurian 377
Murchison and the Cambrian nomenclature 378
He confounds the Longmynd and Bala groups 380
The statements of his Siluria criticised 380
Disagreement as to the Cambrian and Silurian nomenclature . . . 381
Distribution of Lower and Middle Cambrian rocks 382
Crystalline schists of Malvern and of Anglesea 383
Gold-bearing Lingula flags of North Wales 383
Hicks on the classification of lower palæozoic rocks 384
Sedgwick's latest views on classification 384
Tabular view of lower palæozoic rocks 386

PART III. — CAMBRIAN AND SILURIAN ROCKS IN NORTH AMERICA.

The geological survey of New York 387
Hall on the rocks of the New York system 387
The Taconic system equivalent to Lower and Middle Cambrian . . . 388
The paleontological determinations of Hall 389
Stratigraphical errors of the Taconic system 390
The Red sand-rock and the primordial trilobites of Vermont . . . 391
Contributions of Barrande and Billings to the subject 392
Logan on the Taconic rocks of Vermont 394
Hall's determinations and the errors of Hisinger 395
Bigsby on the fossiliferous rocks near Quebec 396
Bayfield and Logan on the same rocks 397

The graptolites of Point Levis 399
Discovery of trilobites at Point Levis 400
Logan describes and defines the Quebec group 401
He supposes a great and continuous dislocation 402
Hall accepts Logan's stratigraphical conclusions 403
Potsdam of the Ottawa basin and of Wisconsin 403
Its relations to the primordial of Europe 404
History of the Paradoxides Harlani of Braintree 405
The primordial fauna in Newfoundland and New Brunswick . . . 406
Murray on the geology of Newfoundland 406
The Lower Potsdam fauna of Billings 407
Fossiliferous rocks of Troy, New York 407
Menevian fauna in New Brunswick 407
Crystalline schists of Nova Scotia 408
Eophyton and its supposed geological relations 409
Hicks and Barrande on the early trilobitic fauna 409
Murray on ancient fossiliferous rocks in Newfoundland 410
Dawson on ancient foraminiferal forms 411
On the Palæotrochis of Emmons 411
Billings on paleontological breaks in the Ottawa basin 412
The true horizon of the Levis limestone 412
Its equivalents in Great Britain and elsewhere 412
Unconformability of Calciferous and Trenton formations 413
Discordance between the Quebec and Trenton groups 413
Lesley on a similar discordance in Pennsylvania 414
The Chazy formation on the Ottawa River 414
Absence of the second fauna to the eastward 415
Distribution of the Lower Helderberg fauna 415
History of the Oneida conglomerate 416
Mingling of second and third faunas on the Saguenay 417
Fossiliferous rocks of Anticosti 417
Middle Silurian division in Great Britain 417
Middle Silurian of Billings different therefrom 418
Two faunas in the Upper Silurian of Murchison 418
The Onondaga and Water-lime formations 418
Introduction of the terms Silurian and Devonian in America 419
Views of De Verneuil and of Hall 419
Names adopted by the geological survey of Canada 420
The geological survey of Pennsylvania 420
The nomenclature adopted by Rogers 421
Rogers on the British equivalents of American rocks 422
Errors of the Silurian nomenclature 422
The Upper Cambrian or Siluro-Cambrian division 423
Jukes and Giekie on the Silurian nomenclature 424
Barrande's downward extension of Silurian 424
Great importance of Sedgwick's geological labors 425

XVI.

THEORY OF CHEMICAL CHANGES AND EQUIVALENT VOLUMES (1853).

The physical and chemical history of matter 426
Generation of chemical species considered 427
Theory of double decomposition 428
On the relations of lower to higher species 428
The significance of combination by volumes 429
The nature of chemical union and of solution 429
Relations of chlorine to hydrogen and hydrocarbons 430
Laurent's law of divisibility in formulas 431
Reasons for doubling the equivalents of oxygen and carbon 431
Extension of the principle of progressive series 432
Relations between density and equivalent weight in gases 432
Relations between density and equivalent weight in solids 433
High equivalent weights of solid species 434
Playfair and Joule on equivalent volumes 434
Equivalent volumes of crystalline solids 435
Equivalent volumes of liquid species 436

XVII.

THE CONSTITUTION AND EQUIVALENT VOLUME OF MINERAL SPECIES (1853-1863).

Progressive or homologous series in chemistry 439
General formula for silica and other oxides 440
Equivalent volume of certain salts 440
Probable constitution of the carbon-spars 441
Illustrations of isomorphism and of homology 442
Relations between the various triclinic feldspars 443
A similar view subsequently adopted by Tschermak 444
The feldspathides; scapolites, beryl, and iolite 445
The grenatides; zoisite or saussurite 446
Polymerism in mineral species illustrated 446
Relations between the jades and the scapolites 447
The allomerism of Professor Cooke 447

XVIII.

THOUGHTS ON SOLUTION AND THE CHEMICAL PROCESS (1854).

Views of various chemists as to the nature of solution 448
Solution maintained to be chemical union 449
Chemical union is identification 450
Chemical decomposition or differentiation 451
Nature of double decomposition 451
Action by pressure or catalysis 452

XIX.

ON THE OBJECTS AND METHOD OF MINERALOGY (1867).

Mineralogy in its relations to chemistry and natural history . . . 453
Mineralogy the natural history of all unorganized matter 454
Objects to be attained in a natural classification 454
Views of Oken and of Stallo 454
The nature of chemical species defined 455
Varying condensation and equivalents of solid species 455
Relations of vapors to liquids and solids 456
Evidences of polymerism in solid species 457

XX.

THEORY OF TYPES IN CHEMISTRY (1848 - 1861).

Kolbe on oxides of carbon as types in chemistry 459
Ad. Wurtz's criticism of Kolbe 460
Importance of the conception of types in chemistry 461
Views of Williamson and of Gerhardt 462
Laurent on water as a type 463
The author's views on the water-type 463
On anhydrous monobasic acids 464
The conception of condensed or polymeric types 464
The nature of sulphur, ozone, and nitrogen 464
Hydrogen the fundamental type 465
Note on the theory of nitrification 465
On the value and significance of rational formulas 465
The hypothesis of radicles and substitution by residues 465
Ad. Wurtz on polyatomic radicles 466
The genesis of the phosphoric acids explained 466
Gerhardt on polybasic and sub-salts 467
The sulphates considered as derived from polyatomic radicles 467
Priority of the author to Williamson and to Gerhardt 468

APPENDIX.

The theory of nitrification 470
Views as to the double nature of nitrogen gas 470
Its conversion into ammonia and nitrous acid 470
The intervention of ozone in the process 471
Experiments of Schönbein on nitrification 471
Nicklès on the priority of the author 472
Schaeffer on the theory of nitrification 473

I.

THEORY OF IGNEOUS ROCKS AND VOLCANOES.

(1858.)

The following Essay, read before the Canadian Institute, at Toronto, March 13, 1858, was printed in the Canadian Journal for May of the same year. It may be regarded as a first contribution to the theoretical notions developed in some of the following papers.

In a note in the American Journal of Science for January, 1858, I have ventured to put forward some speculations upon the chemistry of a cooling globe, such as the igneous theory supposes our earth to have been at an early period. Considering only the crust with which geology makes us acquainted, and the liquid and gaseous elements which now surround it, I have endeavored to show that we may attain to some idea of the chemical conditions of the cooling mass by conceiving these materials to again react upon each other under the influence of an intense heat. The quartz, which is present in such a great proportion in many rocks, would decompose the carbonates and sulphates, and, aided by the presence of water, the chlorides both of the rocky strata and the sea; while the organic matters and the fossil carbon would be burned by the atmospheric oxygen. From these reactions would result a fused mass of silicates of alumina, alkalies, lime, magnesia, iron, etc.; while all the carbon, sulphur, and chlorine, in the form of acid gases, mixed with watery vapor, azote, and a probable excess of oxygen, would form an exceedingly dense atmosphere. When the cooling permitted condensation, an acid rain would fall upon the heated crust of the earth, decomposing the silicates, and giving rise to chlorides and sulphates of the various bases, while the separated silica would probably take the form of crystalline quartz.

In the next stage, the portions of the primitive crust not covered by the ocean undergo a decomposition under the influence of the hot moist atmosphere charged with carbonic acid, and the feldspathic silicates are converted into clays with separation of an alkaline silicate, which, decomposed by the carbonic acid, finds its way to the sea in the form of alkaline bicarbonate, where, having first precipitated any dissolved sesquioxides, it changes the dissolved lime-salts into bicarbonate. This, precipitated chemically or separated by organic agencies, gives rise to limestones, the chloride of calcium being at the same time replaced by common salt.* The separation from the waters of the ocean of gypsum and sea-salt, and of the salts of potash by the agency of marine plants, and by the formation of glauconite, are considerations foreign to our present study.

In this way we obtain a notion of the processes by which, from a primitive fused mass, may be generated the silicious, calcareous, and argillaceous rocks which make up the greater part of the earth's crust, and we also understand the source of the salts of the ocean. But the question here arises whether this primitive crystalline rock, which probably approached to dolerite in its composition, is now anywhere visible upon the earth's surface. It is certain that the oldest known rocks are stratified deposits of limestone, clay, and sands, generally in a highly altered condition; but these, as well as more recent strata, are penetrated by various injected rocks, such as granites, trachytes, syenites, porphyries, dolerites, phonolites, etc. These offer in their mode of occurrence, not less than their composition, so many analogies with the lavas of modern volcanoes, that they also are universally supposed to be of igneous origin, and to owe their peculiarities to slow cooling under pressure. This conclusion being admitted, we proceed to inquire into the sources of these liquid masses which, from the earliest known geological period up to the present day, have been from time

* See in this connection the note appended, page 10.

to time ejected from below. They are generally regarded as evidences both of the igneous fusion of the interior of our planet, and of a direct communication between the surface and the fluid nucleus, which is supposed to be the source of the various ejected rocks.

These intrusive masses, however, offer very great diversities in their composition, from the highly silicious and feldspathic granites, eurites, and trachytes, in which lime, magnesia, and iron are present in very small quantities, and in which potash is the predominant alkali, to the denser basic rocks, dolerite, diorite, trap, and basalt; in these, lime, magnesia, and iron-oxide are abundant, and soda prevails over the potash. To account for these differences in the composition of the injected rocks, Phillips, and after him Durocher, suppose the interior fluid mass to have separated into a denser stratum of the basic silicates, upon which a lighter and more silicious portion floats like oil upon water; and that these two liquids, occasionally more or less modified by a partial crystallization and eliquation, or by a refusion, give rise to the principal varieties of silicious and basic rocks; while from the mingling of the two zones of liquid matter intermediate rocks are formed. (Phillips's Manual of Geology, p. 556, and Durocher, Annales des Mines, 1857, Vol. I. p. 217.)

An analogous view was suggested by Bunsen in his researches on the volcanic rocks of Iceland, and extended by Streng to similar rocks in Hungary and Armenia. These investigators suppose the existence beneath the earth's crust of a trachytic and a pyroxenic magma of constant composition, representing respectively the two great divisions of rocks which we have just distinguished; and have endeavored to calculate from the amount of silica in any intermediate variety, the proportions in which these two magmas must have been mingled to produce it, and consequently the proportions of alumina, lime, magnesia, iron-oxide and alkalies which such a rock may be expected to contain. But the amounts thus calculated, as may be seen from Dr. Streng's results, do not always correspond with the results of analysis. (Streng, Annales de Chimie et de Physique,

Third Series, Vol. XXXIX. p. 52.) Besides, there are intrusive rocks, such as the phonolites, which are highly basic, and yet contain but very small quantities of lime, magnesia, and iron-oxide; being essentially silicates of alumina and alkalies, in part hydrated.

We may here remark that many of the so-called igneous rocks are often of undoubted sedimentary origin. It will scarcely be questioned that this is true of many granites, and it is certain that all the feldspathic rocks coming under the categories of hyperite, labradorite, diorite, and amphibolite, which make so large a part of the Laurentian system in North America, are of sedimentary origin. They are here interstratified with limestones, dolomites, serpentines, crystalline gneisses and quartzites, which latter are often conglomerate. The same thing is true of similar feldspathic rocks in the crystalline strata of the Green Mountains. These metamorphic strata have been exposed to conditions which have rendered some of them quasi-fluid or plastic. Thus, for example, crystalline limestone may be seen in positions which have led many observers to regard it as intrusive rock, although its general mode of occurrence leaves no doubt as to its sedimentary origin. We find in the Laurentian system that the limestones sometimes envelope the broken and contorted fragments of the beds of quartzite, with which they are often interstratified, and penetrate like a veritable trap into fissures in the quartzite and gneiss. A rock of sedimentary origin may then assume the conditions of a so-called igneous rock, and who shall say that any intrusive granites, dolerites, euphotides, or serpentines have an origin distinct from the metamorphic strata of the same kind which make up such vast portions of the older stratified formations? To suppose that each of these sedimentary rocks has also its representative among the ejected products of the central fire, seems a hypothesis not only unnecessary, but, when we consider their varying composition, untenable.

We are next led to consider the nature of the agencies which have produced this plastic condition in various crystalline rocks. Certain facts, such as the presence in them of graphite in contact with carbonate of lime and oxide of iron, not less than the presence of alkaliferous silicates like the feldspars in crystalline limestones, forbid us to admit the ordinary notion of the intervention of an intense heat such as would produce an igneous fusion, and lead us to consider the view first put forward by Poulett Scrope,* and since ably advocated by Scheerer and by Elie de Beaumont, of the intervention of water, aided by heat, which they suppose may communicate a plasticity to rocks at a temperature far below that required for their igneous fusion. The presence of water in the lavas of modern volcanoes led Mr. Scrope to speculate upon the effect which a small portion of this element might exert, at an elevated temperature and under pressure, in giving liquidity to masses of rock, and he extended this idea from proper volcanic rocks to granites.

Scheerer, in his inquiry into the origin of granite, has appealed to the evidence afforded us by the structure of this rock, that the more fusible feldspars and mica crystallized before the almost infusible quartz. He also points to the existence in granite of what he has called pyrognomic minerals, such as allanite and gadolinite, which, when heated to low redness, undergo a peculiar and permanent molecular change, accompanied by an augmentation in density and a change in chemical properties; a phenomenon completely analogous to that offered by titanic acid and chromic oxide in their change by ignition from a soluble to an insoluble condition. These facts seem to exclude the idea of igneous fusion, and point to some other cause of liquidity. The presence of natrolite as an integral part of the zircon-syenites of Norway, and of talc, chlorite, and other hydrous minerals in many granites shows that water was not excluded from the original granitic paste. Scheerer appeals, by way of illustration, to the influence of small portions of carbon and sulphur in greatly reducing the fusing point of iron. He alludes to the experiments of Schafhautl and Wöhler, which show that quartz and apophyllite may be dissolved by heated water, under pressure, and recrystallized on cooling.

* See Journal of Geological Society of London, Vol. XII. p. 326.

He recalls the aqueous fusion of many hydrated salts, and finally suggests that the presence of a small amount of water, perhaps five or ten per cent, may suffice, at a temperature which may approach that of redness, to give to a granitic mass a liquidity partaking at once of the characters of an igneous and an aqueous fusion.

This ingenious hypothesis, sustained by Scheerer in his discussion with Durocher,* is strongly confirmed by the late experiments of Daubrée. He found that common glass, a silicate of lime and alkali, when exposed to a temperature of 400° C., in presence of its own volume of water, swelled up and was transformed into an aggregate of crystals of wollastonite, the alkali, with the excess of silica, separating, and a great part of the latter crystallizing in the form of quartz. When the glass contained oxide of iron, the wollastonite was replaced by crystals of diopside. Obsidian in the same manner yielded crystals of feldspar, and was converted into a mass like trachyte. In these experiments upon vitreous alkaliferous matters, the process of nature in the metamorphosis of sediments is reversed; but Daubrée found still farther that kaolin, when exposed to a heat of 400° C. in the presence of a soluble alkaline silicate, is converted into crystalline feldspar, while the excess of silica separates in the form of quartz. He found natural feldspar and diopside to be extremely stable in the presence of alkaline solutions. These beautiful results were communicated to the French Academy of Sciences on the 16th of November last, and, as the author well remarked, enable us to understand the part which water may play in giving origin to crystalline minerals in lavas and intrusive rocks. The swelling up of the glass also shows that water gives a mobility to the particles of the glass at a temperature far below that of its igneous fusion.

I had already shown in the Report of the Geological Sur-

* See for the arguments on the two sides, Bulletin of the Geol. Soc. of France, Second Series, Vol. IV. pp. 468, 1018; VI. 644; VII. 276; VIII. 500; also, Elie de Beaumont, Ibid., Vol. IV. p. 1312. See also the recent microscopical observations of Mr. Sorby, confirming the theory of the aqueo-igneous origin of granite in the L. E. & D. Phil. Mag., February, 1858.

vey of Canada for 1856, p. 479, that the reaction between alkaline silicates and carbonates of lime, magnesia and iron at a temperature of 100° C. gives rise to silicates of these bases, and enables us to explain their production from a mixture of carbonates and quartz, in the presence of a solution of alkaline carbonate. I there also suggested that the silicates of alumina in sedimentary rocks may combine with alkaline silicates to form feldspars and mica, and that it would be possible to crystallize these minerals from hot alkaline solutions in sealed tubes. In this way I explained the occurrence of these silicates in altered fossiliferous strata. My conjectures are now confirmed by the experiments of Daubrée, which serve to complete the demonstration of my theory of the normal metamorphism of sedimentary rocks by the interposition of heated alkaline solutions.

But to return to the question of intrusive rocks : Calculations based on the increasing temperature of the earth's crust as we descend, lead to the belief that at a depth of about twenty-five miles the heat must be sufficient for the igneous fusion of basalt. The recent observations of Hopkins, however, show that the melting points of various bodies, such as wax, sulphur and resin, are greatly and progressively raised by pressure, so that from analogy we may conclude that the interior portions of the earth are, although ignited, solid from great pressure. This conclusion accords with the mathematical deductions of Mr. Hopkins, who, from the precession of the equinoxes, calculates the solid crust of the earth to have a thickness of 800 or 1,000 miles. Similar investigations by Mr. Hennessey, however, assign 600 miles as the maximum thickness of the crust. The region of liquid fire being thus removed so far from the earth's surface, Mr. Hopkins suggests the existence of lakes or limited basins of molten matter, which serve to feed the volcanoes.

Now the supposed mode of formation of the primitive molten crust of the earth would naturally exclude all combined or intermingled water; while all the sedimentary rocks are necessarily permeated by this liquid, and consequently in a condition to be rendered semi-fluid by the application of heat as supposed in the theory of Scrope and Scheerer. If now we admit that all igneous rocks, ancient plutonic masses as well as modern lavas, have their origin in the liquefaction of sedimentary strata, we at once explain the diversities in their composition. We can also understand why the products of volcanoes in different regions are so unlike, and why the lavas of the same volcano vary at different periods. We find an explanation of the water and carbonic acid which are such constant accompaniments of volcanic action, as well as the hydrochloric acid, sulphuretted hydrogen, and sulphuric acid, which are so abundantly evolved by certain volcanoes. The reaction between silica and carbonates must give rise to carbonic acid, and the decomposition of sea-salt in saliferous strata by silica, in the presence of water, will generate hydrochloric acid; while gypsum in the same way will evolve its sulphur in the form of sulphurous acid mixed with oxygen. The presence of fossil plants in the melting strata would generate carburetted hydrogen gases, whose reducing action would convert the sulphurous acid into sulphuretted hydrogen; or the reducing agency of the carbonaceous matters might give rise to sulphuret of calcium, which would be, in its turn, decomposed by carbonic acid or otherwise. The intervention of such matters in volcanic phenomenon is indicated by the recent investigations of Deville, who has found carburetted hydrogen in the gaseous emanations of the region of Etna and the lagoons of Tuscany. The ammonia and the nitrogen of volcanoes are also in many cases probably derived from organic matters in the strata decomposed by subterranean heat. The carburetted hydrogen and bitumen evolved from mud-volcanoes, like those of the Crimea and of Bakou, and the carbonized remains of plants in the *moya* of Quito, and in the volcanic matters of the Island of Ascension, not less than the infusorial remains found by Ehrenberg in the ejected matters of most volcanoes, all go to show that fossiliferous sediments are very generally implicated in volcanic phenomena.

It is to Sir John F. W. Herschel that we owe, so far as I am aware, the first suggestions of the theory of volcanic action which I have here brought forward. In a letter to Sir Charles Lyell, dated February 20, 1836 (Proceedings Geol. Soc., London, Vol. XI. p. 548), he maintains that with the accumulation of sediment the isothermal lines in the earth's crust must rise, so that strata buried deep enough will be crystallized and metamorphosed, and eventually be raised, with their included water, to the melting-point. This will give rise to evolutions of gases and vapors, earthquakes, volcanic explosions, etc., all of which results must, according to known laws, follow from the fact of a high central temperature; while from the mechanical subversion of the equilibrium of pressure, following upon the transfer of sediments, while the yielding surface reposes upon a mass of matter partly liquid and partly solid, we may explain the phenomena of elevation and subsidence. Such is a summary of the views put forward more than twenty years since by this eminent philosopher, which, although they have passed almost unnoticed by geologists, seem to me to furnish a simple and comprehensive explanation of several of the most difficult problems of chemical and dynamical geology.

To sum up in a few words the views here advanced. We conceive that the earth's solid crust of anhydrous and primitive igneous rock is everywhere deeply concealed beneath its own ruins, which form a great mass of sedimentary strata, permeated by water. As heat from beneath invades these sediments, it produces in them that change which constitutes normal metamorphism. These rocks, at a sufficient depth, are necessarily in a state of igneo-aqueous fusion, and in the event of fracture of the overlying strata, may rise among them, taking the form of eruptive rocks. Where the nature of the sediments is such as to generate great amounts of elastic fluids by their fusion, earthquakes and volcanic eruptions may result, and these, other things being equal, will be most likely to occur under the more recent formations.*

[NOTE to page 2. — I have since pointed out that the evidences of a similar process are still to be seen in the granites and crystalline schists

* See further in this connection Essays VI. and VII.

of eozoic ages which in many regions are decomposed to great depths, the feldspar being converted into kaolin, while the hornblende has lost its protoxide bases, the peroxidized iron and the silica remaining behind. This change has affected the crystalline rocks of the southern United States and of Brazil to depths of a hundred feet or more, and doubtless at one time extended to all such rocks as were above the surface of the ocean. The absence of this decayed material from certain regions of crystalline rocks is to be attributed to its subsequent removal by denudation, a process which in the northern parts of Europe and America terminated at the close of the pliocene period, when the remaining softened material was swept away by the action of water and ice, and the hard and unchanged rocks beneath were exposed and glaciated, since which time the chemical decomposition of the surface has been insignificant. It is this process which was called by Dolomieu the *maladie du granit*, and ascribed by him to the influence of carbonic-acid gas from subterranean sources. It was, however, in my opinion a universal phenomenon, and dependent upon the peculiar composition of the atmosphere in early times. These decomposed strata furnished the great deposits of clay and sand of the paleozoic and later periods; and from them was dissolved the iron which in various forms is found at different horizons in the uncrystalline rocks; while the silica and the alkaline and earthy carbonates, removed in a soluble form from these decaying eozoic rocks, have generated the limestones, dolomites, and various silicious deposits. (See Proceedings Boston Society of Natural History for October 15, 1873.)

In the Proceedings of the same Society for February 18, 1874, I have called attention to the fact that the clay resulting from this decay of rocks remains for many days suspended in pure water, though not in waters even slightly saline, and is therefore readily precipitated in a few hours when the turbid fresh waters mingle with those of the sea, thus forming fine argillaceous sediments. The geological significance of this fact was, it is believed, first pointed out in 1861 by Mr. Sidell in Humphreys and Abbot's Report on the Physics and Hydraulics of the Mississippi River (Appendix A, page xi.), where he applied it to explain the accumulations of mud at this river's mouth. Many chemical precipitates, in like manner, which may be washed on a filter with acid or saline solutions, readily pass through its pores if suspended in pure water. I have sought to explain these phenomena by the principle that saline matters reduce the cohesion between water and the suspended particles, thus allowing gravity and their own cohesion to come into play. Guthrie (Proceedings Royal Society, XIV.) has shown that the addition of small quantities of saline matters to water diminishes the size of its drops, evidently for the same reason.]

II.

ON SOME POINTS IN CHEMICAL GEOLOGY.

(1859.)

A paper with the above title was sent to the Geological Society of London in August, 1858, and read before that body, January, 1859. An abstract of it appeared in the L. E. & D. Philosophical Magazine for February, and it was published in full in the Quarterly Journal of the Geological Society for November, from which it was reprinted, with the addition of a few notes, in the Canadian Naturalist for December, 1859. Such portions of this paper as were but a repetition of the preceding one are here omitted; what follows may be regarded as a supplement to that.

. . . . .

WHEN we examine the waters charged with saline matters which impregnate the great mass of calcareous strata constituting in Canada the base of the palæozoic series, we find that only about one half of the chlorine is combined with sodium; the remainder exists as chlorides of calcium and magnesium, the former predominating, — while sulphates are present only in small amount. If now we compare this composition, which may be regarded as representing that of the palæozoic sea, with that of the modern ocean, we find that the chloride of calcium has been in great part replaced by common salt, — a process involving the intervention of carbonate of soda, and the formation of carbonate of lime. The amount of magnesia in the sea, although diminished by the formation of dolomite and magnesite, is now many times greater than that of the lime; for so long as chloride of calcium remains in the water, the magnesian salts are not precipitated by bicarbonate of soda.*

When we consider that the vast amount of argillaceous sedi-

* See Report Geol. Surv. Canada, 1857, pp. 212-214; Am. Jour. Science (2), XXVIII. pp. 170, 305; and further, Essays VIII. and IX.

mentary matter in the earth's strata has doubtlessly been formed by the same process which is now going on, namely, the decomposition of feldspathic minerals, it is evident that we can scarcely exaggerate the importance of the part which the alkaline carbonates, formed in this process, must have played in the chemistry of the seas. (Page 2.) We have only to recall waters like Lake Van, the natron-lakes of Egypt, Hungary, and many other regions, the great amounts of carbonate of soda furnished by springs like those of Carlsbad and Vichy, or contained in the waters of the Loire, the Ottawa, and probably many other rivers that flow from regions of crystalline rocks, to be reminded that a similar though much slower process of decomposition of alkaliferous silicates is still going on.

A striking and important fact in the history of the sea, and of most alkaline and saline waters, is the small proportion of potash-salts which they contain. Soda is pre-eminently the soluble alkali; while the potash in the earth's crust is locked up in the form of insoluble orthoclase, the soda-feldspars readily undergo decomposition. Hence we find in the analyses of clays and argillites, that of the alkalies which these rocks still retain, the potash almost always predominates greatly over the soda. At the same time these sediments contain silica in excess, and but small portions of lime and magnesia. These conditions are readily explained when we consider the nature of the soluble matters found in the mineral waters which issue from these argillaceous rocks. I have elsewhere shown that (setting aside the waters charged with soluble lime and magnesia-salts, issuing from limestones and from gypsiferous and saliferous formations) the springs from argillaceous strata are marked by the predominance of bicarbonate of soda, often with portions of silicate and borate, besides bicarbonates of lime and magnesia, and occasionally of iron. The atmospheric waters filtering through such strata remove soda, lime and magnesia, leaving behind the silica, alumina and potash, — the elements of granitic and trachytic rocks. The more sandy clays and argillites being most permeable, the action of the infiltrating waters will be more or less complete; while finer and more compact clays and marls, resisting the penetration of this liquid, will retain their soda, lime, and magnesia, and by subsequent alteration will give rise to basic feldspars containing lime and soda, and if lime and magnesia predominate, to hornblende or pyroxene.

The presence or absence of iron in sediments demands especial consideration, since its elimination requires the interposition of organic matters, which, by reducing the peroxide to the condition of protoxide, render it soluble in water, either as bicarbonate or combined with some organic acid. This action of waters holding organic matter upon sediments containing iron-oxide has been described by Bischof and many other writers, particularly by Dr. J. W. Dawson * in a paper on the coloring matters of some sedimentary rocks, and is applicable to all cases where iron has been removed from certain strata and accumulated in others. This is seen in the fire-clays and iron-stones of the coal-measures, and in the white clays associated with great beds of green-sand (essentially a silicate of iron) in the cretaceous series of New Jersey. Similar alternations of white feldspathic beds with others of iron-ore occur in the Green Mountain rocks of Canada, and on a still more remarkable scale in those of the Laurentian series. We may probably look upon the formation of beds of iron-ore as in all cases due to the intervention of organic matters; so that its presence, not less than that of graphite, affords evidence of the existence of organic life at the time of the deposition of these old crystalline rocks.

The agency of sulphuric and muriatic acids, from volcanic and other sources, is not, however, to be excluded in the solution of oxide of iron and other metallic oxides. The oxidation of pyrites, moreover, gives rise to solutions of iron and alumina-salts, the subsequent decomposition of which, by alkaline or earthy carbonates, will yield oxide of iron and alumina; the absence of the latter element serves perhaps to characterize the iron-ores of organic origin.† In this way the deposits of emery,

* Quar. Jour. Geol. Soc., Vol. V. p. 25.

† The occurrence of hydrated mixtures of oxide of iron and alumina, like

which is a mixture of crystallized alumina with oxide of iron, have doubtless been formed.

Waters deficient in organic matters may remove soda, lime, and magnesia from sediments, and leave the granitic elements intermingled with oxide of iron; while on the other hand, by the admixture of organic materials, the whole of the iron may be removed from strata which will still retain the lime and soda necessary for the formation of basic feldspars. The fact that bicarbonate of magnesia is much more soluble than bicarbonate of lime, is also to be taken into account in considering these reactions.

The study of the chemistry of mineral waters, in connection with that of sedimentary rocks, shows us that the result of processes continually going on in nature is to divide the silico-argillaceous rocks into two great classes (mentioned on page 3), — the one characterized by an excess of silica, by the predominance of potash, and by the small amounts of lime, magnesia and soda, and represented by the granites and trachytes; while in the other class silica and potash are less abundant, and soda, lime and magnesia prevail, giving rise to pyroxenes and triclinic feldspars. The metamorphism and displacement of such sediments may thus enable us to explain the origin of the different varieties of plutonic rocks without calling to our aid the ejections of the central fire.

. . . . .

Mr. Babbage * has shown that the horizons or surfaces of equal temperature in the earth's crust must rise and fall, as a consequence of the accumulation of sediment in some parts and its removal from others, producing thereby expansion and contraction in the materials of the crust, and thus giving rise to gradual and wide-spread vertical movements. Sir John Her-

bauxite, serves to show an intimate relation between the origin of these two bases in an uncombined state. Hydrous alumina, gibbsite, is moreover found incrusting limonite, and the existence of compounds like mellite and pigotite, in which alumina is united to organic acids, shows that this base may, under certain conditions, be set free in a soluble condition.

* On the Temple of Serapis, Proc. Geol. Soc., Vol. II. p. 73.

schel* subsequently showed that, as a result of the internal heat thus retained by accumulated strata, sediments deeply enough buried will become crystallized, and ultimately be raised, with their included water, to the melting-point. From the chemical reactions at this elevated temperature gases and vapors will be evolved, and earthquakes and volcanic eruptions will result. At the same time the disturbance of the equilibrium of pressure consequent upon the transfer of sediments, while the yielding surface reposes upon a mass of matter partly liquid and partly solid, will enable us to explain the phenomena of elevation and subsidence.

According, then, to Sir John Herschel's view, all volcanic phenomena have their source in sedimentary deposits; and this ingenious hypothesis, which is a necessary consequence of a high central temperature, explains in a most satisfactory manner the dynamical phenomena of volcanoes, and many other obscure points in their history, as, for instance, the independent action of adjacent volcanic vents, and the varying nature of their ejected products.† Not only are the lavas of different volcanoes very unlike, but those of the same crater vary at different times; the same is true of the gaseous matters, hydrochloric, hydrosulphuric, and carbonic acids. As the ascending heat penetrates saliferous strata, we shall have hydrochloric acid, from the decomposition of sea-salt by silica in the presence of water; while gypsum and other sulphates, by a similar reaction, would lose their sulphur in the form of sulphurous acid and oxygen. The intervention of organic matters, either by direct contact or by giving rise to reducing gases, would convert the sulphates into sulphurets, which would yield sulphuretted hydrogen when decomposed by water and silica or by carbonic acid; the latter being the result of the action of silica upon earthy carbonates. We conceive the ammonia so often found among the products of volcanoes to be evolved from the heated strata, where it exists in part as ready-formed ammonia (which is absorbed from air and water, and pertinaciously re-

* On the Temple of Serapis, Proc. Geol. Soc., Vol. II. pp. 548, 596.

† For a further development of this theory, see Essays VI. and VII.

tained by argillaceous sediments), and is in part formed by the action of heat upon azotized organic matter present in these strata, as already maintained by Bischof.* Nor can we hesitate to accept this author's theory of the formation of boracic acid from the decomposition of borates by heat and aqueous vapor.†

. . . . .

The metamorphism of sediments *in situ*, their displacement in a pasty condition from igneo-aqueous fusion as plutonic rocks, and their ejection as lavas, with attendant gases and vapors are, then, all results of the same cause, and depend upon the differences in the chemical composition of the sediments, the temperature, and the depth to which they are buried: while the unstratified nucleus of the earth, which is doubtless anhydrous, and, according to the calculations of Messrs. Hopkins and Hennessey, probably solid to a great depth, intervenes in the phenomena under consideration only as a source of heat.‡

* Lehrbuch der Geologie, Vol. II. pp. 115-122.

† Ibid., Vol. I. p. 669.

‡ The notion that volcanic phenomena have their seat in the sedimentary formations of the earth's crust, and are dependent upon the combustion of organic matters, is, as Humboldt remarks, one which belongs to the infancy of geognosy. (Cosmos, Vol. V. p. 443. Otté's translation.) In 1834, Christian Keferstein published his Naturgeschichte des Erdkörpers, in which he maintains that all crystalline non-stratified rocks, from granite to lava, are products of the transformation of sedimentary strata, in part very recent, and that there is no well-defined line to be drawn between neptunian and volcanic rocks, since they pass into each other. Volcanic phenomena, according to him, have their origin, not in an igneous fluid centre, nor an oxidizing metallic nucleus, but in known sedimentary formations, where they are the result of a peculiar process of fermentation, which crystallizes and arranges in new forms the elemente of the sedimentary strata, with evolution of heat as an accompaniment of the chemical process. (Naturgeschichte, Vol. I. p. 109; also Bull. Soc. Géol. de France (1), Vol. VII. p. 197.)

These remarkable conclusions were unknown to me at the time of writing this paper, and seem indeed to have been entirely overlooked by geological writers; they are, as will be seen, in many respects an anticipation of the views of Herschel and my own; although in rejecting the influence of an incandescent nucleus as a source of heat, he has, as I conceive, excluded the exciting cause of that chemical change, which he has not inaptly described as a process of fermentation, and which is the source of all volcanic and plutonic phenomena. See in this connection Essays I. and VII. of the present volume.

The volcanic phenomena of the present day appear, so far as I am aware, to be confined to regions covered by the more recent secondary and tertiary deposits, which we may suppose the central heat to be still penetrating (as shown by Mr. Babbage), a process which has long since ceased in the palæozoic regions. Both normal metamorphism and volcanic action are generally connected with elevations and foldings of the earth's crust, all of which phenomena we conceive to have a common cause, and to depend upon the accumulation of sediments and the subsidence consequent thereon, as maintained by Mr. James Hall in his theory of mountains.*

. . . . .

* See, for an exposition of the views of Professor Hall, Essays V. and VII. of the present volume.

III.

THE CHEMISTRY OF METAMORPHIC ROCKS.

(1863.)

This paper was read before the Dublin Geological Society, April 10, 1863, published in the Dublin Quarterly Journal for July, and reprinted in the Canadian Naturalist for the same year. The notions expressed in the first paragraph as to the existence of crystalline strata of all geological ages, the results of a subsequent alteration of palæozoic, meseozoic, and even of cenozoic sediments, are in strict accordance with those which were then (and are even now) maintained by most of the authorities in geology; and at that time had scarcely been questioned. Hence it is that the rocks of what are here designated the third and fourth series were, in conformity with the conclusions generally accepted, referred to the palæozoic age. It will, however, be seen that I had at that time no doubt that the rocks of the third (or Green Mountain) series, then regarded as altered Lower Silurian, were, as Macfarlane had already maintained, the equivalents of a part at least of the Primitive Slate or Urschiefer formation of Norway. He, as is here stated, supposed the Huronian to represent another part of the same formation; while Bigsby soon after expressed the opinion that the Huronian and the Urschiefer are the same. My own extended studies of these rocks in the Green Mountains, in New Brunswick, and on Lakes Superior and Huron, have since convinced me that this view is correct, and that the Green Mountain series is represented in the crystalline strata around the great lakes just mentioned; and, moreover, that both this series and the crystalline rocks of the fourth or White Mountain series existed in their present crystalline form before the deposition of the oldest Cambrian sediments. The further history of these crystalline series will be found in an Essay on the Geognosy of the Appalachians (XIII. of the present volume), and in its Appendix. In this connection the reader is also referred to portions of those on Granitic Rocks (XI.), on Alpine Geology (XIV.), and to the third part of that on Cambrian and Silurian (XV.). See also a note to the present paper (page 33).

These conclusions carry back the origin of these two series of crystalline rocks to a much more remote period in geological history than was formerly supposed; but the chemical principles laid down in this paper I believe to be still true, and of general application, and for this reason it is reprinted with the omission of a few sentences which, by their reference to the supposed palæozoic age of the crystalline rocks above referred to, might serve to mislead the reader.

While retaining the original title, I however regard the name of *metamorphic rocks*, as applied to crystalline strata, an unfortunate one, which it would be well to banish from the science of geology. Although it is not to be questioned that local and exceptional agencies, apparently hydrothermal, have occasionally given rise to crystalline silicated minerals in palæozoic and even in more recent sediments, and may thus help us to form some conception of processes which were universal in eozoic times, the notion that any of the great series of crystalline rocks are the stratigraphical equivalents of formations elsewhere known to us as uncrystalline sediments, will be found to rest on very uncertain evidence. Those crystalline rocks have doubtless, since their deposition, undergone certain molecular modifications (by what has been named diagenesis) which have changed their original aspect; but something of the same sort is to a greater or less extent true of many sedimentary rocks to which we do not give the name of metamorphic. This term has not only come to be familiarly used as a synonyme for all crystalline stratified rocks, but is associated with the notion of a profound epigenic change (pseudomorphism) extended alike to uncrystalline sediments and to crystalline eruptive rocks; a notion has been embodied in the assertion that "regional metamorphism is pseudomorphism on a grand scale." See in this connection Essay XIII. and its Appendix.

At a time not very remote in the history of geology, when all crystalline stratified rocks were included under the common designation of primitive, and were supposed to belong to a period anterior to the fossiliferous formations, the lithologist confined his studies to descriptions of the various species of rocks, without reference to their stratigraphical or geological distribution. But with the progress of geological science a new problem is presented to his investigation. While paleontology has shown that the fossils of each formation furnish a guide to its age and stratigraphical position, it has been found that sedimentary strata of all ages, up to the tertiary inclusive, may undergo such changes as to obliterate the direct evidences of organic life; and to give to the sediments the mineralogical characters once assigned to primitive rocks.* The question here arises, whether in the absence of organic remains, or of stratigraphical evidence, there exists any means of determining, even approximately, the geological age of a given series of crystalline stratified rocks; in other words, whether the chemical conditions which have presided over the formation of sedimentary rocks have so far varied in the course of ages, as to impress upon these rocks marked chemical and mineralogical differences. In the case of unaltered sediments it would be difficult to arrive at any solution of this question without greatly multiplied analyses; but in the same rocks, when altered, the crystalline minerals which are formed, being definite in their composition, and varying with the chemical constitution of the sediments,

* See the remarks on the preceding page.

may perhaps to a certain extent become to the geologist what organic remains are in the unaltered rocks, — a guide to the geological age and succession.

It was while engaged in the investigation of metamorphic rocks of various ages in North America, that this problem suggested itself; and I have endeavored from chemical considerations, conjoined with multiplied observations, to attempt its solution. In the Quarterly Journal of the Geological Society of London for 1859 (Essay II. of the present volume) will be found the germs of the ideas on this subject, which I shall endeavor to explain in the present paper. It cannot be doubted that in the earlier periods of the world's history, chemical forces of certain kinds were much more active than at the present day. Thus the decomposition of earthy and alkaline silicates, under the combined influences of water and carbonic acid, would be greater when this acid was more abundant in the atmosphere, and when the temperature was probably higher (page 2). The larger amounts of alkaline and earthy carbonates then carried to the sea from the decomposition of these silicates would furnish a greater amount of calcareous matter to the sediments; and the chemical effects of vegetation, both on the soil and on the atmosphere, must have been greater during the carboniferous period, for example, than at present. In the spontaneous decomposition of feldspars, which may be described as silicates of alumina combined with silicates of potash, soda and lime, these latter bases are removed, together with a portion of silica; and there remains, as the final result of the process, a hydrous silicate of alumina, which constitutes kaolin or clay. This change is favored by mechanical division; and Daubrée has shown that by the prolonged attrition of fragments of granite under water, the softer and readily cleavable feldspar is in great part reduced to an impalpable powder, while the uncleavable grains of quartz are only rounded, and form a readily subsiding sand; the water at the same time dissolving from the feldspar a certain portion of silica and of alkali. It has been repeatedly observed, where potash and soda-feldspars are associated, that the latter is much the more readily decomposed, becoming friable, and finally being reduced to clay, while the orthoclase is unaltered. The result of combined chemical and mechanical agencies acting upon rocks which contain quartz, with orthoclase and a soda-feldspar such as albite or oligoclase, would thus be a sand, made up chiefly of quartz and potash-feldspar, and a finely divided and suspended clay, consisting for the most part of kaolin and of partially decomposed soda-feldspar, mingled with some of the smaller particles of orthoclase and of quartz. With this sediment will also be included the oxide of iron, and the earthy carbonates set free by the sub-aerial decomposition of silicates like pyroxene and the anorthic feldspars, or formed by the action of the carbonate of soda derived from the latter upon the lime-salts and the magnesia-salts of sea-water. The débris of hornblende and pyroxene will also be found in this finer sediment. This process is evidently the one which must go on in the wearing away of rocks by aqueous agency, and explains the fact that while quartz, or an excess of combined silica, is for the most part wanting in rocks which contain a large proportion of alumina, it is generally abundant in those rocks in which potash-feldspar predominates.

So long as this decomposition of alkaliferous silicates is sub-aerial, the silica and alkali are both removed in a soluble form. The process is often, however, submarine or subterranean, taking place in buried sediments which are mingled with carbonates of lime and magnesia. In such cases the silicate of soda set free reacts either with these earthy carbonates, or with the corresponding chlorides of sea-water, and forms in either event a soluble soda-salt, and insoluble silicates of lime and magnesia which take the place of the removed silicate of soda. The evidence of such a continued reaction between alkaliferous silicates and earthy carbonates is seen in the large amounts of carbonate of soda, with but little silica, which infiltrating waters constantly remove from argillaceous strata; thus giving rise to alkaline springs and to natron-lakes. In these waters it will be found that soda greatly predominates, sometimes almost to the exclusion of potash. This is due not only to the fact that soda-feldspars are more readily decomposed than orthoclase, but to the well-known power of argillaceous sediments to abstract from water the potash-salts which it already holds in solution. Thus when a solution of silicate, carbonate, sulphate or chloride of potassium is filtered through common earth, the potash is taken up, and replaced by lime, magnesia, or soda, by a double decomposition between the soluble potash-salt and the insoluble silicates or carbonates of the latter bases. Soils, in like manner, remove from infiltrating waters, ammonia, and phosphoric and silicic acids, the bases which were in combination with these being converted into carbonates. The drainage-water of soils, like that of most mineral springs, contains only carbonates, chlorides and sulphates of lime, magnesia and soda; the ammonia, potash, phosphoric and silicic acids being retained by the soil.

The elements which the earth retains or extracts from waters are precisely those which are removed from it by growing plants. These, by their decomposition under ordinary conditions, yield their mineral matters again to the soil; but when decay takes place in water, these elements become dissolved, and hence the waters from peat-bogs and marshes contain large amounts of potash and silica in solution, which are carried to the sea, there to be separated, — the silica by protophytes, and the potash by algæ, which latter, decaying on the shore, or in the ooze at the bottom, restore the alkali to the earth. The conditions under which the vegetation of the coal-formation grew and was preserved being similar to those of peat, the soils became exhausted of potash, and are seen in the fire-clays of the carboniferous period.

Another effect of vegetation on sediments is due to the reducing or deoxidizing agency of the organic matters from its decay. These, as is well known, reduce the peroxide of iron to a soluble protoxide, and remove it from the soil, to be afterwards deposited in the forms of iron-ochre and iron-ores, which by subsequent alteration become hard, crystalline and insoluble. Thus, through the agency of vegetation, is the iron-oxide of the sediments withdrawn from the terrestrial circulation; and it is evident that the proportion of this element diffused in the more recent sediments must be much less than in those of ancient times. The reducing power of organic matter is further shown in the formation of metallic sulphurets; the reduction of sulphates having precipitated in this insoluble form the heavy metals, copper, lead and zinc, which, with iron, appear to have been in solution in the waters of early times, but are now by this means also abstracted from the circulation, and accumulated in beds and fahlbands, or by a subsequent process have been redissolved and deposited in veins. All analogies lead us to the conclusion that the primeval condition of the metals, and of sulphur, was, like that of carbon, one of oxidation, and that vegetable life has been the sole medium of their reduction.

The source of the carbonates of lime and magnesia in sedimentary strata is twofold: — first, the decomposition of silicates containing these bases, such as anorthic feldspars and pyroxene; and, second, the action of the alkaline carbonates formed by the decomposition of feldspars, upon the chlorides of calcium and magnesium originally present in sea-water; which have thus, in the course of ages, been in great part replaced by chloride of sodium. The clay, or aluminous silicate which has been deprived of its alkali, is thus at once a measure of the carbonic acid removed from the air, of the carbonates of lime and magnesia precipitated, and of the amount of chloride of sodium added to the waters of the primeval ocean.

The coarser sediments, in which quartz and orthoclase prevail, are readily permeable to infiltrating waters, which gradually remove from them the soda, lime and magnesia which they contain; and, if organic matters intervene, the oxide of iron; leaving at last little more than silica, alumina and potash, — the elements of granite, trachyte, gneiss and mica-schist. On the other hand, the finer marls and clays, resisting the penetration of water, will retain all their soda, lime, magnesia, and oxide of iron; and containing an excess of alumina, with a small amount of silica, will, by their metamorphism, give rise to basic lime-feldspars and soda-feldspars, and to pyroxene and hornblende, — the elements of diorites and dolerites. In this way the operation of the chemical and mechanical causes which we have traced naturally divides all the crystalline silico-aluminous rocks of the earth's crust into two types. These correspond to the two classes of igneous rocks, distinguished first by Professor Phillips, and subsequently by Durocher and by Bunsen, as derived from two distinct magmas which these geologists imagine to exist beneath the solid crust, and which the latter denominates the trachytic and pyroxenic types. I have however elsewhere endeavored to show that all intrusive or exotic rocks are probably nothing more than altered and displaced sediments, and have thus their source within the lower portions of the stratified crust, and not beneath it (pages 4, 8 and 14).

It may be well in this place to make a few observations on the chemical conditions of rock-metamorphism. I accept in its widest sense the view of Hutton and Boué, that all the crystalline stratified rocks have been produced by the alteration of mechanical and chemical sediments. The conversion of these into definite mineral species has been effected in two ways: first, by molecular changes, that is to say, by crystallization, and a rearrangement of particles; and, secondly, by chemical reactions between the elements of the sediments. Pseudomorphism, which is the change of one mineral species into another by the introduction or the elimination of some element or elements, presupposes metamorphism; since only definite mineral species can be the subjects of this process. To confound metamorphism with pseudomorphism, as Bischof, and others after him, have done, is therefore an error. It may be further remarked, that, although certain pseudomorphic changes may take place in some mineral species, in veins, and near to the surface, the alteration of great masses of silicated rocks by such a process is as yet an unproved hypothesis.*

The cases of local metamorphism in proximity to intrusive rocks go far to show, in opposition to the views of certain geologists, that heat has been one of the necessary conditions

* See further on this subject Essay XIII. and its appendix.

of the change. The source of this has been generally supposed to be from below; but to the hypothesis of alteration by ascending heat, Naumann has objected that the inferior strata in some cases escape change, and that, in descending, a certain plane limits the metamorphism, separating the altered strata above from the unaltered ones beneath, there being no apparent transition between the two. This, taken in connection with the well-known fact that in many cases the intrusion of igneous rocks causes no apparent change in the adjacent unaltered sediments, shows that heat and moisture are not the only conditions of metamorphism. In 1857 I showed by experiments that, in addition to these conditions, certain chemical reagents might be necessary; and that water impregnated with alkaline carbonates and silicates would, at a temperature not above that of 212° F., produce chemical reactions among the elements of many sedimentary rocks, dissolving silica, and generating various silicates.* Some months subsequently, Daubrée found that in the presence of solutions of alkaline silicates, at temperatures above 700° F., various silicious minerals, such as quartz, feldspar and pyroxene, could be made to assume a crystalline form; and that alkaline silicates in solution at this temperature would combine with clay to form feldspar and mica.† These observations were the complement of my own, and both together showed the agency of heated alkaline waters to be sufficient to effect the metamorphism of sediments by the two modes already mentioned, — namely, by molecular changes and by chemical reactions. Following upon this, Daubrée observed that the thermal alkaline spring of Plombières, with a temperature of 160° F., had in the course of centuries given rise to the formation of zeolites, and other crystalline silicated minerals, among the bricks and cement of the old Roman baths. From this he was led to suppose that the metamorphism of great regions

* Proc. Royal Soc. of London, May 7, 1857; and Philos. Mag. (4), XV. 68; also Amer. Jour. Science (2), XXII. and XXV. 435.

† Comptes Rendus de l'Acad., Nov. 16, 1857; also Bull. Soc. Geol de France (2), XV. 103.

might have been effected by hot springs; which, rising along certain lines of dislocation, and thence spreading laterally, might produce alteration in strata near to the surface, while those beneath would in some cases escape change.* This ingenious hypothesis may serve in some cases to meet the difficulty pointed out by Naumann; but while it is undoubtedly true in certain instances of local metamorphism, it seems to be utterly inadequate to explain the complete and universal alteration of areas of sedimentary rocks, embracing many hundred thousands of square miles. On the other hand, the study of the origin and distribution of mineral springs shows that alkaline waters (whose action in metamorphism I first pointed out, and whose efficient agency Daubrée has since so well shown) are confined to certain sedimentary deposits, and to definite stratigraphical horizons; above and below which saline waters wholly different in character are found impregnating the strata. This fact seems to offer a simple solution of the difficulty advanced by Naumann, and a complete explanation of the theory of metamorphism of deeply buried strata by the agency of ascending heat; which is operative in producing chemical changes only in those strata in which soluble alkaline salts are present.†

When the sedimentary strata have been rendered crystalline by metamorphism, their permeability to water, and their alterability, become greatly diminished; and it is only when again broken down by mechanical agencies to the condition of soils and sediments, that they once more become subject to the chemical changes which have just been described. Hence,

* It should be remembered that normal or regional metamorphism is in no way dependent upon the proximity of unstratified or igneous rocks, which are rarely present in metamorphic districts. The ophiolites, amphibolites, euphotides, diorites, and granites of such regions, which it has been customary to regard as exotic or intrusive rocks, are in most cases indigenous.

† See Report of the Geological Survey of Canada, 1853 - 56, pp. 479, 480; also Canadian Naturalist, Vol. VII. p. 262. For a consideration of the relations of mineral waters to geological formations, see General Report on the Geology of Canada, 1863, p. 561, and also Chap. XIX. of the same Report, on Sedimentary and Metamorphic Rocks; where most of the points touched in the present paper are discussed at greater length.

the mean composition of the argillaceous sediments of any geological epoch, or, in other words, the proportion between the alkalies and the alumina, will depend not only upon the age of the formation, but upon the number of times which its materials have been broken up, and the periods during which they have remained unmetamorphosed, and exposed to the action of infiltrating waters. . . . . The proportion between the alkalies and the alumina in the argillaceous sediments of any given formation is not therefore in direct relation to its age; but indicates the extent to which these sediments have been subjected to the influences of water, carbonic acid, and vegetation. If, however, it may be assumed that this action, other things being equal, has on the whole been proportionate to the newness of the formation, it is evident that the chemical and mineralogical composition of different systems of rocks must vary with their antiquity; and it now remains to find in their comparative study a guide to their respective ages.

It will be evident that silicious deposits and chemical precipitates, like the carbonates and silicates of lime and magnesia, may exist with similar characters in the geological formations of any age; not only forming beds apart, but mingled with the impermeable silico-aluminous sediments of mechanical origin. Inasmuch as the chemical agencies giving rise to these compounds were then most active, they may be expected in greatest abundance in the rocks of the earlier periods. In the case of the permeable and more highly silicious class of sediments already noticed, whose chief elements are silica, alumina, and alkalies, the deposits of different ages will be marked chiefly by a progressive diminution in the amount of potash and more especially of the soda which they contain. In the oldest rocks the proportion of alkali will be nearly or quite sufficient to form orthoclase and albite with the whole of the alumina present; but as the alkali diminishes, a portion of the alumina will crystallize, on the metamorphism of the sediments, in the form of a potash-mica, such as muscovite or margarodite. While the oxygen-ratio between the alumina and the alkali in the feldspars just named is 3 : 1, it becomes 6 : 1 in margarodite, and 12 : 1 in muscovite. The appearance of these micas in a rock denotes, then, a diminution in the amount of alkali, until in some strata the feldspar almost entirely disappears, and the rock becomes a quartzose mica-schist. In sediments still further deprived of alkali, metamorphism gives rise to schists filled with crystals of kyanite or of andalusite, which are simple silicates of alumina, into whose composition alkalies do not enter; or in case the sediment still retains oxide of iron, staurolite and iron-alumina garnet take their place. The matrix of all of these minerals is generally a quartzose mica-schist. The last term in this exhaustive process appears to be represented by the disthene and pyrophyllite rocks, which occur in some regions of crystalline schists.

In the second class of sediments we have alumina in excess, with a small proportion of silica, and a deficiency of alkalies, besides a variable proportion of silicates or carbonates of lime, magnesia, and oxide of iron. The result of the processes already described will produce a gradual diminution in the amount of alkali, which is chiefly soda. So long as this predominates, the metamorphism of these sediments will give rise to feldspars like oligoclase, labradorite, or scapolite (a dimetric feldspar); but in sediments where lime replaces a great proportion of the soda, there appears a tendency to the production of denser silicates, like lime-alumina garnet, and epidote, or zoisite, which replace the soda-lime feldspars. Minerals like the chlorites, dichroite and chloritoid are formed when magnesia and iron replace lime. In all of these cases the excess of the silicates of earthy protoxides over the silicate of alumina is represented in the altered strata by hornblende, pyroxene, olivine, and similar species; which give rise, by their admixture with the double aluminous silicates, to diorite, diabase, euphotide, eklogite, and similar compound rocks.

In eastern North America, the crystalline strata, so far as yet studied, may be conveniently classed in five groups, corresponding to as many different geological series, four of which will be considered in the present paper.

I. The Laurentian system represents the oldest known rocks of the globe, and is supposed to be the equivalent of the Primitive Gneiss formation of Scandinavia, and that of the Western Islands of Scotland, to which also the name of Laurentian is now applied. It has been investigated in Canada along a continuous outcrop from the coast of Labrador to Lake Superior, and also over a considerable area in northern New York.

II. Associated with this system is a series of strata characterized by a great development of anortholites, of which the hypersthenite or opalescent feldspar-rock of Labrador may be taken as a type. These strata overlie the Laurentian gneiss, and are regarded as constituting a second and more recent group of crystalline rocks, to which the name of the Labrador series may be provisionally given. [Since called Norian; see note to page 31.] From evidence recently obtained, Sir William Logan conceives it probable that this series is uncomformable with the older Laurentian system, and is separated from it by a long interval of time.

III. In the third place is a great series of crystalline schists (the Green Mountain series), which are in Canada referred to the Quebec group, an inferior part of the Lower Silurian system. They appear to correspond both lithologically and stratigraphically with the Schistose group of the Primitive Slate formation of Norway, as recognized by Naumann and Keilhau, and to be there represented by the strata in the vicinity of Drontheim, and those of the Dofrefeld. The Huronian series of Canada in like manner appears to correspond to the Quartzose group of the same Primitive Slate formation.* It consists of quartzites, varieties of imperfect gneiss, diorites, silicious and feldspathic schists passing into argillites, with limestones, and great beds of hematite. . . . . The Huronian series is as yet but imperfectly studied, and for the present will not be further considered.†

* See Macfarlane, — Primitive Formations of Norway and Canada compared, — Canadian Naturalist, VII. 113, 162.

[† It will be seen above that I have indicated *five* groups of crystalline rocks,

IV. In the fourth place are to be noticed the metamorphosed strata of Upper Silurian and Devonian age, with which may also be included those of the Carboniferous system in eastern New England. This group has as yet been imperfectly studied, but presents interesting peculiarities.

In the oldest of these, the Laurentian system, the first class of aluminous rocks takes the form of granitoid gneiss, which is often coarse-grained and porphyritic. Its feldspar is frequently a nearly pure potash-orthoclase, but sometimes contains a considerable proportion of soda. Mica is often almost entirely wanting, and is never abundant in any large mass of this gneiss, although small bands of mica-schist are occasionally met with. Argillites, which from their general predominance of potash and silica are related to the first class of sediments, are, so far as known, wanting throughout the Laurentian series; nor is any rock here met with, which can be regarded as derived from the metamorphism of sediments like the argillites of more modern series. Chloritic and chiastolite-schists and kyanite are, if not altogether wanting, extremely rare in the Laurentian system. The aluminous sediments of the second class are, however, represented in this system by a diabase made up of dark green pyroxene and bluish labradorite, often associated with a red alumino-ferrous garnet. This latter mineral also sometimes constitutes small beds, often with quartz, and occasionally with a little pyroxene. These basic aluminous minerals form, however, but an insignificant part of the mass of strata. This system is further remarkable by the small amount of ferruginous matter diffused through the strata; from which the greater part of the iron seems to have been removed, and accumulated in the form of immense beds of hematite and magnetic iron. Beds and veins of crystalline plumbago also characterize this series, and are generally found with the limestones, which are here developed to an extent

while attempting to describe but *four;* the fifth being the Huronian series, which from its close resemblance to the third series (from which it was by Logan regarded as geologically distinct), was to me a source of great perplexity. For further considerations touching this question, see the remarks on page 18.]

unknown in more recent formations, and are associated with veins of crystalline apatite, which sometimes attain a thickness of several feet. The serpentines of this series, so far as yet studied in Canada, are generally pale-colored, and contain an unusual amount of water, a small proportion of oxide of iron, and neither chrome nor nickel; both of which are almost always present in the serpentines of the third series.*

The second or Labrador series is characterized, as already remarked, by the predominance of great beds of anortholite, composed chiefly of triclinic feldspars, which vary in composition from anorthite to andesine. These feldspars sometimes form mountain masses, almost without any admixture, but at other times include portions of pyroxene, which passes into hypersthene. Beds of nearly pure pyroxenite are met with in this series, and others which would be called hyperite and diabase. These anortholite rocks are frequently compact, but are more often granitoid in structure. They are generally grayish, greenish, or bluish in color, and become white on the weathered surfaces. The opalescent labradorite-rock of Labrador is a characteristic variety of these anortholites; which often contain small portions of red garnet and brown mica, and more rarely, epidote, olivine, and a little quartz. They are sometimes slightly calcareous. Magnetic iron and ilmenite are often disseminated in these rocks, and occasionally form masses or beds of considerable size. These anortholites constitute the predominant part of the Labrador series, so far as yet examined. They are, however, associated with beds of quartzose orthoclase-gneiss, which represent the first class of aluminous sediments, and with crystalline limestones; and they will probably be found, when further studied, to offer a complete lithological series. These rocks have been observed in several areas among the Laurentide Mountains, from the coast of Labrador to Lake Huron, and are also met with among the

* See in this connection the author on the History of Ophiolites, Am. Jour. Science (2), XXV. 117, and XXVI. 234.

Adirondack Mountains; of which, according to Emmons, they form the highest summits.*

In the third (or Green Mountain) series, which we have referred to the Lower Silurian age, the gneiss is sometimes granitoid, but less markedly so than in the first; and it is much more frequently micaceous, often passing into micaceous schist, a common variety of which contains disseminated a large quantity of chloritoid. Argillites abound, and under the influence of metamorphism sometimes develop crystalline orthoclase. At other times they are converted into a soft micaceous mineral, and form a kind of mica-schist. Chiastolite and staurolite are never met with in the schists of this series, at least in its northern portions, throughout Canada and New England. The anortholites of the Labrador series are here represented by fine-grained diorites, in which the feldspar varies from albite to very basic varieties, which are sometimes associated with an aluminous mineral allied to chlorite in composition. Chloritic schists, frequently accompanied by epidote, abound in this series. The great predominance of magnesia in the forms of dolomite, magnesite, steatite, and serpentine, is also characteristic of portions of this series. The latter, which forms great beds (ophiolites), is marked by the almost constant presence of small portions of the oxides of chrome and nickel. These metals are also common in the other magnesian rocks of the series; green chrome-garnets, and chrome-mica occur; and beds of chromic iron ore are found in the ophiolites of the series. It is also a gold-bearing formation in eastern North America, and contains large quantities of copper ores in interstratified beds. . . . . The only graphite which has been found in the third series is in the form of impure plumbaginous shales.

The metamorphic rocks of the fourth (or White Mountain) series, as seen in southeastern Canada, are for the greater part quartzose and micaceous schists, more or less feldspathic; which in certain portions become remarkable for a great

* A further description of this Labrador or Norian series is given in Essay XIII.

development of crystals of staurolite and of red garnet. A large amount of argillite occurs in this series; and when altered, whether locally by the proximity of intrusive rock, or by normal metamorphism, exhibits a micaceous mineral, and crystals of andalusite; so that it becomes known as chiastolite-slate in parts of its distribution. Granitoid gneiss is abundantly associated with these crystalline schists. . . . . The crystalline limestones and ophiolites of eastern Massachusetts, which are probably of this series, resemble those of the Laurentian system; and the coal beds in that region are in some parts changed into graphite.* . . . .

Large masses of intrusive granite occur among the crystalline strata of the fourth series, but the so-called granites of the Laurentian appear to be in every case indigenous rocks; that is to say, strata altered *in situ*, and still retaining evidences of stratification. The same thing is true with regard to the ophiolites and the anortholites of both series. No evidences of the hypothetical granitic substratum are met with in the Laurentian system, although this is in one district penetrated by great masses of syenite, orthophyre, and dolerite. Granitic veins, with minerals containing the rarer elements, such as boron, fluorine, lithium, zirconium, and glucinum, are met with alike in the oldest and the newest gneiss in North America. These, however, I regard as having been formed, like metalliferous veins, by aqueous deposition in fissures in the strata.

The above observations upon the metamorphic strata of a wide region seem to be in conformity with the chemical principles already laid down in this paper; which it remains for geologists to apply to the rocks of other regions, and thus determine whether they are susceptible of a general application. I have found that the blue crystalline labradorite of the Labrador series of Canada is exactly represented by speci-

[* See in this connection the prefatory note to this essay, and also Essays XIII. and XV. The carboniferous age of the graphite of eastern Massachusetts has been generally assumed by geologists, though without any good reason. The crystalline rocks of this region, embracing New Hampshire and eastern Massachusetts, include representatives of the second, third, and fourth, and probably also of the first series.]

mens from Scarvig, in Skye; and the ophiolites of Iona resemble those of the Laurentian series in Canada. Many of the rocks of Donegal appear to me lithologically identical with those of the Laurentian period; while the serpentines of Aghadoey, containing chrome and nickel, and the andalusite and kyanite-schists of other parts of Donegal, cannot be distinguished from those which characterize the altered palæozoic strata of Canada. It is to remarked that chrome and nickel bearing serpentines are met with in the same geological horizon in Canada and Norway; and that those of the Scottish Highlands, which contain the same elements, belong to the newer gneiss formation; which, according to Sir Roderick Murchison, would be of similar age.* The serpentines of Cornwall, the Vosges, Mount Rosa, and many other regions, agree in containing chrome and nickel; which, on the other hand, seem to be absent from the serpentines of the Primitive Gneiss formation of Scandinavia. It remains to be determined how far chemical and mineralogical differences, such as those which have been here indicated, are geological constants. Meanwhile it is greatly to be desired that future chemical and mineralogical investigations of crystalline rocks should be made with this question in view; and that the metamorphic strata of the British Isles, and of southern and central Europe, be studied with reference to the important problem which it has been my endeavor, in the present paper, to lay before the Society.

* See in this connection the Essays XIII. and XV.

IV.

THE CHEMISTRY OF THE PRIMEVAL EARTH.

(1867.)

The following paper is an abstract of a Friday-evening lecture, given before the Royal Institution of Great Britain, London, May 31, 1867, and here reprinted from the Proceedings of the Institution. As an attempt to bring together in a connected form some of the latest conclusions of chemical and geological science, it attracted at the time considerable attention, having been frequently reprinted, several times translated, and adversely criticised both in the Chemical News and the Geological Magazine. My replies to these criticisms the reader will find in these same journals for February, 1868.

As bearing upon the subject of the lecture, an Appendix is subjoined including a note on the relation of the atmosphere of early times to climate, and to the temperature near the sea-level. For further discussion upon the origin and mode of formation of dolomites and gypsum, and their relation to the composition of the atmosphere, the reader is referred to Paper VIII. in this volume.

THE natural history of our planet, to which we give the name of geology, is necessarily a very complex science, including, as it does, the concrete sciences of mineralogy, botany, and zoölogy, and the abstract sciences, chemistry and physics. These latter sustain a necessary and very important relation to the whole process of development of our earth from its earliest ages, and we find that the same chemical laws which have presided over its changes apply also to those of extra-terrestrial matter. Recent investigations show the presence in the sun, and even in the fixed stars, — suns of other systems, — the same chemical elements as in our own planet. The spectroscope, that marvellous instrument, has, in the hands of modern investigators, thrown new light upon the composition of the farthest bodies of the universe, and has made clear many points which the telescope was impotent to resolve. The results of extra-terrestrial spectroscopic research have lately been set forth in an admirable manner by one of its most successful students, Mr.

Huggins. We see, by its aid, mattter in all its stages, and trace the process of condensation and the formation of worlds. It is long since Herschel, the first of his illustrious name, conceived the nebulæ, which his telescope could not resolve, to be the uncondensed matter from which worlds are made. Subsequent astronomers, with more powerful glasses, were able to show that many of these nebulæ are really groups of stars, and thus a doubt was thrown over the existence of nebulous luminous matter in space; but the spectroscope has now placed the matter beyond doubt. By its aid, we find in the heavens, planets, bodies like our earth, shining only by reflected light; suns, self-luminous, radiating light from solid matter; and, moreover, true nebulæ, or masses of luminous gaseous matter. These three forms represent three distinct phases in the condensation of the primeval matter from which our own and other planetary systems have been formed.

This nebulous matter is conceived to be so intensely heated as to be in the state of true gas or vapor, and, for this reason, feebly luminous when compared with the sun. It would be out of place, on the present occasion, to discuss the detailed results of spectroscopic investigation, or the beautiful and ingenious methods by which modern science has shown the existence in the sun, and in many other luminous bodies in space, of the same chemical elements that are met with in our earth, and even in our own bodies.

Calculations based on the amount of light and heat radiated from the sun show that the temperature which reigns at its surface is so great that we can hardly form an adequate idea of it. Of the chemical relations of such intensely heated matter, modern chemistry has made known to us some curious facts, which help to throw light on the constitution and luminosity of the sun. Heat, under ordinary conditions, is favorable to chemical combination, but a higher temperature reverses all affinities. Thus, the so-called noble metals, gold, silver, mercury, etc., unite with oxygen and other elements; but these compounds are decomposed by heat, and the pure metals are regenerated. A similar reaction was many years since shown by Mr. Grove with regard to water, whose elements, — oxygen and hydrogen, — when mingled and kindled by flame, or by the electric spark, unite to form water, which, however, at a much higher temperature, is again resolved into its component gases. Hence, if we had these two gases existing in admixture at a very high temperature, cold would actually effect their combination precisely as heat would do if the mixed gases were at the ordinary temperature, and literally it would be found that "frost performs the effect of fire." The recent researches of Henry Ste.-Claire Deville and others go far to show that this breaking up of compounds, or dissociation of elements by intense heat, is a principle of universal application; so that we may suppose that all the elements which make up the sun or our planet would, when so intensely heated as to be in that gaseous condition which all matter is capable of assuming, remain uncombined, — that is to say, would exist together in the condition of what we call chemical elements, whose further dissociation in stellar or nebulous masses may even give us evidence of matter still more elemental than that revealed by the experiments of the laboratory, where we can only conjecture the compound nature of many of the so-called elementary substances.

The sun, then, is to be conceived as an immense mass of intensely heated gaseous and dissociated matter, so condensed, however, that, notwithstanding its excessive temperature, it has a specific gravity not much below that of water; probably offering a condition analogous to that which Cagniard de la Tour observed for volatile bodies when submitted to great pressure at temperatures much above their boiling point. The radiation of heat going on from the surface of such an intensely heated mass of uncombined gases will produce a superficial cooling, permitting the combination of certain elements and the production of solid or liquid particles, which, suspended in the still dissociated vapors, become intensely luminous and form the solar photosphere. The condensed particles, carried down into the intensely heated mass, again meet with a heat of dissociation; so that the process of combination at the surface is incessantly renewed, while the heat of the sun may be supposed to be maintained by the slow condensation of its mass; a diminution by $\frac{1}{1000}$th of its present diameter being sufficient, according to Helmholtz, to maintain the present supply of heat for 21,000 years.

This hypothesis of the nature of the sun and of the luminous process going on at its surface is the one lately put forward by Faye, and, although it has met with opposition, appears to be that which accords best with our present knowledge of the chemical and physical conditions of matter such as we must suppose it to exist in the condensing gaseous mass, which, according to the nebular hypothesis, should form the centre of our solar system. Taking this, as we have already done, for granted, it matters little whether we imagine the different planets to have been successively detached as rings during the rotation of the primal mass, as is generally conceived, or whether we admit with Chacornac a process of aggregation or concretion operating within the primal nebular mass, resulting in the production of sun and planets. In either case we come to the conclusion that our earth must at one time have been in an intensely heated gaseous condition, such as the sun now presents, self-luminous, and with a process of condensation going on at first at the surface only, until by cooling it must have reached the point where the gaseous centre was exchanged for one of combined and liquefied matter.

Here commences the chemistry of the earth, to the discussion of which the foregoing considerations have been only preliminary. So long as the gaseous condition of the earth lasted, we may suppose the whole mass to have been homogeneous; but when the temperature became so reduced that the existence of chemical compounds at the centre became possible, those which were most stable at the elevated temperature then prevailing would be first formed. Thus, for example, while compounds of oxygen with mercury, or even with hydrogen, could not exist, oxides of silicon, aluminum, calcium, magnesium, and iron might be formed and condense in a liquid form at the centre of the globe. By progressive cooling, still other elements would be removed from the gaseous mass, which would form the atmosphere of the non-gaseous nucleus. We may suppose an arrangement of the condensed matters at the centre according to their respective specific gravities, and thus the fact that the density of the earth as a whole is about twice the mean density of the matters which form its solid surface may be explained. Metallic or metalloidal compounds of elements, grouped differently from any compounds known to us, and far more dense, may exist in the centre of the earth.

The process of combination and cooling having gone on until those elements which are not volatile in the heat of our ordinary furnaces were condensed into a liquid form, we may here inquire what would be the result, upon the mass, of a further reduction of temperature. It is generally assumed that in the cooling of a liquid globe of mineral matter, congelation would commence at the surface, as in the case of water; but water offers an exception to most other liquids, inasmuch as it is denser in the liquid than in the solid form. Hence, ice floats on water, and freezing water becomes covered with a layer of ice, which protects the liquid below. With most other matters, however, and notably with the various mineral and earthy compounds analogous to those which may be supposed to have formed the fiery-fluid earth, numerous and careful experiments show that the products of solidification are much denser than the liquid mass; so that solidification would have commenced at the centre, whose temperature would thus be the congealing point of these liquid compounds. The important researches of Hopkins and Fairbairn on the influence of pressure in augmenting the melting point of such compounds as contract in solidifying are to be considered in this connection.

It is with the superficial portions of the fused mineral mass of the globe that we have now to do; since there is no good reason for supposing that the deeply seated portions have intervened in any direct manner in the production of the rocks which form the superficial crust. This, at the time of its first solidification, presented probably an irregular, diversified surface from the result of contraction of the congealing mass, which at last formed a liquid bath of no great depth, surrounding the solid nucleus. It is to the composition of this crust that we must direct our attention, since therein would be found all the elements (with the exception of such as were still in the gaseous form) now met with in the known rocks of the earth. This crust is now everywhere buried beneath its own ruins, and we can only from chemical considerations attempt to reconstruct it. If we consider the conditions through which it has passed, and the chemical affinities which must have come into play, we shall see that they are just what would now result if the solid land, sea, and air were made to react upon each other under the influence of intense heat. To the chemist it is at once evident that from this would result the conversion of all carbonates, chlorides, and sulphates into silicates, and the separation of the carbon, chlorine, and sulphur in the form of acid gases, which, with nitrogen, watery vapor, and a probable excess of oxygen, would form the dense primeval atmosphere. The resulting fused mass would contain all the bases as silicates, and must have much resembled in composition certain furnace-slags or volcanic glasses. The atmosphere, charged with acid gases, which surrounded this primitive rock, must have been of immense density. Under the pressure of such a high barometric column, condensation would take place at a temperature much above the present boiling point of water, and the depressed portions of the half-cooled crust would be flooded with a highly heated solution of hydrochloric and sulphuric acids, whose action in decomposing the silicates is easily intelligible to the chemist. The formation of chlorides and sulphates of the various bases, and the separation of silica, would go on until the affinities of the acids were satisfied, and there would be a separation of silica, taking the form of quartz, and the production of a sea-water holding in solution, besides the chlorides and sulphates of sodium, calcium, and magnesium, salts of aluminum and other metallic bases. The atmosphere, being thus deprived of its volatile chlorine and sulphur compounds, would approximate to that of our own time, but differ in its greater amount of carbonic acid.

We next enter into the second phase in the action of the

atmosphere upon the earth's crust. This, unlike the first, which was subaqueous, or operative only on the portion covered with the precipitated water, is subaerial, and consists in the decomposition of the exposed parts of the primitive crust under the influence of the carbonic acid and moisture of the air, which convert the complex silicates of the crust into a silicate of alumina, or clay; while the separated lime, magnesia, and alkalies, being converted into carbonates, are carried down into the sea in a state of solution.

The first effect of these dissolved carbonates would be to precipitate the dissolved alumina and the heavy metals, after which would result a decomposition of the chloride of calcium of the sea-water, resulting in the production of carbonate of lime or limestone, and chloride of sodium or common salt. This process is one still going on at the earth's surface, slowly breaking down and destroying the hardest rocks, and, aided by mechanical processes, transforming them into clays; although the action, from the comparative rarity of carbonic acid in the atmosphere, is far less energetic than in earlier times, when the abundance of this gas, and a higher temperature, favored the chemical decomposition of the rocks. But now, as then, every clod of clay formed from the decay of a crystalline rock corresponded to an equivalent of carbonic acid abstracted from the atmosphere, and to equivalents of carbonate of lime and common salt formed from the chloride of calcium of the sea-water.

It is very instructive, in this connection, to compare the composition of the waters of the modern ocean with that of the sea in ancient times, whose composition we learn from the fossil sea-waters which are still to be found in certain regions, imprisoned in the pores of the older stratified rocks. These are vastly richer in salts of lime and magnesia than those of the present sea, from which have been separated, by chemical processes, all the carbonate of lime of our limestones, with the exception of that derived from the subaerial decay of calcareous and magnesian silicates belonging to the primitive crust.

The gradual removal, in the form of carbonate of lime, of the carbonic acid from the primeval atmosphere, has been connected with great changes in the organic life of the globe. The air was doubtless at first unfit for the respiration of warm-blooded animals, and we find the higher forms of life coming gradually into existence as we approach the present period of a purer air. Calculations lead us to conclude that the amount of carbon thus removed in the form of carbonic acid has been so enormous, that we must suppose the earlier forms of air-breathing animals to have been peculiarly adapted to live in an atmosphere which would probably be too impure to support modern reptilian life. The agency of plants in purifying the primitive atmosphere was long since pointed out by Brongniart, and our great stores of fossil fuel have been derived from the decomposition, by the ancient vegetation, of the excess of carbonic acid of the early atmosphere, which through this agency was exchanged for oxygen gas. In this connection the vegetation of former periods presents the curious phenomenon of plants allied to those now growing beneath the tropics flourishing within the polar circles. Many ingenious hypotheses have been proposed to account for the warmer climate of earlier times, but are at best unsatisfactory, and it appears to me that the true solution of the problem may be found in the constitution of the early atmosphere, when considered in the light of Dr. Tyndall's beautiful researches on radiant heat. He has found that the presence of a few hundredths of carbonic-acid gas in the atmosphere, while offering almost no obstacle to the passage of the solar rays, would suffice to prevent almost entirely the loss, by radiation, of obscure heat, so that the surface of the land beneath such an atmosphere would become like a vast orchard-house, in which the conditions of climate necessary to a luxuriant vegetation would be extended even to the polar regions.

This peculiar condition of the early atmosphere cannot fail to have influenced in many other ways the processes going on at the earth's surface.* To take a single example: one of the processes by which gypsum may be produced at the earth's surface involves the simultaneous production of bicarbonate of magnesia. This, being more soluble than the gypsum, is not

* See Appendix to this paper.

always now found associated with it; but we have indirect evidence that it was formed and subsequently carried away, in the case of many gypsum deposits, whose thickness indicates a long continuance of the process under conditions much more perfect and complete than we can attain under our present atmosphere. While studying this reaction I was led to inquire whether the carbonic acid of the earlier periods might not have favored the formation of gypsum; and I found, by repeating the experiments in an artificial atmosphere impregnated with carbonic acid, that such was really the case.* We may thence conclude that the peculiar composition of the primeval atmosphere was the essential condition under which the great deposits of gypsum, generally associated with magnesian limestones, were formed.

The reactions of the atmosphere, which we have considered, would have the effect of breaking down and disintegrating the surface of the primeval globe, covering it everywhere with beds of stratified rock of mechanical or of chemical origin. These now so deeply cover the partially cooled surface that the amount of heat escaping from below is inconsiderable, although in earlier times it was very much greater, and the increase of temperature met with in descending into the earth must then have been many times more rapid than now. The effect of this heat upon the buried sediments would be to soften them, producing new chemical reactions between their elements, and converting them into what are known as crystalline or metamorphic rocks, such as gneiss, greenstone, granite, etc. We are often told that granite is the primitive rock or substratum of the earth; but this is not only unproved, but extremely improbable. As I endeavored to show in the early part of this discourse, the composition of this primitive rock, now everywhere hidden, must have been very much like that of a slag or lava; and there are excellent chemical reasons for maintaining that granite is in every case a rock of sedimentary origin, — that is to say, it is made up of materials which were deposited from water, like beds of modern sand and gravel, and includes in its com-

* See Paper VIII.

position quartz, which, so far as we know, can only be generated by aqueous agencies, and at comparatively low temperatures.

The action of heat upon many buried sedimentary rocks, however, not only softens or melts them, but gives rise to a great disengagement of gases, such as carbonic and hydrochloric acids, and sulphur compounds, all of which are results of the reaction of the elements of sedimentary rocks, heated in presence of the water which everywhere filled their pores. In the products thus generated we have a rational explanation of the chemical phenomena of volcanoes, which are vents through which these fused rocks and confined gases find their way to the surface of the earth. In some cases, as where there is no disengagement of gases, the fused or half-fused rocks solidify *in situ*, or in rents or fissures in the overlying strata, and constitute eruptive or plutonic rocks, such as granite and basalt.

This theory of volcanic phenomena was put forward in germ by Sir John F. W. Herschel thirty years since, and, as I have during the past few years endeavored to show, it is the one most in accordance with what we know both of the chemistry and the physics of the earth. That all volcanic and plutonic phenomena have their seat in the deeply buried and softened zone of sedimentary deposits of the earth, and not in its primitive nucleus, accords with the conclusions already arrived at relative to the solidity of that nucleus; with the geological relations of these phenomena, as I have elsewhere shown; and also with the remarkable mathematical and astronomical deductions of the late Mr. Hopkins of Cambridge, based upon the phenomena of precession and nutation, those of Archdeacon Pratt, and those of Professor Thompson on the theory of the tides, — all of which lead to the same conclusion, namely, that the earth, if not solid to the centre, must have a crust several hundred miles in thickness, which would practically exclude it from any participation in the plutonic phenomena of the earth's surface, except such as would result from its high temperature communicated by conduction to the sedimentary strata reposing upon it.

The old question between the plutonists and the neptunists, which divided the scientific world in the last generation, was, in brief, this: whether fire or water has been the great agent in giving origin and form to the rocks of the earth's crust. While some maintained the direct igneous origin of such rocks as gneiss, mica-schist, and serpentine, and ascribed to fire the filling of metallic veins, others — the neptunian school — were disposed to shut their eyes to the evidences of igneous action on the earth, and even sought to derive all rocks from a primal aqueous magma. In the light of the exposition which I have laid before you this evening, we can, I think, render justice to both of these opposing schools. We have seen how reactions dependent on water and acid solutions have transformed the primitive plutonic mass, and how the resulting aqueous sediments, when deeply buried, come again within the domain of fire, to be transformed into crystalline and so-called plutonic or volcanic rocks.

The scheme which I have thus sought to put before you in the short time allotted this evening is, as I have endeavored to show, in strict conformity with known chemical laws and the facts of physical and geological science. Did time permit, I would gladly have attempted to demonstrate at greater length its adaptation to the explanation of the origin of the various classes of rocks, of metallic veins and deposits, of mineral springs, and of gaseous exhalations. I shall not, however, have failed in my object, if, in the hour which we have spent together, I shall have succeeded in showing that chemistry is able to throw a great light upon the history of the formation of our globe, and to explain in a satisfactory manner some of the most difficult problems of geology; and I feel that there is a peculiar fitness in bringing such an exposition before the members of this Royal Institution, which has been for so many years devoted to the study of pure science, and whose glory it is, through the illustrious men who have filled, and those who now fill, its professorial chairs, to have contributed more than any other school in the world to the progress of modern chemistry and physics.

APPENDIX.

ON THE CLIMATE OF THE EARTH IN FORMER GEOLOGICAL PERIODS.

The following note appeared in the London, Edinburgh, and Dublin Philosophical Magazine for October, 1863. I subsequently found that this consequence of his discoveries had not escaped Tyndall, who, in his Bakerian lecture for 1861 (Ibid., October, 1861), after showing that from its influence on terrestrial radiation all variation in the amount of aqueous vapor must produce changes in climate, added, "Similar remarks would apply to the carbonic acid diffused through the air, while an almost inappreciable admixture of any of the hydro-carbon vapors would produce great effects on the terrestrial rays, and corresponding changes in climate. It is not therefore necessary to assume alterations in the density and height of the atmosphere, to account for different amounts of heat being preserved to the earth at different times; a slight change in its variable constituents would account for this. Such changes, in fact, may have produced all the mutations of climate which the researches of geologists reveal." A letter from the author to Dr. Tyndall, in which this passage was cited, appeared in the above-named magazine for March, 1864.

THE late researches of Dr. John Tyndall on the relation of gases and vapors to radiant heat are important in their bearing upon the temperature of the earth's surface in former geological periods. He has shown that heat, from whatever source, passes through hydrogen, oxygen, and nitrogen gases, or through dry air, with nearly the same facility as through a vacuum. These gases are thus to radiant heat what rock-salt is among solids. Glass, and some other solid substances which are readily permeable to light and to solar heat, offer, as is well known, great obstacles to the passage of radiant heat from non-luminous bodies; and Tyndall has recently shown that many colorless vapors and gases have a similar effect, intercepting the heat from such sources, by which they become warmed and in their turn radiate heat. Thus, while for a vacuum the absorption of heat from a body at 212° F. is represented by 0, and that for dry air is 1, the absorption by an atmosphere of carbonic-acid gas equals 90, by marsh gas 403, by olefiant gas 970, and by ammonia 1,195. The diffusion of olefiant gas of one-inch tension in a vacuum produces an absorption of 90, and the same amount of carbonic-acid gas an absorption of 5.6. The small quantities of ozone present in electrolytic oxygen were found to raise its absorptive power from 1 to 85, and even to 136; and the watery vapor present in the air at ordinary temperatures in like manner produces an absorption of heat represented by 70 or 80. Air saturated with moisture at the ordinary temperature absorbs more than five hundredths of the heat radiated from a metallic vessel filled with boiling water, and Tyndall calculates that of the heat radiated from the earth's surface warmed by the sun's rays, one tenth is intercepted by the aqueous vapor within ten feet of the surface. Hence the powerful influence of moist air upon the climate of the globe. Like a covering of glass, it allows the sun's rays to reach the earth, but prevents to a great extent the loss by radiation of the heat thus communicated.

When, however, the supply of heat from the sun is interrupted during long nights, the radiation which goes on into space causes the precipitation of a great part of the watery vapor from the air, and the earth, thus deprived of this protecting shield, becomes more and more rapidly cooled. If now we could suppose the atmosphere to be mingled with some permanent gas, which should possess an absorptive power like that of the vapor of water, this cooling process would be in a great measure arrested, and an effect would be produced similar to that of a screen of glass ; which keeps up the temperature beneath it, directly, by preventing the escape of radiant heat, and indirectly by hindering the condensation of the aqueous vapor in the air confined beneath.

Now we have only to bear in mind that there are the best of reasons for believing that, during the earliest geological periods, all of the carbon since deposited in the forms of limestone and of mineral coal existed in the atmosphere in the state of carbonic acid, and we see at once an agency which must have aided greatly to maintain the elevated temperature that then prevailed at the earth's surface.* Without doubt the great extent of sea, and the absence

* [The carbonic acid contained in a layer of pure carbonate of lime or marble, covering the entire surface of the globe, and having a thickness of 8.61 metres, would, if set free, double the weight of the atmosphere. (Canadian Naturalist (2), III. 119.) It is probable that the amount of carbonic acid thus fixed in the earth's crust must surpass this many times, but from the activity of chemical forces then prevailing, the greater part of this was doubtless fixed in the form of carbonate of lime at a very early period in the history of the globe, so that the atmosphere in the palæozoic age may not have contained more than a few hundredths of carbonic acid. It must not be supposed that the whole of the vast deposits of limestone which have since been formed are directly and immediately due to the reaction of carbonic acid on the alkaline and earthy silicates of the rocks. A large part of the carbonate of lime deposited in later times was doubtless derived from the solution of the limestones of pre-existing formations. It nevertheless remains true that a reaction between the carbonic acid of the atmosphere and mineral silicates, similar to that of early times, though small in amount, is still going on at the earth's surface. (*Ante*, pages 10 and 20.)]

or rarity of high mountains, contributed much towards the mild climate of later ages, when a vegetation as luxuriant as that now found in the tropics flourished within the Arctic circle ; but to these causes must be added the influence of a portion of carbon which was afterwards condensed in the forms of coal and carbonate of lime, and which then existed in the condition of a transparent and permanent gas, mingled with the atmosphere, surrounding the earth, and protecting it like a dome of glass. To this effect of carbonic acid it is possible that other gases may have contributed. The ozone, which is mingled with the oxygen set free from growing plants, and the marsh gas, which is now evolved from decomposing vegetation under conditions similar to those then presented by the coal fields, may, by their great absorptive power, have very well aided to maintain at the earth's surface that high temperature the cause of which has been one of the enigmas of geology.

V.

THE ORIGIN OF MOUNTAINS.

(1861.)

The following pages are from a review entitled Some Points in American Geology, which appeared in the American Journal of Science for May, 1861, and was devoted in part to a notice of the remarkable essay which forms the Introduction to the third volume of Hall's Paleontology of New York, from which numerous extracts are given below. Read in connection with Paper VII. of the present volume, on Dynamical Geology, it will serve to give a notion of the views of Professor Hall and the author on the nature and origin of mountains.

The sediments of the carboniferous period, like those of earlier formations, exhibit, towards the east, a great amount of coarse sediments, evidently derived from a wasting continent, and are nearly destitute of calcareous beds. In Nova Scotia, Sir William Logan found, by careful measurement, 14,000 feet of carboniferous strata; and Professor Rogers gives their thickness in Pennsylvania as 8,000 feet, including at the base 1,400 feet of a conglomerate, which disappears before reaching the Mississippi. In Missouri, Professor Swallow finds but 640 feet of carboniferous strata, and in Iowa their thickness is still less, the sediments composing them being at the same time of finer materials. In fact, as Mr. Hall remarks, throughout the whole palæozoic period we observe a greater accumulation and a coarser character of sediments along the line of the Appalachian chain, with a gradual thinning westward, and a deposition of finer and farther-transported matter in that direction. To the west, as this shore-derived material diminishes in volume, the amount of calcareous matter rapidly augments. Mr. Hall concludes, therefore, that the coal-measure sediments were driven westward into an ocean where there already existed a marine fauna. At length, the marine limestones predominating, the coal-measures come to be of little importance, and we have a great limestone formation of marine origin, which in the Rocky Mountains and New Mexico occupies the horizon of the coal, and, itself unaltered, rests on crystalline strata like those of the Appalachian range. In truth, Mr. Hall observes, the carboniferous limestone is one of the most extensive marine formations of the continent, and is characterized over a much greater area by its marine fauna than by its terrestrial vegetation.

"The accumulations of the coal-period were the last that gave form and contour to the eastern side of our continent, from the Gulf of St. Lawrence to the Gulf of Mexico; and as we have shown that the great sedimentary deposits of successive periods have followed essentially the same course, parallel to the mountain ranges, we naturally inquire: What influence this accumulation has had upon the topography of our country, and whether the present line of mountain-elevation from northeast to southwest is in any way connected with the original accumulation of sediments." (Hall's Paleontology, Vol. III.; Introduction, p. 66.)

The total thickness of the palæozoic strata along the Appalachain chain is about 40,000 feet, while the same formations in the Mississippi Valley, including the carboniferous limestone, which is wanting in the east, have, according to Mr. Hall, a thickness of scarcely 4,000 feet. In many places in this valley we find the palæozoic formations exposed, exhibiting hills of 1,000 feet, made up of horizontal strata, with the Potsdam sandstone for their base, and capped by the Niagara limestone; while the same strata in the Appalachians would give from ten to sixteen times that thickness. Still, as Mr. Hall remarks, we have there no mountains of corresponding altitude, that is to say, none whose height, like those of the Mississippi valley, equals the actual vertical thickness of the strata. In the west there has been little or no disturbance, and the highest elevations mark essentially the aggregate thickness of the strata composing them. In the disturbed regions of the east, on the contrary, though we can prove that certain formations of known thickness are included in the mountains, the height of these is never equal to the aggregate amount of the formations. "We thus find that in a country not mountainous, the elevations correspond to the thickness of the strata, while in a mountainous country, where the strata are immensely thicker, the mountain heights bear no comparative proportion to the thickness of the strata. . . . . While the horizontal strata give their whole elevation to the highest parts of the plain, we find the same beds folded and contorted in the mountain region, and giving to the mountian elevations not one sixth of their actual measurement."

Both in the east and west the valleys exhibit the lower strata of the palæozoic series, and it is evident that, had the eastern region been elevated, without folding of the strata, so as to make the base of the series correspond nearly with the sea-level, as in the Mississippi Valley, the mountains exposed between these valleys, and including the whole palæozoic series, would have a height of 40,000 feet; so that the mountains evidently correspond to depressions of the surface, which have carried down the bottom-rocks below the level at which we meet them in the valleys. In other words, the synclinal structure of these mountains depends upon an actual subsidence of the strata along certain lines.

"We have been taught to believe that mountains are produced by upheaval, folding, and plication of the strata, and that, from some unexplained cause, these lines of elevation extend along certain directions, gradually dying out on either side, and subsiding at the extremities. We have, however, here shown that the line of the Appalachian chain is the line of the greatest accumulation of sediments, and that this great mountain-barrier is due to original deposition of materials, and not to any subsequent forces breaking up or disturbing the strata of which it is composed."

We have given Mr. Hall's reasonings on this subject for the most part in his own words, and with some detail, for we conceive that the views which he is here urging are of the highest importance to a correct understanding of the theory of mountains. In the Canadian Naturalist for December, 1859, p. 425, and in the American Journal of Science (2), XXX. 137, will be found an allusion to the rival theories of upheaval and accumulation as applied to volcanic mountains, the discussion between which we conceive to be settled in favor of the latter theory by the reasonings and observations of Constant-Prevost, Scrope, and Lyell. A similar view to the former applied to mountain-chains like those of the Alps, Pyrenees, and Alleghanies, which are made up of aqueous sediments, has been imposed upon the world by the authority of Humboldt, Von Buch, and Elie de Beaumont, with scarcely a protest. Buffon, it is true, when he explained the formation of continents by the slow accumulation of detritus beneath the ocean, conceived that the irregular action of the water would give rise to great banks or ridges of sediments, which when raised above the waves must assume the form of mountains. Later, in 1832, we find De Montlosier protesting against the elevation-hypothesis of Von Buch, and maintaining that the great mountain-chains of Europe are but the remnants of continental elevations which have been cut away by denudation, and that the foldings and inversions to be met with in the structure of mountains are to be looked upon only as local and accidental.

In 1856, Mr. J. P. Lesley published a little volume entitled Coal and its Topography, in the second part of which he has, in a few brilliant and profound chapters, discussed the principles of topographical science with the pen of a master. He there tells us that the mountain lies at the base of all topographical geology. Continents are but congeries of mountains, or rather the latter are but fragments of continents, separated by valleys which represent the absence or removal of mountain-land; and again, "mountains terminate where the rocks thin out."

The arrangement of the sedimentary strata of which mountains are composed may be either horizontal, synclinal, anticlinal, or vertical, but from the greater action of diluvial forces upon anticlinals in disturbed strata it results that great mountain-chains are generally synclinal in their structure, being in fact but fragments of the upper portion of the earth's crust lying in synclinals, and thus preserved from the destruction and translation which have exposed the lower strata in the anticlinal valleys, leaving the intermediate mountains capped with lower strata. The effects of those great and mysterious denuding forces which have so powerfully modified the surface of the globe become less apparent as we approach the equatorial regions, and accordingly we find that in the southern portions of the Appalachian chain many of the anticlinal folds have escaped erosion, and appear as hills of an anticlinal structure. The same thing is occasionally met with farther north; thus Sutton Mountain in eastern Canada, lying between two anticlinal valleys, has an anticlinal centre, with two synclinals on its opposite slopes. Its form appears to result from three anticlinals, the middle one of which has to a great extent escaped denudation.

The error of the prevailing ideas upon the nature of mountain chains may be traced to the notion that a disturbed condition of the rocky strata is not only essential to the structure of a mountain, but an evidence of its having been formed by local upheaval; and the great merit of De Montlosier and Lesley (the latter altogether independently) is to have seen that the upheaval has been in all cases not local but continental, and that the disturbance so often seen in the strata is neither dependent upon elevation nor essential to the formation of a mountain.

. . . . .

Such was the state of the question when Mr. Hall came forward, bringing his great knowledge of the sedimentary formations of North America to bear upon the theory of continents and mountains. These were first advanced in his address delivered before the American Association for the Advancement of Science, as its president, at Montreal, in August, 1857. This address was never published, but the author's views were brought forward in the first volume of his Report on the Geology of Iowa, p. 41, and with more detail in the Introduction to the third volume of his Paleontology of New York, from which we have taken the abstract already given. He has shown that the difference between the geographical features of the eastern and central parts of North America is directly connected with the greater accumulation of sediment along the Appalachians. He has further shown that so far from local elevation being concerned in the formation of these mountains, the strata which form their base are to be found beneath their foundations at a much lower horizon than in the undisturbed hills of the Mississippi Valley, and that to this depression chiefly is due the fact that the mountains of the Appalachian range do not, like those hills, exhibit in their vertical height above the sea the whole accumulated thickness of the palæozoic strata which lie buried beneath their summits. . . . .

The lines of mountain-elevation of De Beaumont are, according to Hall, simply those of original accumulations, which took place along current or shore lines, and have subsequently, by continental elevations, produced mountain-chains. "They were not then due to a later action upon the earth's crust, but the course of the chain and the source of the materials were predetermined by forces in operation long anterior to the existence of the mountains or of the continent of which they form a part." (p. 86.)

It will be seen from what we have said of Buffon, De Montlosier, and Lesley, that many of the views of Mr. Hall are not new, but old; it was, however, reserved to him to complete the theory and give to the world a rational system of orographic geology. He modestly says: "I believe I have controverted no established fact or principle beyond that of denying the influence of local elevating forces, and the intrusion of ancient or plutonic formations beneath the lines of mountains, as ordinarily understood and advocated. In this I believe I am only going back to the views which were long since entertained by geologists relative to continental elevations." (p. 82.)

The nature of the palæozoic sediments of North America clearly shows that they were accumulated during a slow progressive subsidence of the ocean's bed, lasting through the palæozoic period, and this subsidence, which would be greatest along the line of greatest accumulation, was doubtless, as Mr. Hall considers, connected with the transfer of sediment and the variations of local pressure acting upon the yielding crust of the earth, agreeably to the view of Sir John Herschel. This subsidence of the ocean's bottom would, according to Mr. Hall, cause plications in the soft and yielding strata. Lyell, in speculating upon the results of a cooling and contracting sea of molten matter, such as he imagined might have once underlaid the Appalachians, had already suggested that the incumbent flexible strata, collapsing in obedience to gravity, would be forced, if this contraction took place along narrow and parallel zones of country, to fold into a smaller space as they conformed to the circumference of a smaller arc, "thus enabling the force of gravity, though originally exerted vertically, to bend and squeeze the rocks as if they had been subjected to lateral pressure." *

Admitting thus Herschel's theory of subsidence and Lyell's theory of plication, Mr. Hall proceeds to inquire into the great system of foldings presented by the Appalachians. The sinking along the line of greatest accumulation produces a vast synclinal, which is that of the mountain ranges, and the result of such a sinking of flexible beds will be the production within the greater synclinal of numerous smaller synclinal and anticlinal axes, which must gradually decline toward the margin of the great synclinal axis. This process, the author observes, appears to furnish a satisfactory explanation of the difference of slope observed on the two sides of the Appalachian anticlinals, where the dips on one side are uniformly steeper than on the other. (p. 71.)

An important question here arises, which is this: while admitting with Lyell and Hall that parallel foldings may be the result of the subsidence which accompanied the deposition of the Appalachian sediments, we inquire whether the cause is adequate to produce the vast and repeated flexures presented by the Alleghanies. Mr. Billings, in a recent paper in the Canadian Naturalist (Jan., 1860), has endeavored to show that the foldings thus produced must be insignificant when compared with the great undulations of strata; whose origin

* Travels in North America, First Visit, Vol. I. p. 78.

Professor Rogers has endeavored to explain by his theory of earthquake-waves propagated through the igneous fluid mass of the globe, and rolling up the flexible crust. We shall not stop to discuss this theory, but call attention to another agency hitherto overlooked, which must also cause contraction and folding of the strata, and to which we have already elsewhere alluded. (Am. Jour. Sci. (2), XXX. 138.) It is the condensation which must take place when porous sediments are converted into crystalline rocks like gneiss and mica-slate, and still more when the elements of these sediments are changed into minerals of high specific gravity, such as pyroxene, garnet, epidote, staurolite, chiastolite, and chloritoid. This contraction can only take place when the sediments have become deeply buried and are undergoing metamorphism, and is, as many attendant phenomena indicate, connected with a softened and yielding condition of the lower strata.

We have now in this connection to consider the hypothesis which ascribes the corrugation of portions of the earth's crust to the gradual contraction of the interior. An able discussion of this view will be found in the American Journal of Science (2), III. 176, from the pen of Mr. J. D. Dana, who, in common with all others who have hitherto written on the subject, adopts the notion of the igneous fluidity of the earth's interior. We have, however, elsewhere given our reasons for accepting the conclusion of Hopkins and Hennessey that the earth, instead of being a liquid mass covered with a thin crust, is essentially solid to a great depth, if not indeed to the centre, so that the volcanic and igneous phenomena generally ascribed to a fluid nucleus have their seat, as Keferstein and, after him, Sir John Herschel long since suggested, not in the anhydrous solid nucleus, but in the deeply buried layers of aqueous sediments, which, permeated with water, and raised to a high temperature, become reduced to a state of more or less complete igneo-aqueous fusion. So that beneath the outer crust of sediments, and surrounding the solid nucleus, we may suppose a zone of plastic sedimentary material adequate to explain all the phenomena hitherto ascribed to a fluid nucleus. (Quar.

Jour. Geol. Society, Nov., 1859; Canadian Naturalist, Dec., 1859; Amer. Jour. Sci. (2), XXX. 136; and *ante*, page 9.)

This hypothesis, as we have endeavored to show, is not only completely conformable with what we know of the behavior of aqueous sediments impregnated with water and exposed to a high temperature, but offers a ready explanation of all the phenomena of volcanoes and igneous rocks, while avoiding the many difficulties which beset the hypothesis of a nucleus in a state of igneous fluidity. At the same time any changes in volume resulting from the contraction of the nucleus would affect the outer crust through the medium of the more or less plastic zone of sediments, precisely as if the whole interior of the globe were in a liquid state.

The accumulation of a great thickness of sediment along a given line would, by destroying the equilibrium of pressure, cause the somewhat flexible crust to subside; the lower strata becoming altered by the ascending heat of the nucleus would crystallize and contract, and plications would thus be determined parallel to the line of deposition. These foldings, not less than the softening of the bottom strata, establish lines of weakness or of least resistance in the earth's crust, and thus determine the contraction which results from the cooling of the globe to exhibit itself in those regions and along those lines where the ocean's bed is subsiding beneath the accumulating sediments. Hence we conceive that the subsidence invoked by Mr. Hall (and by Lyell), although not the sole nor even the principal cause of the corrugations of the strata, is the one which determines their position and direction, by making the effects produced by the contraction not only of sediments, but of the earth's nucleus itself, to be exerted along the lines of greatest accumulation. . . . .

On the subject of igneous rocks and volcanic phenomena, Mr. Hall insists upon the principles which we were, so far as we know, the first to point out, namely, their connection with great accumulations of sediment, and that of active volcanoes with the newer deposits. We have elsewhere said: "The volcanic phenomena of the day appear, so far as we are aware, to be confined to regions of newer secondary and tertiary deposits, which we may suppose the central heat to be still penetrating (as shown by Mr. Babbage), a process which has long since ceased in the palæozoic regions." * To the accumulation of sediments then we referred both modern volcanoes and ancient plutonic rocks; these latter, like lavas, we regard in all cases as but altered and displaced sediments, for which reason we have called them exotic rocks. (Am. Jour. Sci. (2), XXX. 133.) Mr. Hall reiterates these views, and calls attention moreover to the fact that the greatest outbursts of igneous rock in the various formations appear to be in all cases connected with rapid accumulation over limited areas, causing perhaps disruptions of the crust, through which the semi-fluid stratum may have risen to the surface. He cites in this connection the traps with the palæozoic sandstones of Lake Superior, and with the mesozoic sandstones of Nova Scotia and the Connecticut and Hudson Valleys.

* *Ante*, pp. 9 and 17.

VI.

THE PROBABLE SEAT OF VOLCANIC ACTION.

(1869.)

The following paper was published in the Geological Magazine for June, 1869, and reprinted, with an additional paragraph, in the Am. Jour. Science, from which it is here reproduced. It is, as will be seen, to some extent a reinforcement of the views advanced in Papers I. and II.; but, notwithstanding the repetitions involved, it has been thought proper to reprint it entire for the sake of the context. In further elucidation of the subject I have appended some extracts from a lecture given in April, 1869, before the American Geographical Society in New York, and published in its Proceedings, in which the distribution of volcanic and plutonic phenomena are considered.

THE igneous theory of the earth's crust, which supposes it to have been at one time a fused mass, and to still retain in its interior a great degree of heat, is now generally admitted. In order to explain the origin of eruptive rocks, the phenomena of volcanoes, and the movements of the earth's crust, all of which are conceived by geologists to depend upon the internal heat of the earth, three principal hypotheses have been put forward. Of these the first supposes that in the cooling of the globe a solid crust of no great thickness was formed, which rests upon the still uncongealed nucleus. The second hypothesis, maintained by Hopkins and by Poulett Scrope, supposes solidification to have commenced at the centre of the liquid globe, and to have advanced towards the circumference. Before the last portions became solidified, there was produced, it is conceived, a condition of imperfect liquidity, preventing the sinking of the cooled and heavier particles, and giving rise to a superficial crust, from which solidification would proceed downwards. There would thus be enclosed, between the inner and outer solid parts, a portion of uncongealed matter, which, according to Hopkins, may be supposed still to retain its liquid condition, and to be the seat of volcanic action, whether existing in isolated reservoirs or subterranean lakes; or whether, as suggested by Scrope, forming a continuous sheet surrounding the solid nucleus, whose existence is thus conciliated with the evident facts of a flexible crust, and of liquid ignited matters beneath.

Hopkins, in the discussion of this question, insisted upon the fact, established by his experiments, that pressure favors the solidification of matters which, like rocks, pass in melting to a less dense condition, and hence concludes that the pressure existing at great depths must have induced solidification of the molten mass at a temperature at which, under a less pressure, it would have remained liquid. Mr. Scrope has followed this up by the ingenious suggestion that the great pressure upon parts of the solid igneous mass may become relaxed from the effect of local movements of the earth's crust, causing portions of the solidified matter to pass immediately into the liquid state, thus giving rise to eruptive rocks in regions where all before was solid.*

Similar views have been put forward in a note by Rev. O. Fisher, and in an essay on the formation of mountain-chains, by N. S. Shaler, in the Proceedings of the Boston Society of Natural History, both of which appear in the Geological Magazine for November last. As summed up by Mr. Shaler, the second hypothesis supposes that the earth "consists of an immense solid nucleus, a hardened outer crust, and an intermediate region of comparatively slight depth, in an imperfect state of igneous fusion." In this connection it is curious to remark that, as pointed out by Mr. J. Clifton Ward, in the same Magazine for December (p. 581), Halley was led, from the study of terrestrial magnetism, to a similar hypothesis. He supposed the existence of two magnetic poles situated in the earth's outer crust, and two others in an interior mass, separated from the solid envelope by a fluid medium, and revolving,

* See Scrope On Volcanoes, and his communication to the Geological Magazine for December, 1868.

by a very small degree, slower than the outer crust.* The same conclusion was subsequently adopted by Hansteen.

The formation of a solid layer at the surface of the viscid and nearly congealed mass of the cooling globe, as supposed by the advocates of the second hypothesis, is readily admissible. That this process should commence when the remaining envelope of liquid was yet so deep that the refrigeration from that time to the present has not been sufficient for its entire solidification, is, however, not so probable. Such a crust on the cooling superficial layer would, from the contraction consequent on the further refrigeration of the liquid stratum beneath, become more or less depressed and corrugated, so that there would probably result, as I have elsewhere said, "an irregular diversified surface from the contraction of the congealing mass, which at last formed a liquid bath of no great depth, surrounding the solid nucleus." † Geological phenomena do not, however, in my opinion, afford any evidence of the existence of yet unsolidified portions of the originally liquid material, but are more simply explained by the third hypothesis. This, like the last, supposes the existence of a solid nucleus and of an outer crust, with an interposed layer of partially fluid matter; which is not, however, a still unsolidified portion of the once liquid globe, but consists of the outer part of the congealed primitive mass, disintegrated and modified by chemical and mechanical agencies, impregnated with water, and in a state of igneo-aqueous fusion.

The history of this view forms an interesting chapter in geology. As remarked by Humboldt, a notion that volcanic phenomena have their seat in the sedimentary formations, and are dependent on the combustion of organic substances, belongs

* The elevated temperature of the interior of the globe would probably offer no obstacle to the development of magnetism. In a recent experiment of M. Trève, communicated by M. Faye to the French Academy of Sciences, it was found that molten cast-iron when poured into a mould, surrounded by a helix which was traversed by an electric current, became a strong magnet when liquid at a temperature of 1300° C., and retained its magnetism while cooling. (Comptes Rendus de l'Acad. des Sciences, February, 1869.)

† *Ante*, page 39.

to the infancy of geology. To this period belong the theories of Lémery and Breislak. (Cosmos, V. 443; Otté's translation.) Keferstein, in his Naturgeschichte des Erdkörpers, published in 1834, maintained that all crystalline non-stratified rocks, from granite to lava, are products of the transformation of sedimentary strata, in part very recent, and that there is no well-defined line to be drawn between neptunian and volcanic rocks, since they pass into each other. Volcanic phenomena, according to him, have their origin not in an igneous fluid centre, nor in an oxidizing metallic nucleus (Davy, Daubeny), but in known sedimentary formations, where they are the result of a peculiar kind of fermentation, which crystallizes and arranges in new forms the elements of the sedimentary strata, with an evolution of heat as a result of the chemical process. (Naturgeschichte, Vol. I. p. 109; also Bull. Soc. Geol. de France (1), Vol. VII. p. 197.) In commenting upon these views (Am. Jour. Science, July, 1860), I have remarked that, by ignoring the incandescent nucleus as a source of heat, Keferstein has excluded the true exciting cause of the chemical changes which take place in the buried sediments. The notion of a subterranean combustion or fermentation, as a source of heat, is to be rejected as irrational.

A view identical with that of Keferstein, as to the seat of volcanic phenomena, was soon after put forth by Sir John Herschel, in a letter to Sir Charles Lyell, in 1836. (Proc. Geol. Soc. London, II. 548.) Starting from the suggestions of Scrope and Babbage, that the isothermal horizons in the earth's crust must rise as a consequence of the accumulation of sediments, he insisted that deeply buried strata will thus become crystallized by heat, and may eventually, with their included water, be raised to the melting point, by which process gases would be generated, and earthquakes and volcanic eruptions follow. At the same time the mechanical disturbance of the equilibrium of pressure, consequent upon a transfer of sediments while the yielding surface reposes on matters partly liquefied, will explain the movements of elevation and subsidence of the earth's crust. Herschel was probably ignorant of the extent to which his views had been anticipated by Keferstein; and the suggestions of the one and the other seemed to have passed unnoticed by geologists until, in March, 1858, I reproduced them in a paper read before the Canadian Institute (Toronto), being at that time acquainted with Herschel's letter, but not having met with the writings of Keferstein. I there considered the reaction which would take place under the influence of a high temperature in sediments permeated with water, and containing, besides silicious and aluminous matters, carbonates, sulphates, chlorides and carbonaceous substances. From these, it was shown, might be produced all the gaseous emanations of volcanic districts, while from aqueo-igneous fusion of the various admixtures might result the great variety of eruptive rocks. To quote the words of my paper just referred to: "We conceive that the earth's solid crust of anhydrous and primitive igneous rock is everywhere deeply concealed beneath its own ruins, which form a great mass of sedimentary strata, permeated by water. As heat from beneath invades these sediments, it produces in them that change which constitutes normal metamorphism. These rocks, at a sufficient depth, are necessarily in a state of igneo-aqueous fusion; and in the event of fracture in the overlying strata, may rise among them, taking the form of eruptive rocks. When the nature of the sediments is such as to generate great amounts of elastic fluids by their fusion, earthquakes and volcanic eruptions may result, and these — other things being equal — will be most likely to occur under the more recent formations." (Canadian Journal, May, 1858, Vol. III. p. 207; and *ante*, page 9.)

The same views are insisted upon in a paper On some Points in Chemical Geology (Quar. Jour. Geol. Soc., London, Nov., 1859, Vol. XV. p. 594), and have since been repeatedly put forward by me, with further explanations as to what I have designated above, *the ruins of the crust of anhydrous and primitive igneous rock*. This, it is conceived, must, by contraction in cooling, have become porous and permeable, for a considerable depth, to the waters afterwards precipitated upon its surface. In this way it was prepared alike for mechanical disintegration, and for the chemical action of the acids, which, as shown in the two papers just referred to, must have been present in the air and the waters of the time. It is, moreover, not improbable that a yet unsolidified sheet of molten matter may then have existed beneath the earth's crust, and may have intervened in the volcanic phenomena of that early period, contributing, by its extravasation, to swell the vast amount of mineral matter then brought within aqueous and atmospheric influences. The earth, air, and water thus made to react upon each other, constitute the first matter from which, by mechanical and chemical transformations, the whole mineral world known to us has been produced.

It is the lower portions of this great disintegrated and water-impregnated mass which form, according to the present hypothesis, the semi-liquid layer supposed to intervene between the outer solid crust and the inner solid and anhydrous nucleus. In order to obtain a correct notion of the condition of this mass, both in earlier and later times, two points must be especially considered, — the relation of temperature to depth, and that of solubility to pressure. It being conceded that the increase of temperature in descending in the earth's crust is due to the transmission and escape of heat from the interior, Mr. Hopkins showed mathematically that there exists a constant proportion between the effect of internal heat at the surface and the rate at which the temperature increases in descending. Thus, at the present time, while the mean temperature at the earth's surface is augmented only about one twentieth of a degree Fahrenheit, by the escape of heat from below, the increase is found to be equal to about one degree for each sixty feet in depth. If, however, we go back to a period in the history of our globe when the heat passing upwards through its crust was sufficient to raise the superficial temperature twenty times as much as at present, that is to say, one degree of Fahrenheit, the augmentation of heat in descending would be twenty times as great as now, or one degree for each three feet in depth. (Geol. Journal, VIII. 59.) The conclusion is inevitable that a condition of things must have existed during long periods in the history of the cooling globe when the accumulation of comparatively thin layers of sediment would have been sufficient to give rise to all the phenomena of metamorphism, vulcanicity, and movements of the crust, whose origin Herschel has so well explained.

Coming, in the next place, to consider the influence of pressure upon the buried materials derived from the mechanical and chemical disintegration of the primitive crust, we find that, by the presence of heated water throughout them, they are placed under conditions very unlike those of the original cooling mass. While pressure raises the fusing point of such bodies as expand in passing into the liquid state, it depresses that point for those which, like ice, contract in becoming liquid. The same principle extends to that liquefaction which constitutes solution; where, as is with few exceptions the case, the process is attended with condensation or diminution of volume, pressure will, as shown by the experiments of Sorby, augment the solvent power of the liquid.* Under the influence of the elevated temperature and the great pressure which prevail at considerable depths, sediments should, therefore, by the effect of the water which they contain, acquire a certain degree of liquidity; rendering not improbable the suggestion of Scheerer, that the presence of five or ten per cent of water may suffice, at temperatures approaching redness, to give to a granitic mass a liquidity partaking at once of the character of an igneous and an aqueous fusion. The studies by Mr. Sorby of the cavities in crystals have led him to conclude that the constituents of granitic and trachytic rocks have crystallized in the presence of liquid water, under great pressure, at temperatures not above redness, and consequently very far below that required for simple igneous fusion. The intervention of water in giving liquidity to lavas has, in fact, long been taught by Scrope, and, notwithstanding the opposition of plutonists like Durocher, Fournet, and Rivière, is now very generally admitted. In this connection, the reader is referred to the Geological Magazine for February, 1868, page 57, where the history of this question is discussed.

* Sorby, Bakerian Lecture, Royal Society, 1863.

It may here be remarked, that if we regard the liquefaction of heated rocks under great pressure, and in presence of water, as a process of solution rather than of fusion, it would follow that diminution of pressure, as supposed by Mr. Scrope, would cause, not liquefaction, but the reverse. The mechanical pressure of great accumulations of sediment is to be regarded as co-operating with heat to augment the solvent action of the water, and as being thus one of the efficient causes of the liquefaction of deeply buried sedimentary rocks.

That water intervenes not only in the phenomena of volcanic eruptions, but in the crystallization of the minerals of eruptive rocks, which have been formed at temperatures far below that of igneous fusion, is a fact not easily reconciled with either the first or the second hypothesis of volcanic action, but is in perfect accordance with the one here maintained, which is also strongly supported by the study of the chemical composition of igneous rocks. These are generally referred to two great divisions, corresponding to what have been designated the trachytic and pyroxenic types, and to account for their origin, a separation of a liquid igneous mass beneath the earth's crust into two layers of acid and basic silicates was imagined by Phillips, Durocher, and Bunsen. The latter, as is well known, has calculated the normal composition of these supposed trachytic and pyroxenic magmas, and conceives that from them, either separately or by admixture, the various eruptive rocks are derived; so that the amounts of alumina, lime, magnesia, and alkalies sustain a constant relation to the silica in the rock. If, however, we examine the analyses of the eruptive rocks in Hungary and Armenia, made by Streng, and put forward in support of this view, there will be found such discrepancies between the actual and the calculated results as to throw grave doubts on Bunsen's hypothesis.

Two things become apparent from a study of the chemical nature of eruptive rocks: first, that their composition presents such variations as are irreconcilable with the simple origin generally assigned to them; and, second, that it is similar to that of sedimentary rocks whose history and origin it is, in most cases, not difficult to trace. I have elsewhere pointed out how the natural operation of mechanical and chemical agencies tends to produce among sediments a separation into two classes, corresponding to the two great divisions above noticed. From the mode of their accumulation, however, great variations must exist in the composition of the sediments, corresponding to many of the varieties presented by eruptive rocks. The careful study of stratified rocks of aqueous origin discloses, in addition to these, the existence of deposits of basic silicates of peculiar types. Some of these are in great part magnesian, others consist of compounds like anorthite and labradorite, highly aluminous basic silicates in which lime and soda enter, to the almost complete exclusion of magnesia and other bases; while in the masses of pinite or agalmatolite-rock we have a similar aluminous silicate, in which lime and magnesia are wanting, and potash is the predominant alkali. In such sediments as these just enumerated we find the representatives of eruptive rocks like peridotite, phonolite, leucitophyre, and similar rocks, which are so many exceptions in the basic group of Bunsen. As, however, they are represented in the sediments of the earth's crust, their appearance as exotic rocks, consequent upon a softening and extravasation of the more easily liquefiable strata of deeply buried formations, is readily and simply explained.

APPENDIX.

DISTRIBUTION OF VOLCANOES.

WE regard the extravasation of igneous matter, whether as lava or ashes at the surface, or as plutonic rock in the midst of strata, as, in its wider sense, a manifestation of vulcanicity; and, for the elucidation of our subject, consider both those regions characterized by great outbursts of plutonic rock in former geologic periods, and those now the seats of volcanic activity, which, in these cases, can generally be traced back some distance into the tertiary age. To begin with the latter, the first and most important is the great continental region which may be described as including the Mediterranean and Aralo-Caspian basins, extending from the Iberian peninsula eastward to the Thian-Chan Mountains of central Asia. In this great belt, extending over about ninety degrees of longitude, are included all the historic volcanoes of the ancient world, to which we must add the extinct volcanoes of Murcia, Catalonia, Auvergne, the Vivarais, the Eifel, Hungary, etc., some of which have probably been active during the human period.

It is a most significant fact that this region is nearly coextensive with that occupied for ages by the great civilizing races of the world. From the plateau of central Asia, throughout their westward migration to the pillars of Hercules, the Indo-European nations were familiar with the volcano and the earthquake ; and that the Semitic race were not strangers to the same phenomena, the whole poetic imagery of the Hebrew Scriptures bears ample evidence. In the language of their writers, the mountains are molten ; they quake and fall down at the presence of the Deity, when the melting fire burneth. The fury of his wrath is poured forth like fire ; he toucheth the hills and they smoke ; while fire and sulphur come down to destroy the doomed cities of the plain, whose foundation is a molten flood. Not less does the poetry and the mythology of Greece and of Rome bear the impress of that nether realm of fire in which the volcano and the earthquake have their seat. The influence of these is conspicuous throughout the imaginative literature and the religious systems of the Indo-European nations, whose contact with terrible manifestations of unseen forces beyond their foresight or control could not fail to act strongly on their moral and intellectual development ; which would have doubtless presented very different phases had the early home of these races been in Australia or on the eastern side of the American continent, where volcanoes are unknown, and earthquakes are scarcely felt.

Besides the great region just indicated, must be mentioned that of our own Pacific slope, from Fuegia to Alaska, whence, along the eastern shore of Asia, a line of volcanic activity extends to the burning mountains of the Indian archipelago. Volcanic islands are widely scattered over the Pacific basin, and volcanoes are conspicuous in the Antarctic continent. The Atlantic area is in like manner marked by volcanic islands from Jan Mayen and Iceland to the Canaries, the Azores, and the Caribbean Islands, and southward to Ascension, St. Helena, and Tristan d'Acunha.

. . . . .

If we look at the North American continent, we find along its northeastern portion evidences of great subsidence, and an accumulation of not less than 40,000 feet of sediment along the line of the Appalachians from the Gulf of St. Lawrence southwards, during the palæozoic period. This region is precisely that characterized by considerable eruptions of plutonic rocks during this period and for some time after its close. To the westward of the Appalachians, the deposits of palæozoic sediments were much thinner, and in the Mississippi valley are probably less than 4,000 feet in thickness. Conformably with this, there are no traces of plutonic or volcanic outbursts from the northeast region just mentioned throughout this vast palæozoic basin, with the exception of the shores of Lake Superior, where we find the early portion of the palæozoic age marked by a great accumulation of sediments, comparable to that occurring at the same time in the region of New England, and followed or accompanied by similar plutonic phenomena. Across the plains of northern Russia and Scandinavia, as in the Mississippi valley, the palæozoic period was represented by not more than 2,000 feet of sediments, which still lie undisturbed, while in the British Islands 50,000 feet of palæozoic strata, contorted and accompanied by igneous rocks, attest the connection between great accumulation and plutonic phenomena.

Coming now to modern volcanoes, we find them in their greatest activity in oceanic regions where subsidence and accumulation are still going on. Of the two continental regions already pointed out, that along the Mediterranean basin is marked by an accumulation of mesozoic and tertiary sediments, 20,000 feet or more in thickness. It is evident that the great mountain zone which includes the Pyrenees, the Alps, the Caucasus, and the Himalaya was, during the later secondary and tertiary periods, a basin in which vast depositions were taking place, as in the Appalachian belt during the palæozoic times. Turning to the other continental region, the American Pacific slope, similar evidences of great accumulations during the same periods are found throughout its whole extent, showing that the great Pacific mountain-belt of North and South America, with its attendant volcanoes, is, in the main, the geological equivalent or counterpart of the great east and west belt of the eastern world. (Proceedings of the American Geographical Society, April, 1869.)

VII.

ON SOME POINTS IN DYNAMICAL GEOLOGY.

The following paper, which appeared in the American Journal of Science for April, 1873, may be read as a supplement to Essays I., II., V., and VI. in the present volume. Nearly all of the views which I have maintained since 1858-1861 in my endeavors to reconstruct dynamical geology on a new basis, as set forth in the essays just referred to, have of late been appropriated, without recognition, perhaps unconsciously, by LeConte, Mallet, and others; and therefore some assertion of priority on my part seemed not out of place. The reader may also consult in this connection Professor J. D. Dana's essay on The Results of the Earth's Contraction on Cooling, in the American Journal of Science for June-September, 1873, and further a note in the same Journal for November, 1873 (page 381). containing his acknowledgment of my claims to priority on important points which he had denied me in the essay in question.

In his late essay on The Formation of the Features of the Earth's Crust, in the American Journal of Science for November and December, 1872, Professor Joseph LeConte has discussed a wide range of subjects in geological dynamics, in a manner for which the student cannot but be grateful. After a consideration of the arguments with regard to the nature of the earth's interior, he arrives at the conclusion that "the whole theory of igneous agencies — which is little less than *the whole foundation of theoretic geology — must be reconstructed on the basis of a solid earth*"; a conclusion which forms the starting-point of his essay. It is here to be noted, that the late William Hopkins, to whom we owe one of our great arguments in favor of a solid globe, did not take this ground, but sought to explain the phenomena of igneous action by the hypothesis of portions of matter still remaining unsolidified at no great depth between the solid nucleus and the superficial crust. Dissenting from this view, though accepting the general conclusions of Hopkins and others as to a solid globe, I have been endeavoring, since 1858, to reconstruct, in the language of Professor LeConte, "*the theory of igneous agencies on the basis of a solid earth.*" Alone up to this time, so far as I am aware, I have labored to expand, complete, and give geological and chemical consistency to the suggestion long since put forth, both by Keferstein and by Sir John Herschel, that the deeply buried and water-impregnated strata between the superficial crust of the earth and the solid nucleus constitute a region "of plastic material adequate to explain all the phenomena hitherto ascribed to a fluid nucleus," since "any changes in volume resulting from the contraction of the (solid) nucleus would affect the outer crust through the medium of the more or less plastic zone of sediments precisely as if the whole interior of the globe were liquid."

A softening by heat of previously solid porous sediments, filled with water, was maintained (in accordance with the views of Babbage as to the rise of the isogeothermal horizons from the deposition of newer strata) to depend upon the accumulation of large thicknesses of sediment, the results of which heat and softening were declared by me to offer a "*ready explanation of all the phenomena of volcanoes and igneous rocks.*" This relation of volcanic phenomena to great acccumulation, and of those of recent times to more modern sedimentary deposits, which was also maintained by me, was subsequently insisted upon and enforced by Professor James Hall in the introduction to the third volume of the Paleontology of New York. A summing up of these views as put forth by me in March, 1858, and in November, 1859, will be found in the American Journal of Science for May, 1861. (See Essays I., II., and V. of the present volume.) In this last it was shown, in opposition to the notion of Babbage (who had speculated upon the *expansion* and consequent *elevation* of the deeply buried strata by heat), that one of the effects of heat and water upon the buried sediments would be *condensation*, from the diminution of porosity and still more from the conversion of the earthy materials into crystalline species of higher specific gravity, thus causing *con-*

traction of the mass. A further and very important result of this accumulation there pointed out was by the softening of the underlying floor, or *the "bottom strata to establish lines of weakness or of least resistance in the earth's crust, and thus determine the contraction which results from the cooling of the globe to exhibit itself in those regions, and along those lines where the ocean's bed is subsiding beneath the accumulated sediments.*" Hence, I added, "We conceive the subsidence invoked by Mr. Hall, though not the sole nor even the principal cause of the corrugations of the earth's strata, is the one which determines their position and direction by making the effects produced by the contraction, not only of sediments but of the earth's nucleus itself, to be exerted along the lines of greatest accumulation." (*Ante*, page 57.) As further results of this process of accumulation, I also asserted "the metamorphism of sediments *in situ*, their displacement in a pasty condition from igneo-aqueous fusion as plutonic rocks, and their ejection as lavas, with attendant gases and vapors." (*Ante*, page 16.)

With these conclusions, enunciated in 1858 – 1861, we may compare those arrived at by Professor LeConte in his recent essay, where he recognizes as consequences of the heating of great thicknesses of sediments accumulated along the shores of a continent, a process of *condensation* in the lower strata, resulting in "contraction and subsidence *pari passu* with the deposit," followed by "aqueo-igneous softening or even melting, not only of the lower portion of the sediments themselves, but of *the underlying strata upon which they were deposited;* the subsidence probably continues during this process. Finally, this *softening determines a line of yielding to horizontal pressure*, and a consequent upswelling of the line into a chain. Thus are accounted for, first, the *subsidence*, then the *subsequent upheaval*, and also the *metamorphism* of the lower strata." Beneath every great line of sediments there will moreover be found, according to him, a reservoir of sedimentary material in a state of more or less complete fusion, in which volcanic phenomena have their seat. The reader cannot fail to see that these views are identical with those which I have so long advocated.

The views of Professor James Hall, as to the relation between great accumulations of strata and mountain-elevations, are cited with approval by LeConte, who, following him, asserts that "mountain-chains are masses of immensely thick sediments." I venture, however, to remark in this connection, that the views both of Mr. Hall and of myself, as his expounder, have as yet been but imperfectly understood either by LeConte or our other critics. Thus they have been defined as "a theory of mountains with the origin of mountains left out"; while LeConte says, "Hall and Hunt leave the sediments just after the whole preparation has been made, but before the actual mountain-formation has taken place." Now, in fact, so far as I am aware, neither Hall nor yet myself in my exposition of his views (*ante*, page 49) has proposed any theory to explain this latter part of the process, that is to say, the uplifting of the deposited sediments which LeConte calls "the actual mountain-formation." Hall's contribution to the problem, which, as our author well says, forms "an era in geological science," was to show the relation between mountain-chains and great accumulations of sediments; to illustrate this by the history of the palæozoic rocks of North America; and moreover to protest against the generally received theory that mountain elevations are due to local upthrusts; he, to use his own language, "going back to the theories long since entertained by geologists relative to continental elevations." That mountains were the remnants of eroded continental areas had already been taught by Lesley, and long before by Buffon and De Montlosier. It was left for Hall, through a new way, to lead us back to these views; but the whole theory of the cause of continental elevations was left by him where he found it. In my exposition of his views, I have only endeavored, in addition, to show in what manner a contracting globe and a solid nucleus may be related to the great facts of local subsidence and accumulation.

I shall not attempt to follow LeConte in his objections to the views of Dana and Whitney with regard to the uplifting of mountains, but proceed to notice briefly his own, according to which the horizontal thrust resulting from the slow contraction of the nucleus is brought, in the manner which I long since explained, to act upon the great accumulations of sediment, so that they are "crushed together horizontally and swelled up vertically," thus producing not only plications and slaty cleavage, but an amount of vertical extension "*fully adequate to account for the upheaval of the greatest mountain-chains*"; the ridges, peaks, gorges, and valleys of mountain-regions being, however, the results of subsequent erosion. This theory of the plications of strata, and their relations both to great accumulations and to a contracting nucleus, is fully set forth in my paper of 1861, already quoted; where I have also insisted upon the results of "the lateral pressure brought to bear upon the strata in an elongated basin (of subsidence) by the contraction of the globe."

But while admitting that the process here described must cause elevations of the compressed strata, it must be said that it fails to solve the problem of the uplifting of mountain-regions, the strata of which have, in many cases, undergone neither folding nor lateral compression, but are nearly or quite horizontal. Foldings, contortions, and slaty cleavage, though met with in many mountain-regions, are, in fact, accidents which are to be left out of view in considering the origin of mountains. The student of physical geography may learn from the great elevated plateaus of the globe the truth of De Montlosier's statement, that the great mountain-chains of Europe are but the remains of continental elevations which have been cut away by denudation, and that the foldings and inversions to be met with in the structure of mountains are to be looked upon only as local and accidental. (*Ante*, page 52.) In a similar spirit Jukes remarks that we learn "how completely the present surface of the earth is a sculptured surface carved out by denudation, and how little, as a rule, it is effected by the dislocations, upheavals and convolutions of the rocks beneath it." (Manual of Geology, 3d ed., 449.) In the case of the uplifted palæozoic basin of eastern North America, as Hall has well shown, the process of elevation was the same for the thicker and corrugated sediments of the eastern portion and for the thinner and undisturbed strata of the valley of the upper Mississippi. The hills in the latter region, built up of 1,000 feet of horizontal beds, having the Potsdam sandstone at the base, and capped by the Niagara limestone, show us the production of mountains by erosion, uncomplicated by the accidents which make their study more difficult in regions where contortion of the strata has supervened. Hall has also noted in this connection the nearly horizontal strata of the Catskill Mountains.

The question of the structure and the origin of the Appalachians has been complicated by the assumption that the crystalline strata which constitute their higher portions are altered sediments of palæozoic age, rather than parts of an ancient continent of eozoic rocks which formed the eastern border of the palæozoic sea, corresponding to the Rocky Mountains on the west. The former view has been very generally held by American geologists, and was maintained by the present writer until 1870, when he endeavored to show that the crystalline rocks of New England, and their lithological representatives both to the southwest and the northwest, are of pre-palæozoic age, and in part Laurentian. (Amer. Jour. Science (2), L. 83; also Address before the American Association, Indianapolis, 1871, Paper XIII. of this volume.) This view, already before maintained by Credner and by Emmons, is now advocated by LeConte, who conceives that the gneissic region of the Atlantic slope is Laurentian, and was probably "land during the palæozoic times," constituting an "eastern continental mass," which, judging from the immense thickness of sediments in the eastern parts of the palæozoic area, must have been of great extent. A similar view was put forward by H. D. Rogers in 1842, and again by James Hall in 1859, when, after describing these palæozoic sediments, he said, "We may have had a coast-line nearly parallel to and coextensive with the Appalachians" (Paleontology, Vol. III. p. 96, note); commenting upon which, in 1861, I asserted that these coarse sediments "were evidently derived from a wasting continent." In a paper read before the American Geographical Society,

New York, November 12, 1872, I adduced a further argument in favor of such a pre-palæozoic continent to the eastward, from the climatic conditions of great dryness which gave rise in the palæozoic region of North America to deposits of salt, gypsum, and dolomite over considerable areas from Nova Scotia to Michigan and Ohio, and from the time of the Calciferous formation to the Lower Carboniferous. (Engineering and Mining Journal, January 14 and 23, 1873.)

In concluding his essay, Professor LeConte declares that an important problem in geological dynamics remains, in his opinion, unsolved, namely, the cause of those "great and wide-spread *oscillations* which have marked the great divisions of *time*, and have left their impress in the general unconformability of the strata"; the last being that of the post-pliocene period. Now, it is precisely the *upward* movements of this kind which constitute the continental elevations of De Montlosier, Hall, and myself, giving rise to plateaus, and by the partial erosion and denudation of these to mountains. The cause of these continental elevations was not discussed by Hall, and is by LeConte declared to be unexplained; while such is the case, "the actual mountain-formation," to use his words, is still unaccounted for. That these gentle and widespread movements of oscillation are, however, in some way not yet clearly explained, connected with the contracting of the nucleus and the consequent conforming thereto of the envelope, we can scarcely doubt; or that the latter, from its nature and origin, must present great differences in constitution and in flexibility in its various parts. From this it might be expected that the movements imparted to the envelope alike by the process of secular cooling and contraction of the nucleus, and by the disturbance of the equilibrium of pressure consequent on the processes of erosion and sedimentation, would give rise to seemingly irregular oscillations, resulting in the depression or the elevation of considerable areas, constituting continental movements.

The grave question here arises as to whether the heat which plays such an important part in the phenomena under consideration is a cause or an effect of the activities beneath the earth's surface. Starting from the notion of an igneous centre, Babbage and Herschel adopted the first view, in which I have followed them, maintaining that the heat from a yet uncooled nucleus is the efficient cause of all igneous and volcanic manifestations. According to Keferstein, on the other hand, the hypothesis of an incandescent nucleus is unnecessary, and the internal heat results from what he called a fermentation among the deeply buried sedimentary layers. A view which unites these two is proposed by LeConte, who suggests that heat from a central source invading the buried sediments may there excite chemical action, which will, in its turn, evolve heat, and thus greatly augment their temperature. It is, however, I think, probable that any chemical processes which may be set up in the buried sediments for their conversion into igneous rocks and volcanic products would absorb rather than generate heat.

In his remarkable study of the Secular Cooling of the Earth (Trans. Royal Soc. Edinb., XXIII. pt. 1, p. 157), Sir William Thomson arrives at the conclusion that the observed mean rate of increase in descending in the earth's crust will continue with but little variation for 100,000 feet, but will gradually diminish at greater depths, from an increase of conducting power. Estimating with him the rate of increase at one degree of Fahrenheit for fifty-one feet, it would require the depth just named, or about nineteen miles, to give a temperature of 2,000° F., which may be supposed sufficient to produce the chemical actions required. But it is probable that the seat of volcanic activity may be much less profound than above supposed, in which case the central heat would be inadequate. *Chemical* action, as suggested by Keferstein and LeConte, being rejected as a source of heat, there however remained the hypothesis that thermal effects might result from local *physical* causes, and that the immense mechanical force which is exerted in the movements of the earth's crust might be converted into heat. This view was, so far as I am aware, first suggested by George L. Vose, whose review of Orographic Geology, a very valuable contribution to the literature of the subject, was published in 1866. In it, while recognizing with Sorby the conversion of mechanical force into chemical action, he insists that "the enormous pressure generated in the folding of masses of rocks the depth of which is measured by miles" is an agent potent to produce changes both mechanical and chemical. The causes of the conversion of sediments into plutonic rocks like granite, he conceives to be "*mechanical compression, with the heat and chemical action* which proceed therefrom," and adds in a note, alluding to the view which explains their conversion by the action of heat from beneath, "we should prefer to get the heat needed by the compression which accompanies the disturbance of the strata where metamorphism occurs." (Orographic Geology, pp. 129, 130.)* This suggestion of Vose is sustained by the late researches of Robert Mallet, who concludes that, "as the solid crust sinks together to follow down the shrinking nucleus, the *work* expended in mutual crushing and dislocation of its parts is *transformed into heat*, by which, at the places where the crushing sufficiently takes place, the material of the rock so crushed and that adjacent to it are heated even to fusion. The access of water at such points determines volcanic eruption." (American Journal of Science, III. iv, 411.) To this it may be added, that, inasmuch as the crushing process takes place in strata which, from their depth, are already at an elevated temperature, the heat developed by the mechanical process comes in to supplement that derived by conduction from the igneous centre.

* It was not until after the publication of this paper that I became aware that Professor Henry Wurtz had previously enunciated the view suggested by Vose and adopted and applied by Mallet. In a paper read before the American Association for the Advancement of Science, at Buffalo, in August, 1866, and published in the American Journal of Mining for January 25, 1868, under the title of Gold-Genetic Metamorphism, etc., Professor Wurtz concludes that "the tremendous dynamic agencies whose effects of upheaval, subsidence, disruption and displacement we find so widely manifest, while doubtless themselves engendered of the pent-up heat-energy of the interior, must have given birth to or been in part transmuted into heat-motion," and proceeds to say that in the heat which must be evolved in these movements we may find an explanation of metamorphic changes, and of "thermal springs and many like phenomena."

Moreover, these strata already include, not only water, but the compounds of chlorine, sulphur, and carbon necessary for the generation of the various gases which are the frequent accompaniments of volcanic eruptions.* With the contributions of Vose and Mallet, the theory of volcanic action advocated by Keferstein, Herschel and myself would seem to be wellnigh complete.

* That the view so fully set forth in papers I. and II. of the present volume, in the years 1858 and 1859, as to the origin of volcanic products, is the one now adopted by Mallet, appears from the following extract of a letter by him in Nature, for March 20, 1873: "There is just that general similarity in the constitution of volcanic products which we should expect to result from the heating or the fusion together of the beds, more or less silicious, aluminous, magnesian, or calcareous, mixed with metallic oxides or other compounds, and often with carbon, boron, and other elements, all variously superposed or mixed, which constitute the known crust of the earth."

VIII.

ON LIMESTONES, DOLOMITES AND GYPSUMS.

(1858 - 1866.)

The results of the author's researches on the chemistry of the salts of lime and magnesia, undertaken with reference to the theory of mineral-waters and the origin of calcareous and magnesian rocks, were first announced in the American Journal of Science for July, 1858, and subsequently more at length in an essay in that journal for September and November, 1859. This paper, which extended over thirty-six pages, was divided into five parts, of which the first treats of the action of solutions of bicarbonate of soda on the soluble salts of lime and magnesia; the second on the reactions between solutions of bicarbonate of lime and the sulphates of soda and magnesia; the third describes the production of the double carbonate of lime and magnesia (dolomite); the fourth discusses various facts in the history of gypsums, dolomites, magnesites, and limestones; and the fifth treats of the mode of formation of these rocks. The continuation of the subject in the same journal for July, 1866, occupies nineteen pages, and includes researches on the hydrated double carbonates of lime and magnesia, on supersaturated solutions of these two carbonates, and on the alleged decomposition of gypsum by dolomite, besides further experiments on the artificial production of dolomite.

Allusions to some of the results obtained are made in paper IV., and many more of the results are embodied in the one on Natural Waters (IX.), while in XIII. will be found a brief summary of the results so far as the origin of dolomites and magnesian limestones is concerned. I have thought it well to reproduce in the present collection some few selections from the fifth part of the essay of 1859, and to preface them by a translation of parts of a letter written to Elie de Beaumont and printed in the Comptes Rendus of the French Academy of Science for June 9, 1862, and subsequently in the Canadian Naturalist.

FROM THE COMPTES RENDUS OF THE FRENCH ACADEMY OF SCIENCES, JUNE 9, 1862.

On the 28th of October, 1844, a memoir was deposited with the Academy by the illustrious Cordier. Being in a sealed packet, its contents remained unknown until after his death, when, at the request of his widow, the seal was broken, and the paper, which bears the date of October 22, 1844, was first made public in the Comptes Rendus of the Academy for February 17, 1862. In this remarkable memoir, which has for its title, On the Origin of the Calcareous Rocks which do not belong to the Primordial Crust (*De l'origine des roches calcaires,* etc.), the author gives his views upon the formation of limestone and dolomite. He rejects Von Buch's theory of dolomitization, which still finds some supporters, and which supposes that the magnesia was introduced subsequent to the deposition of the sediments, by a "certain mysterious action of intrusive pyroxenic rocks" which have been ejceted in the vicinity of deposits of pure limestone. Mr. Cordier also combats the idea that these last have been formed entirely of the débris of testacea and zoöphytes, which, according to him, form but a small proportion of limestone-formations. Going back still further, he finds the source alike of the carbonate of lime of these organic remains, and of the great mass of calcareous rocks, in certain chemical reactions. The pure limestones, according to him, pass into magnesian limestones by an admixture of dolomite, and form thus a transition to the pure dolomites, so that we must admit a common origin for all these rocks. The source of the two carbonates which compose them, according to Mr. Cordier, is to be found in the reaction of carbonate of soda upon the chlorides of calcium and magnesium in sea-water; the carbonate of soda being derived from the decomposition of feldspars, from alkaline springs, and from plutonic emanations. This alkaline salt, reacting upon the salts of sea-water, would give rise to chloride of sodium and carbonate of lime, and, under certain conditions, to calcareo-magnesian precipitates. From this reaction must result a secular variation in the composition of the ocean, which corresponds to the progressive changes in the marine fauna of successive geological epochs.

Such is the theory of Mr. Cordier, which is now published, for the first time, in 1862; and which I have thus noticed in order to call the attention of the Academy to my own published papers, in which I have maintained similar views for the last four years. [See, for the origin of carbonate of lime, the first paper in the present volume, an abstract of which was given in the letter of which this is a part.]

In the American Journal of Science for 1859 will be found the results of a series of investigations of the reactions of solutions of bicarbonate of soda with sea-water, and upon the conditions required for the precipitation of carbonate of magnesia and the formation of dolomite. I have there also shown the mutual decomposition, at ordinary temperatures, of solutions of bicarbonate of lime and sulphate of magnesia, resulting in the formation of gypsum and of a soluble bicarbonate of magnesia, which becomes the source of dolomite or of magnesite. A notice of the first part of these researches will be found in the Comptes Rendus for May 23, 1859. In the continuation of them, as cited above, it is shown that the association of magnesian and pure limestones establishes the fact that these rocks have both been deposited as sediments, and that the hypothesis which explains the origin of dolomites by a subsequent alteration of pure limestones is inadmissible. It is also shown that great portions of limestone, even in fossiliferous formations, have the characters of precipitates resulting from chemical reactions, and have never formed part of organized beings; which last, moreover, owe their carbonate of lime to similar reactions.

My views upon the composition of the primitive ocean were further supported by the analyses of numerous saline waters from lower palæozoic limestones. In these waters, the bases of which are almost wholly in the condition of chlorides, about one half of the chlorine is combined with sodium, and the other half is nearly equally divided between calcium and magnesium.

The Academy will perceive, from the short analysis above given, the extent and the importance of my generalizations, with which the ideas of Mr. Cordier are, for the most part, in perfect accordance. It will further be observed, that the publication of Mr. Leymerie, in which similar views are, to a certain extent, indicated (see the Comptes Rendus of March 10, 1862), dates only from 1861, while my own papers appeared in 1858 and 1859.

My researches upon the origin of dolomites and limestones fully justify the previsions of Mr. Cordier. He, however, in his theory, excepted the limestones of primitive formations, but these are regarded by modern geologists as also sedimentary formations, and consequently offer no exception to the general view. The different sources of carbonate of soda indicated by Mr. Cordier may moreover be reduced to a single one, inasmuch as both the salts of alkaline springs, and those of what he calls plutonic emanations, are probably derived from the decomposing feldspathic minerals of sedimentary rocks. The argillaceous rocks, deprived of a large proportion of the alkali which they once contained in the form of feldspars, are the equivalents of the limestones which have been formed at the expense of the chloride of calcium of the primitive ocean.

EXTRACTS FROM A PAPER IN THE AMERICAN JOURNAL OF SCIENCE FOR NOVEMBER, 1859, ON THE SALTS OF LIME AND MAGNESIA AND THE FORMATION OF GYPSUMS AND MAGNESIAN ROCKS.

The theory of the formation of magnesian sediments will be readily understood from the experiments which have been described in the earlier parts of this paper; but before proceeding to its consideration I wish to call attention to the results of the concentration by evaporation of natural waters in basins without an outlet. If such a basin contain sea-water, the gypsum, being insoluble in a saturated brine, will be entirely deposited before the crystallization of the sea-salt, and there will remain a liquid containing no lime-salts, but chlorides of sodium and magnesium, with a large amount of sulphate of magnesia. Such are the waters of Lake Elton and many of the brine-pools of the Russian steppes; while on the contrary the saturated brines of the Dead Sea and some other salt lakes contain little sulphate, but abundance of chloride of calcium, and if they are the residues of sea-water, have been modified by additions of this salt, which has converted the sulphate of magnesia into chloride of magnesium and gypsum, the calcareous chloride remaining in excess.

But while some of these saline lakes may be supposed to be basins of sea-water, modified by evaporation either alone or conjoined with the influx of foreign saline matters, others were evidently once fresh-water lakes in which, the loss of water by evaporation being equal to the supply, have gradually accumulated the soluble salts of all the rivers and springs flowing into the lake. We may arrive at some notion of the diverse natures of the different saline lakes which would be formed in this way if we suppose the waters of different European rivers to be subjected to evaporation under conditions like those of the salt lakes of western Asia. In the waters of the Elbe and Thames chlorides greatly predominate (in the latter with gypsum), with small amounts of magnesian salts, and the evaporation of these waters would give rise to lakes containing a large proportion of common salt. In the Seine, on the contrary, sulphate of lime predominates, while the waters of the Rhine, the Danube, the Arr, and the Arve contain but small amounts of chlorides and large proportions of sulphates of lime and magnesia.

In other rivers we find alkaline salts. The Loire at Orleans, according to Deville, contains in 100,000 parts, 13.46 of solid matters, of which 35.0 p. c. is carbonate of lime and 30.0 p. c. silica; while two thirds of the more soluble salts consist of carbonate of soda. In the waters of the Garonne, with as large a proportion of silica and more carbonate of lime, the carbonate of soda equals one fourth of the soluble salts; while 100,000 parts of the water of the Ottawa, according to my analysis, contain 6.11 parts of solid matters, consisting of carbonate of lime 2.48, carbonate of magnesia 0.69, silica 2.06, sulphates and chlorides of potassium and sodium 0.47, and carbonate of soda 0.41. Silica, although more abundant in alkaline river-waters, is not wanting in waters containing neutral earthy salts like the Seine and the Rhone, of the solid matters of which, according to Deville, it forms respectively 10.0 and 13.0 p. c. (Annales de Chimie et de Physique (3), XXIII. 32.) The waters which rise from the lower palæozoic shales of the St. Lawrence valley are, as I have shown, remarkable for the predominance of alkaline salts, which sometimes amount to one thousandth, or to more than one half the solid matters present. These waters are distinguished from the river-waters just mentioned by their comparatively small amounts of silica and earthy carbonates, and by the presence of a notable proportion of borates.*

We may here refer to the strongly alkaline waters furnished by the artesian wells of Paris and London as evidences of the abundance of alkaline carbonates in natural waters, and to the springs of Vichy and Carlsbad, the latter of which, according to the calculations of Gilbert, furnish annually more than thirteen millions of pounds of carbonate of soda. The evaporation of these alkaline waters, whether of rivers or of springs, must give rise to natron-lakes like Lake Van and those of the plains of Araxes, Lower Egypt, and Hungary. (Bischof, Lehrbuch, II. 1143.)

The carbonate of soda contained in these waters has its source in the decomposition of feldspathic minerals, and shows the continuance in our time of a process whose great activity in former geologic ages is attested, as I have elsewhere maintained, by vast accumulations of argillaceous sediments deprived of a large portion of their alkali, and also by the carbonate of lime which, by the intervention of carbonate of soda, has been formed from the chloride of calcium of the primeval ocean and deposited as limestone (*ante*, pages 2 and 10.)

An indispensable condition for the precipitation of carbonate of magnesia is the absence of chloride of calcium from the solutions, and this, in the presence of an excess of sulphates, is attained simply by evaporating to the point where gypsum becomes insoluble. In nearly all river and spring waters bicarbonate of lime is present in a large proportion, and is often the most abundant salt. We have shown that, when mingled with a solution containing sulphate of magnesia, it gives rise, by double decomposition, to bicarbonate of magnesia and sulphate of lime. By the evaporation of such a solution, the latter salt,

* For descriptions of these various waters of Canada, see the following essay in this volume.

being the less soluble, is first deposited in the form of gypsum, while the magnesian carbonate is only separated after further evaporation; when, provided the supply of bicarbonate of lime still continues, the two carbonates may fall down in a state of intermixture. In this way sediments will be formed containing the elements of dolomite or of magnesite.

The solution of magnesian bicarbonate remaining after the deposition of gypsum from the solution possesses, as we have seen, the power of decomposing chloride of calcium, and, when deprived of a portion of its carbonic acid by evaporation, reacts in a similar manner with a solution of sulphate of lime. In this way, an influx of sea-water into the basin from which gypsum, and perhaps a portion of magnesian carbonate, has already been deposited, would give rise to a precipitate of carbonate of lime, like the tufaceous limestones whose occurrence with gypsum and dolomites has been already noticed. In basins which, like the salt-lagoons of Bessarabia on the shores of the Black Sea, receive occasional additions of sea-water, and deposit every summer large amounts of salt (Bischof, Lehrbuch, II. 1717), the influx of waters containing bicarbonate of lime would give rise to the formation of beds of gypsum, alternating with dolomites or magnesian marls and rock-salt.

We have already referred to the analyses of certain rivers in which the sulphates are more abundant than the chlorides. Thus, in the Rhine, near Bonn, according to Bischof, we have for 100,000 parts of the water, 17.08 of solid matters, of which 1.23 are sulphate of lime, and 1.81 sulphate of magnesia, with only 1.45 of chloride and 8.37 of carbonate of lime; in the Danube, near Vienna, the predominance of sulphates is still more marked. The waters of the Arve, in the month of February, gave to Tingry, for 100,000 parts, 24.5 of solid matters, of which 6.5 were sulphate of lime, 6.2 sulphate of magnesia, and 8.3 carbonate of lime, with only 1.5 of chlorides. Now, as in river-waters there is always present an excess of carbonic acid, and as bicarbonate of lime and sulphate of magnesia in solution are mutually decomposed, these waters, which are to be regarded as solutions of sulphate of lime and bicarbonate of magnesia, would, by their evaporation, yield gypsum and magnesian carbonate, which would appear as portions of a fresh-water formation, like those of Aix and Auvergne.

The similar decomposition of soluble sulphates by bicarbonates of baryta and strontia will explain the formation of heavy spar and celestine, and their frequent association with gypsiferous rocks.

As to the native sulphur which is often associated both with epigenic and sedimentary gypsums, it has doubtless in every case been formed, as Breislak long since indicated, by the decomposition of sulphuretted hydrogen. It is well known that alkaline and earthy sulphates are reduced to sulphurets by organic matters, with the aid of heat, or even at ordinary temperatures, in presence of water. To the decomposition of these sulphurets by water and carbonic acid we are to ascribe not only the sulphuretted hydrogen of solfataras (which, by its oxidation under different conditions, gives rise either to free sulphur or to sulphuric acid and to gypsum by epigenesis), but also the sulphuretted hydrogen which appears in springs and in stagnant waters, where the sulphur produced by the decomposition of the gas is often mingled with sedimentary gypsums.* (Bischof, Lehrbuch, II. 139 – 185.) Bischof has also suggested the decomposition of chloride of magnesium by alkaline or earthy sulphurets as a source of sulphuretted hydrogen and hydrate of magnesia, into which sulphuret of magnesium is readily resolved in the presence of water. (Chemical Geology, I. 16.) If a salt of calcium were present, this reaction could only take place in the absence of carbonic acid, for carbonate of magnesia is incompatible with chloride of calcium. The direct reduction and decomposition of sulphate of magnesia by organic matter and carbonic acid may, however, yield sulphuretted hydrogen and carbonate of magnesia, and thus, in certain cases, give rise to magnesian sediments.

In the preceding sections, we have supposed the waters mingling with the solution of sulphate of magnesia to contain

* On certain modes of decomposition of the sulphates, see Jacquemin, Comptes Rendus, June 14, 1858.

no other bicarbonate than that of lime; but bicarbonate of soda is often present in large proportion in natural waters, and the addition of this salt to sea-water or other solutions containing chlorides and sulphates of lime and magnesia will, as we have shown, separate the lime as carbonate, and give rise to liquids, which, without being concentrated brines, as in the previous case, will contain sulphate of magnesia, but no lime-salts. A further portion of bicarbonate of soda will produce bicarbonate of magnesia, by the evaporation of whose solutions, as before, hydrated carbonate of magnesia would be deposited, mingled with the carbonate of lime which accompanies the alkaline salt, and in the case of the waters of alkaline springs, the compounds of iron, manganese, zinc, nickel, lead, copper, arsenic, chrome, and other metals, which springs of this kind still bring to the surface. In this way the metalliferous character of many dolomites is explained.

As the separation of magnesian carbonate from saline waters by the action of bicarbonate of soda does not suppose a very great degree of concentration, we may conceive this process to go on in basins where animal life exists, and thus explain the origin of fossiliferous magnesian limestones like those of Dudswell and the palæozoic formations of the western United States, whose organic remains, as I am informed by Professor James Hall of Albany, are generally such as indicate a shallow sea. To the intervention of carbonate of soda is, I conceive, to be referred the origin of all those dolomites which are not accompanied by gypsums, and which make up by far the larger part of the magnesian limestones; nor will the dolomites thus derived be necessarily marine, for the same reagent, with waters like those of the Danube and Arve, would give rise to dolomites and magnesites in fresh-water formations, which would not be accompanied by gypsums.

To the first stage of the reaction between alkaline bicarbonates and sea-water I am disposed to ascribe the formation of certain deposits of carbonate of lime which, although included in fossiliferous formations, are, unlike most of their associated limestones, not of organic origin, but have the characters of a chemical precipitate of nearly pure carbonate of lime, in which are often imbedded silicified shells and corals.* It is not perhaps easy in all cases to distinguish between such precipitates, which may assume a concretionary structure (see on this question Bischof, Chemical Geology, I. 428), and those deposits which, like travertines, have been formed from subterranean springs. In neither case, however, should they be confounded with the tufaceous limestones mentioned above.

The union of the mingled carbonates of lime and magnesia to form dolomite is attended with contraction, which, in case the sediment was already somewhat consolidated, would give rise to fissures and cavities in the mass. Should the dolomitic strata be afterwards exposed to the action of infiltrating carbonated waters, the excess of carbonate of lime and any calcareous fossils would be removed, leaving the mass still more porous, and with only the moulds of the fossils. Insoluble, however,

* The large proportion of dissolved silica which many river-waters contain appears in sedimentary deposits, not only replacing fossils and forming concretions and even beds of flint, chert, and jasper, but also in a crystalline state, as is seen in the crystallized quartz often associated with these amorphous varieties, and in some beds of sandstone which are made up entirely of small crystals of quartz. Elie de Beaumont long since called attention to the crystalline nature of certain sandstones which, as Daubrée has remarked, could not have been derived from the disintegration of any known rock; and Mr. J. Brainard, at the meeting of the American Association for the Advancement of Science, held at Cleveland in 1860, insisted upon the crystalline character of the grains composing certain sandstones in Ohio, as evidence that these were chemical deposits. He however fell into the error of supposing that all sandstones and even quartzose conglomerates have had a like origin, while the latter and the greater part of the former are undoubtedly mechanical deposits from the ruins of pre-existing quartzose and granitic rocks.

These crystallized sands, according to Daubrée, are met with in beds in the sandstone of the Vosges, the variegated sandstone (Triassic and Permian), and in the tertiary of the Paris basin and elsewhere. Other sands are made up of globules of chalcedony, apparently, like the crystallized sands, a chemical deposit, and associated with oölitic iron ores in the lias, and with glauconite grains in the green-sand. (Daubrée, Recherches sur le Striage des Roches, etc., Ann. des Mines, 1857.) We may here mention the so-called gaize from the green-sand of the Ardennes, which gave to Sauvage 56.0 p. c. of amorphous soluble silica mixed with quartz-sand and glauconite. (Bischof, Lehrbuch, I. 768-811.)

as it appears to be at ordinary temperatures, the filling up by it of such cavities both in magnesian and in pure limestones, not less than its deposition in veins and druses, indicates that dolomite is under certain conditions soluble in water.

Conclusions.

1. The action of solutions of bicarbonate of soda upon sea-water separates in the first place the whole of the lime in the form of carbonate, and then gives rise to a solution of bicarbonate of magnesia, which by evaporation deposits hydrous magnesian carbonate.

2. The addition of solutions of bicarbonate of lime to sulphate of soda or sulphate of magnesia gives rise to bicarbonates of these bases, together with sulphate of lime, which latter may be thrown down by alcohol. By the evaporation of a solution containing bicarbonate of magnesia and sulphate of lime, either with or without sea-salt, gypsum and hydrous carbonate of magnesia are successively deposited.

3. When the hydrous carbonate of magnesia is heated alone under pressure, it is converted into magnesite; but if carbonate of lime be present, a double salt is formed, which is dolomite.

4. Solutions of bicarbonate of magnesia decompose chloride of calcium, and, when deprived of their excess of carbonic acid by evaporation, even solutions of gypsum, with separation of carbonate of lime.

5. Dolomites, magnesites, and magnesian marls have had their origin in sediments of magnesian carbonate formed by the evaporation of solutions of bicarbonate of magnesia. These solutions have been produced either by the action of bicarbonate of lime upon solutions of sulphate of magnesia, in which case gypsum is a subsidiary product, or by the decomposition of solutions of sulphate or chloride of magnesium by the waters of rivers or springs containing bicarbonate of soda. The subsequent action of heat upon such magnesian sediments, either alone or mingled with carbonate of lime, has changed them into magnesite or dolomite.

SUPPLEMENT.

[In reference to the formation of dolomite, as indicated above, in 3, it was shown in the continuation of these researches, published, as already mentioned, in 1866, that the result was best attained when a mixture of the two carbonates, precipitated together and still amorphous, was gradually heated with water, under pressure, to a temperature of 120° C.; and the question was raised, whether all the deposits of dolomite in nature have been thus heated, or "whether there are yet unknown conditions under which the double carbonate, dolomite, can be formed at lower temperatures."

As to the reaction noticed in 2, it was found that the partial loss of carbonic acid during the spontaneous evaporation of solutions holding gypsum and bicarbonate of magnesia often renders the results of experiments very imperfect, but that the conditions of complete success in the separation of gypsum from such solutions are attained when the evaporation is conducted in an atmosphere containing a large proportion of carbonic-acid gas (*ante*, page 43). This result was first described in a note to the American Association for the Advancement of Science, in August, 1866. (Canadian Naturalist, III. 123.)

I have found in solutions prepared with sulphate of magnesia and bicarbonate of lime, as in 2, the proportion of sulphate of lime to the water as 1 : 404 and 1 : 413; while a saturated solution of gypsum in pure water at 16° C. contained 1 : 372, which nearly agrees with Giese's determination of 1 : 380. Taking, as a mean of these, 1 : 400, we have for a litre of water 2.50 grammes of anhydrous sulphate of lime, equal to 3.16 of gypsum. From a solution prepared as above I have, in a successful experiment, separated by evaporation in the open air 1.18 grammes of gypsum to the litre, leaving in solution an equivalent of magnesian bicarbonate. Of the latter compound, Bineau obtained solutions in which 11.2 grammes of magnesia (equal to 23.5 of monocarbonate) were dissolved in a litre of water with very nearly two equivalents of carbonic acid; and I have found it easy to prepare solutions, permanent in the air, holding, as bicarbonate, 21.0 grammes of monocarbonate of magnesia to the litre; so that the solubility of carbonate of magnesia under these conditions is about nine times as great as that of sulphate of lime. (Amer. Jour. Science (2), XXVIII. pages 170–178.

The fact that the separation of the carbonate of magnesia necessary for the production of dolomites and magnesites requires the absence of chloride of calcium from the waters in which it is deposited, — whether this carbonate is generated by the reaction of bicarbonate of lime on sulphate of magnesia (with simultaneous production of gypsum), or by the intervention of bicarbonate of soda, — and that in both cases isolated and evaporating basins are indispensable conditions of the formation and deposition of this magnesian carbonate, was clearly pointed out, as above, in 1859. The legitimate deductions from this as to the geographical and climatic conditions of regions during the formation of magnesian limestones were further insisted upon in a paper upon the Geology of Southwestern Ontario, in 1868, and again in 1871, in paper XIII. of the present volume. (See also *ante*, page 74.)

It was not, however, I believe, till 1871 that these views of mine found recognition, when Professor A. C. Ramsay, by the investigation of the magnesian limestone of the Permian in England, was led to reject as untenable the notion held by Sorby (and by others), that this was once an ordinary limestone of organic origin subsequently impregnated with magnesian carbonate under conditions not explained; and to conclude that the carbonates of lime and magnesia of which it is composed had been "deposited simultaneously by the concentration of solutions due to evaporation," "in an inland salt lake." To this view, as he informs us, he was led by physical considerations, and by the depauperated condition of the organic remains contained in these strata, without being, at the time, aware that I had twelve years previously announced the same conclusions for all magnesian limestones, and established them on chemical grounds. (Quar. Geol. Jour., 1871, page 249.)]

IX.

THE CHEMISTRY OF NATURAL WATERS.

This paper appeared in the American Journal of Science for March, July, and September, 1865. In reprinting it, the original tables of analyses have been omitted, and in their place a few typical analyses only are given. Some other omissions have been made for the sake of brevity, and a few notes added. In a Supplement I have given the results of the examination of additional waters of the first class, some of them remarkable for the great amount of soluble sulphides present; and in an Appendix, details and results of experiments on the porosity of rocks. Both the Supplement and the Appendix are from the Report of the Geological Survey of Canada for 1863 – 66.

It is proposed to divide this essay into three parts, in the first of which will be considered some general principles which must form the basis of a correct chemical history of natural waters. The second part will embrace a series of chemical analyses of mineral waters from the palæozoic rocks of the Champlain and St. Lawrence and Ottawa basins, together with some river-waters; and the third part will consist chiefly of deductions and generalizations from these analyses.

I.

Contents of Sections. — 1. Atmospheric waters; 2, 3. Results of vegetable decay; 4 – 7. Action on rocky sediments; 8. Action on iron-oxide; 9. Solution of alumina; 10. Reduction of sulphates; 11. Kaolinization; 12. Decay of silicates; 13. Origin of carbonate of soda; 14. Bischof's view rejected; 15, 16. Porosity of rocks, and their contained saline waters; 17. Saliferous strata; 18. Action of carbonate of soda on saline waters; 19. Origin of sulphate of magnesia; 20, 21. Mitscherlich's view rejected; 22, 23. Salts from evaporating sea-water; composition of ancient seas; origin of carbonate of lime; 24 – 27. Origin of gypsum, carbonate of magnesia, and dolomite; 28. Waters from oxidized sulphurets; 29. Origin of free sulphuric and hydrochloric acids; 30. Of hydrosulphuric and boric acids; 31. Of carbonic-acid gas; 32. Of ammoniacal salts; 33 – 35. Classification of mineral waters.

§ 1. The solvent powers of water are such that this liquid is never met with in nature in a perfectly pure state; even meteoric waters hold in solution, besides nitrogen, oxygen, carbonic acid, ammonia, and nitrous compounds, small quantities of solid matters which were previously suspended in the form of dust in the atmosphere. After falling to the earth, these same waters become still further impregnated with foreign elements of very variable nature, according to the conditions of the surface on which they fall.

§ 2. Atmospheric waters, coming in contact with decaying vegetable matters at the earth's surface, take from them two classes of soluble ingredients, organic and inorganic. The waters of many streams and rivers are colored brown with dissolved organic matter, and yield, when evaporated to dryness, colored residues, which carbonize by heat. This organic substance, in some cases at least, is azotized, and similar, if not identical, in composition and properties with the apocrenic acid of Berzelius. The decaying vegetation, at the same time that it yields a portion of its organic matter in a soluble form, parts with the mineral or cinereal elements which it had removed from the soil during life. The salts of potassium, calcium, and magnesium, the silica and phosphates, which are so essential to the growing plant, are liberated during the process of decay; and hence we find these elements almost wanting in peat and coal. (See on this point the analyses by Vohl of peat, peat-moss, and the soluble matters set free during its decay; Ann. der Chem. und Pharm., CIX. 185. Also Liebig, analysis of bog-water, Letters on Modern Agriculture, p. 44; and in the second part of this paper, the analysis of the water of the Ottawa River.)

§ 3. At the same time an important change is effected in the gaseous contents of the atmospheric waters. The oxygen which they hold in solution is absorbed by the decaying organic matter, and replaced by carbonic acid; while any nitrates or nitrites which may be present are by the same means reduced to the state of ammonia (Kuhlmann). By thus losing oxygen, and taking up a readily oxidizable organic matter, these waters become reducing instead of oxidizing media in their further progress.

§ 4. We have thus far considered the precipitated atmospheric waters as remaining at the earth's surface; but a great portion of them, sooner or later in their course, come upon permeable strata, by which they are absorbed, and in their subterranean circulation undergo important changes. The effect of ordinary argillaceous strata destitute of neutral soluble salts may be first examined. Between such sedimentary strata and the waters charged with organic and mineral matters from decaying vegetation there are important reactions. The composition of these waters is peculiar; they contain, relatively to the sodium, a large amount of potassium salts, besides notable quantities of silica and phosphates, in addition to the dissolved organic matters and the earthy carbonates, and in many cases ammoniacal salts and nitrates or nitrites. The sulphuric acid and chlorine are moreover not sufficient to neutralize the alkalies, which are perhaps in part combined with silica or with an organic acid.

§ 5. The experiments of Way, Voelcker, and others have shown that when such waters are brought into contact with argillaceous sediments, they part with their potash, ammonia, silica, phosphoric acid, and organic matter, which remain in combination with the soil; while, under ordinary conditions at least, neither soda, lime, magnesia, sulphuric acid, nor chlorine are retained. This power of the soil appears, from the experiments of Eichhorn, to be in part due to the action of hydrated double aluminous silicates; and the process is one of double exchange, an equivalent of lime or soda being given up for the potash and ammonia retained. The phosphates are probably retained in combination with alumina or peroxide of iron, and the silica and organic matters also enter into insoluble combinations. It follows from these reactions that the surface-waters charged with the products of vegetable decay, after having been brought in contact with argillaceous sediments, retain little else than sulphates, chlorides, or carbonates of soda, lime, and magnesia. In this way the mineral matters required for the growth of plants, and by them removed from the soil, are again restored to it; and from this reaction results the small proportion of potash-salts in the waters of ordinary springs and wells as compared with river-waters. From the waters of rivers, lakes, and seas, aquatic plants again take up the dissolved potash, phosphates, and silica; and the subsequent decay of these plants in contact with the ooze of the bottom, or on the shores, again restores these elements to the earth. See a remarkable essay, by Forchhammer, on the composition of fucoids, and their geological relations, Jour. fur Prakt. Chem., XXXVI. 388.

§ 6. The observations of Eichhorn upon the reaction between solutions of chlorides and pulverized chabazite, which, as a hydrated silicate of alumina and lime, may perhaps be taken as a representative of the hydrous double silicates in the soil, show that these substitutions of protoxide bases are neither complete nor absolute. It would appear, on the contrary, that there takes place a partial exchange or a partition of bases according to their respective affinities. Thus the normal chabazite, in presence of a solution of chloride of sodium, exchanges a large portion of its lime for soda; but if the resulting soda-compound be placed in a solution of chloride of calcium, an inverse substitution takes place, and a portion of lime enters again into the silicate, replacing an equivalent of soda; while, by the action of a solution of chloride of potassium, both lime and soda are, to a large extent, replaced by potash. In like manner, chabazite, in which, by the action of a solution of sal-ammoniac, a part of the lime has been replaced by ammonia, will give up a portion of the ammonia, not only to solutions of chlorides of potassium and sodium, but even to chloride of calcium. It results from these mutual decompositions that there is a point where a chabazite containing both lime and soda, or lime and ammonia, would remain unchanged in mixed solutions of the corresponding chlorides, the affinities of the rival bases being balanced.* Inasmuch, however, as the proportions of ammonia and potash in natural waters are usually small when compared with the amounts of lime and soda existing in the form of hydro-silicates in the

* Amer. Jour. Science (2), XXVIII. 72.

soil, the result of these affinities is an almost complete elimination of the ammonia and potash from infiltrating waters.

§ 7. That the replacement of one base by another in this way is not complete is shown moreover by the experiments of Liebig, Dehérain, and others, who have observed that a solution of gypsum removes from soils a certain amount of potash-salt, which was insoluble in pure water. In this way gypseous waters may also acquire portions of sulphate of soda from silicates.

It is not certain that all the above reactions observed for chabazite are applicable without modification to the double hydro-aluminous silicates of sedimentary strata. Were such the case, important changes might, in certain conditions, be effected in the composition of saline waters. Thus in presence of a great amount of a hydrous silicate of lime and alumina, solutions of chloride of sodium might acquire a considerable amount of chloride of calcium; but it is probable that these reactions, however important they may be in relation to the soil, and to surface-waters with their feeble saline impregnation, have at present but little influence on the composition of the stronger saline waters. It is however not impossible that the action of the ancient sea-waters, holding a large amount of chloride of calcium, upon the hydrated and half-decomposed feldspars which constituted the clays of the period, may have given rise to those double silicates which formed the lime-soda feldspars so abundant in the Labrador series.

§ 8. The reactions just described assume an importance in the case of waters impregnated with soluble matters from vegetable decay; and in this event, another and not less important class of phenomena intervenes, due to the deoxidizing power of the dissolved organic matter. By the action of this upon the insoluble peroxide of iron set free from the decomposition of ferruginous minerals and disseminated in the sediments, protoxide of iron is formed, which is soluble both in carbonic acid and in the excess of the organic (acid) matter. By this means not only are great quantities of iron dissolved, but masses of sediments are sometimes entirely deprived of iron-oxide, and thus beds of white clay and sand are formed. The waters thus charged with proto-salts of iron absorb oxygen when exposed to the air, and then deposit the metal as hydrated peroxide, which, when the organic matter is in excess, carries down a greater or less proportion of it in combination. Such organic matters are rarely absent from limonite, and in some specimens of ochre amount to as much as fifteen per cent.* The conditions under which hydrous peroxide of manganese is often found are very similar to those of hydrous peroxide of iron with which it is so frequently associated; and there is little doubt that oxide of manganese may be dissolved by a process like that just pointed out. A portion of manganese has been observed in the soluble matters from decaying peat-moss; and it seems to be generally present in small quantities with iron in surface-waters.

§ 9. There is reason to believe that alumina is also, under certain conditions, dissolved by waters holding organic acids. The existence of pigotite, a native compound of alumina with an organic acid, and the occasional association of gibbsite with limonite, point to such a reaction. That it is not more abundant in solution, is due to the fact, that, unlike most other metallic oxides, alumina, instead of being separated in a free state by the slow decomposition of its silicious compounds, remains in combination with silica. The formation of bauxite, a mixture of hydrate of alumina with variable proportions of hydrous peroxide of iron, which forms extensive beds in the tertiary sediments of the great Mediterranean basin, indicates a solution of alumina on a grand scale, and perhaps owes its origin to the decomposition of solutions of native alum by alkaline or earthy carbonates. Emery, a crystalline anhydrous form of alumina, has doubtless been formed in a similar manner. (American Journal Science (2), XXXII. 287, and *ante*, page 13.) The existence in many localities of an insoluble subsulphate of alumina, websterite, in layers and concretionary masses in tertiary clays, evidently points to such a process. Compounds consisting chiefly of hydrated alumina are frequently

* Geology of Canada, p. 512.

found in fissures of the chalk in England. On the absence of free hydrated alumina from soils, see Müller, cited as above (2), XXXV. 292.

§ 10. The organic matter dissolved by the surface-waters serves to reduce to the condition of sulphurets the various soluble sulphates which it takes up at the same time or meets with in its course. These sulphurets, decomposed by carbonic acid, which is in part derived from the atmosphere, and in part from the oxidation of the carbon of the organic matter, give rise to alkaline and earthy carbonates on the one hand, and to sulphuretted hydrogen on the other. In this way, under the influence of a somewhat elevated temperature, are generated sulphurous waters, whether of subterranean springs, or of tropical sea-marshes and lagoons. The reaction between the sulphurets thus formed and the salts or oxides of iron, copper, and similar metals which may be present, gives rise to metallic sulphurets. The decomposition of sulphuretted hydrogen by the oxygen of the air produces native sulphur; with which are generally found associated sulphates of lime and strontia. By virtue of these reactions, soluble sulphates of lime and magnesia may be completely eliminated from waters, the bases as insoluble carbonates, and the sulphur as sulphuretted hydrogen, free sulphur, or a metallic sulphuret. Moreover, as Forchhammer has pointed out in the paper already cited, sulphuret of potassium in the presence of ferruginous clays is also completely separated from solution, the sulphur as sulphuret of iron, and the alkali as a double aluminous silicate.

§ 11. We have thus far considered the composition of surface-waters as modified by the decay of vegetation, or by the reactions between the matters derived from this source and the permeated sediments. Not less important however than the elements thus removed by substitution from sedimentary strata are those which are liberated by the slow decomposition of the minerals composing these sediments.

It has long been known that in the transformation of a feldspar into kaolin, the double silicate of alumina and alkali takes up a portion of water, and is resolved into a hydrous silicate of alumina; while the alkali, together with a definite portion of silica, is separated in a soluble state. The feldspar, an anhydrous double salt formed at an elevated temperature, has a tendency under certain conditions to combine, at a lower temperature, with a portion of water, and break up into two simpler silicates. Daubrée has moreover shown that when kaolin is exposed to a heat of 400° C. in presence of a soluble silicate of potash, the two silicates unite and regenerate feldspar. These reactions are completely analogous to those presented by very many other double salts, ethers, amides, and similar compounds. The preliminary conditions of this conversion of feldspar into kaolin and a soluble alkaline silicate, however, still require investigation. It is known that while some feldspathic rocks appear almost unalterable, others containing the same species of feldspar are found converted to a depth of many feet from the surface into kaolin. This chemical alteration, according to Fournet, is always preceded by a mechanical change of the feldspar, which first becomes opaque and friable, and is thus rendered permeable to water. He conceives this alteration to be molecular, and to be connected with the passage of the silicate into a dimorphous or allotropic condition.*

§ 12. The researches of Ebelman on the alterations of various rocks and minerals have thrown considerable light on the relations of sediments and natural waters.† From the analyses of basaltic and similar rocks, which include silicates of lime, magnesia, iron, and manganese in the forms of pyroxene, hornblende, and olivine, and which undergo a slow and superficial decomposition under atmospheric influences, it appears that during the process of decay the greater part of the lime and magnesia is removed, together with a large proportion of silica.

* [Annales de Chimie (2), LV. 225. It is a subject for inquiry how far such changes are recent, and whether all feldspars found thus decomposed are not portions which have been preserved to us from a remote antiquity, when atmospheric agencies more potent than those of the present day were at work. *Ante*, page 10.]

† Ebelman, Recueil des Travaux, II. 1–79.

It was found, moreover, that in the case of a rock apparently composed of labradorite and pyroxene, the removal of the lime and magnesia from the decomposed portion was much more complete than that of the alkalies ; showing thus the comparatively greater stability of the feldspathic element. The decomposition of the feldspar in these mixed rocks is however at length effected, and the final result approximates to a hydrous silicate of alumina or clay. This slow decomposition of silicates of protoxide-bases appears to be due to the action of carbonic acid, which, removing the lime and magnesia as carbonates, liberates the silica in a soluble form ; while the iron and manganese, passing to a state of higher oxidation, remain behind, unless the action of organic matters intervenes to give them solubility.

§ 13. Notwithstanding the complete decomposition, resulting in the production of kaolin, to which orthoclase, in common with the triclinic feldspars (and some other feldspathides, such as the scapolites, beryl, and leucite), is subject, it is to be noticed, that under ordinary atmospheric conditions orthoclase appears less liable to change than the lime-soda feldspars such as albite, oligoclase, and labradorite. Weathered surfaces of these become covered with a thin, soft, white, and opaque crust from decomposition, while the surfaces of orthoclase under similar conditions still preserve their hardness and translucency. A gradual process of this kind is constantly going on in the feldspathic matters which form a large proportion of the mechanical sediments of all formations ; and in deeply buried strata is not improbably accelerated by the elevation of temperature. The soluble alkaline silicate resulting from this process is in most cases decomposed by carbonates of lime and magnesia in the sediments, giving rise to silicates of these bases (which are for the greater part separated in an insoluble state), and to carbonate of soda. Only in rare cases does potash appear in large proportion among the soluble salts thus liberated from sediments, partly because soda-feldspars are more subject to change, and partly from the fact that potash-salts would be separated from the percolating waters in virtue of the reactions mentioned in § 5. Hence it happens that apart from the neutral soda-salts of extraneous origin, waters permeating sediments containing alkaliferous silicates generally bring to the surface little more than soda combined with carbonic and sometimes with boric acid, and carbonates of lime and magnesia with small portions of silica.

§ 14. This explanation of the decomposition of alkaliferous silicates and of the origin of carbonate of soda is opposed to the view of Bischof, who conceives that carbonic acid is the chief agent in decomposing feldspathic minerals.* The solvent action of waters charged with carbonic acid is undoubted, as shown by various experimenters, especially by the Messrs. Rogers,† but this acid is not always present in the quantities required. The proportion of it in atmospheric waters is so inadequate that it becomes necessary to suppose some subterranean source of the gas, which is by no means a constant accompaniment of natron-springs. A copious evolution of carbonic acid is observed in the vicinity of the lake of Laach, where the alkaline waters studied by Bischof occur.‡ The same thing is met with in many other localities of such springs, among which may be mentioned the region around Saratoga, where saline waters containing carbonate of soda, and highly charged with carbonic acid, rise in abundance from the lower palæozoic strata; but farther northward, along the valleys of Lake Champlain and the St. Lawrence, similar alkaline-saline waters, which abound in the continuation of the same geological formations, are not at all acidulous. From this the conclusion seems justifiable that the production of carbonate of soda is a process, in some cases at least, independent of the presence of free carbonic acid. In this connection, it is well to recall the solvent action of pure water on alkaliferous silicates, as shown more especially by Bunsen, and also by Damour, who found that distilled water at temperatures much below 212° takes up from silicates like palagonite and calcined mesotype

* Bischof, Chem. Geol., II. 131.
† Amer. Jour. Sci. (2), V. 401.
‡ Bischof, Lehrbuch, I. 357–363.

comparatively large amounts both of silica and alkalies. (Ann. Chim. et Phys. (3), XIX. 481.)

[The view advanced in § 13 is to be understood of the subterranean decomposition of alkaliferous silicated minerals, the results of which appear in waters like some noted further on in § 67, where, from a deficiency of carbonic acid, parts of the bases are present as silicates, as in the solutions prepared by Damour. At the same time it is clear that carbonic acid greatly favors the process, as seen in the experiments of Rogers, and has played a most important part in the subaerial decomposition of crystalline rocks, from which, as Ebelman showed, have been removed not only the alkalies of the feldspar, but the lime and magnesia of the hornblende. The absence of any excess of carbonic acid in many alkaline-saline waters, as noticed in § 66, appears a conclusive argument for the view set forth in § 13, that the subterranean decomposition of alkaliferous sediments takes place independent of the intervention of carbonic acid.]

§ 15. Another and an important source of mineral impregnation to waters exists in the soluble salts enclosed in sedimentary strata, both in the solid state and in aqueous solution, and for the most part of marine origin. In order to form some conception of the amount of saline matters which may be contained in a dissolved state in the rocky strata of the earth, I have made numerous experiments to determine the porosity of various rocks; a few of the results, so far as regards the lower palæozoic formations of the New York system (in which occur the mineral waters named in the last section), are here given. For further details, and for a table of results, the reader is referred to the Appendix to this paper in the present volume. The volume of water enclosed in 100 volumes of the various rocks having been determined, it was found for three specimens of the Potsdam sandstone to equal 2.26 – 2.71, and for three others of the same rock, much more porous, 6.94 – 9.35. For four specimens of the crystalline dolomite which makes up the so-called Calciferous sand-rock the volume was equal to 1.89 – 2.53, and for two other varieties of the same rock, 5.90 – 7.22.

§ 16. If we take for the Potsdam sandstone the mean of the first three trials, giving 2.5 per cent for the volume of water which it is capable of holding in its pores, we find that a thickness of 100 feet of it would contain in every square mile, in round numbers, 70,000,000 cubic feet of water; an amount which would supply a cubic foot (over seven gallons) a minute for more than thirteen years. The observed thickness of the Potsdam sandstone in the district of Montreal varies from 200 to 700 feet, and a mean of 500 feet may be assumed. To this are to be added 300 feet for the Calciferous sand-rock, whose capacity for water may be taken, like the Potsdam sandstone, at 2.5 per cent. We have thus in each square mile of these formations, wherever they lie below the water-level, a volume of 490,000,000 cubic feet of water, equal to a supply of a cubic foot per minute for 106 years. The capacity of the 800 feet of Chazy and Trenton limestones which succeed these lower formations may be fairly taken at one half that of those just named. But it is unnecessary to multiply such calculations: enough has been said to show that these sedimentary strata include in their pores great quantities of water, which was originally that of the palæozoic ocean. These strata, throughout the palæozoic basin of the St. Lawrence, are now for the greater part beneath the sea-level; nor is there any good reason for supposing them to have ever been elevated much above their present horizon. Wells and borings sunk in various places in these rocks show them to be still filled with bitter saline waters; but in regions where these rocks are inclined and dislocated, surface-waters gradually replace these saline waters, which, in a mixed and diluted state, appear as mineral springs. These saline solutions, other things being equal, will be better preserved in limestones or argillaceous rocks than in the more porous and permeable sandstones.

§ 17. But besides the saline matters thus disseminated in a dissolved state in ordinary sedimentary rocks, there are great volumes of saliferous strata, properly so called, charged with the results of the evaporation of ancient sea-basins. These strata enclose not only gypsum and rock-salt, but in some regions large quantities of the double chloride of potassium and magnesium, carnallite; and in others sulphate of soda, sulphate of magnesia, and complex sulphates like bloedite and polyhallite. Besides these crystalline salts, the mother-liquors containing the more soluble and uncrystallizable compounds may also be supposed to impregnate, in some cases, the sediments of these saliferous formations. The conditions under which these various salts are deposited from sea-water, and their relations to the composition of the ocean in earlier geological periods, are reserved for consideration in § 22. Infiltrating waters remove from these saliferous strata their soluble ingredients; which, together with the ancient sea-waters of other sedimentary rocks, give rise to the various neutral saline waters; while the mingling of these in various proportions with the alkaline waters whose origin has been described in § 13, produces intermediate classes of waters of much interest.

§ 18. I have elsewhere described the results of a series of experiments on the mutual action of the waters of these two classes.* When a dilute solution of bicarbonate of soda is gradually added to a solution which, like sea-water, contains, besides chloride of sodium, the chlorides and sulphates of calcium and magnesium, the greater part of the lime separates as carbonate, carrying down with it only from one to three hundredths of carbonate of magnesia; a portion of lime however remaining in solution as bicarbonate. When the chloride of calcium is wholly decomposed, the magnesian salt is attacked in its turn, and there finally results a solution in which the whole of the earthy chlorides are replaced by chloride of sodium. A further addition of the solution of bicarbonate of soda gives them the character of alkaline-saline waters; which moreover contain an abundance of earthy carbonates.

§ 19. In the saline waters just considered, chlorides generally predominate, the sulphates being small in amount, and often altogether wanting. Some exceptions to this are however met with; for apart from waters impregnated with gypsum, whose origin is readily understood, there are others in which sulphate

* American Journal Science (2), XXVIII. 170.

of soda or sulphate of magnesia enter largely. The soda-salt may sometimes be formed by the reaction between solution of gypsum and natriferous silicates referred to in § 7, or by the decomposition of gypsum by solution of carbonate of soda; while in other cases its origin will probably be found in the natural deposits of sulphates, such as glauberite, thenardite, and glauber-salt, which occur in saliferous rocks. A similar origin is probable for many of those springs in which sulphate of magnesia predominates. This salt also effloresces abundantly in a nearly pure form upon certain limestones, and is in some cases due to the action of sulphates from decomposing pyrites upon magnesian carbonate or silicate. In by far the greater number of cases, however, its appearance is unconnected with any such process; and is, according to Mitscherlich, due to a reaction between dolomite and dissolved gypsum.

§ 20. In support of this view, it was found by the chemist just named that when a solution of sulphate of lime was made to filter for some time through pulverized magnesian limestone, it was decomposed with the formation of carbonate of lime and sulphate of magnesia. This reaction I have been unable to verify. A solution of gypsum in distilled water was made to percolate slowly through a column of several inches of finely powdered dolomite, and after ten filtrations, occupying as many days, no perceptible amount of sulphate of magnesia had been formed. Solutions of gypsum were then digested for many months with pulverized dolomite, and also with crystalline carbonate of magnesia, but with similar negative results; nor did the substitution of a solution of chloride of calcium lead to the formation of any soluble magnesian salt. Solutions of gypsum were then impregnated with carbonic acid, and allowed to remain in contact with pulverized dolomite and with magnesite, as before, during six months of the warm season, when only inappreciable traces of magnesia were taken into solution. These experiments show that no decomposition of dissolved gypsum is effected by native carbonate of magnesia, or by the double carbonate of lime and magnesia, at ordinary temperature.

§ 21. I find, however, that hydrated carbonate of magnesia readily and completely decomposes a solution of gypsum when agitated with it, with formation of carbonate of lime and sulphate of magnesia; and the same result is produced by the native hydrate of magnesia when mingled with a solution of gypsum in presence of carbonic acid. Now there may be dolomites which contain an admixture of hydro-carbonate of magnesia, as there certainly are others which, like predazzite, are penetrated with hydrate of magnesia.* The reaction between solutions of gypsum and such magnesian limestones (with the intervention, in the case of predazzite, of atmospheric carbonic acid) would suffice to explain the results obtained by Mitscherlich, and the appearance in certain cases of sulphate of magnesia as an efflorescence on dolomites. In the experiments above described, the nearly pure crystalline dolomites from the Guelph and Niagara formations were made use of.

§ 22. When sea-water is exposed to spontaneous evaporation, the lime which it contains separates in the form of sulphate, gypsum being but sparingly soluble in a concentrated brine, and the greater portion of the chloride of sodium crystallizes out in a nearly pure state. The mother-liquor of specific gravity 1.24, having lost about four fifths of its chloride of sodium, still contains dissolved a large proportion of sulphate of magnesia. If the evaporation is continued at the ordinary temperature till a density of 1.32 is attained, about one half of the magnesian sulphate separates, mixed with common salt; and by reducing the temperature to 6° C., a large portion of pure sulphate of magnesia now crystallizes out. The further evaporation of the remaining liquor by the heat of summer causes the potassium-salt to separate in the form of a hydrous double chloride of potassium and magnesium, an artificial carnallite.†

[* In subsequent experiments it was found that certain dolomites contain a little hydrous carbonate of magnesia capable of decomposing a limited amount of solution of gypsum. See the author, On Lime and Magnesia Salts, American Journal of Science for July, 1866, §§ 96-101.]

† The hydrous double chloride of potassium and magnesium, to which the name of carnallite has been given, occurs in large quantities in the upper

By varying somewhat the conditions of temperature, the sulphate of magnesia and the chloride of sodium of the mother-liquor undergo mutual decomposition, with the production of sulphate of soda and chloride of magnesium. Hydrated sulphate of soda crystallizes out from such a mixed solution at 0° C., and by reducing the temperature to — 18° C. the greater part of the sulphates may be separated in this form from the mother-liquor of 1.24, previously diluted with one tenth of water; without which addition a mixture of hydrated chloride of sodium would separate at the same time. If, on the other hand, the temperature of the mixed solution be raised above 50° C., the sulphate of soda crystallizes out in the anhydrous form, as thenardite. By the spontaneous evaporation during the heats of summer of the mother-liquors of density 1.35, a double sulphate of potassium and magnesium separates. These reactions are taken advantage of on a great scale in Balard's process, as modified by Merle,* for extracting salts from sea-water.

§ 23. The results of the evaporation of sea-water would however be widely different if an excess of lime-salt were present. In this case the whole of the sulphates present would be deposited in the form of gypsum at an early stage of the evaporation, and the mother-liquor, after the separation of the greater part of the common salt, would contain little else than the chlorides of sodium, potassium, calcium, and magnesium.

§ 24. A consideration of the conditions of the ocean in earlier geological periods will show that it must have contained a much larger quantity of lime-salts than at present. The alkaline carbonates, whose origin has been described in § 13, and which from the earliest times have been flowing

strata of the saliferous formation of Stassfurth in Germany; where it is associated with a hydrous double chloride of calcium and magnesium, tachydrite, and also with a sparingly soluble sulphate of magnesia, kieserite, which contains a small and variable amount of water, and is supposed to be, in its normal condition, an anhydrous salt. When heated to redness in a current of steam this sulphate loses all its acid, which passes off undecomposed.

* See my paper in Amer. Jour. Science (2), XXV. 361; also Report of the Juries of the Exhibition of 1862, Class II. page 48.

into the sea, have gradually modified the composition of its waters, separating the lime as carbonate, and thus replacing the chloride of calcium by chloride of sodium, as I have long since pointed out (*ante*, page 2). This reaction has doubtless been the source of all the carbonate of lime in the earth's crust, if we except that derived from the decomposition of calcareous silicates. (§ 12.) In this decomposition by carbonate of soda, as already described in § 18, it results from the incompatibility of chloride of calcium with hydrous carbonate of magnesia, that the lime is first precipitated, with a little adhering carbonate of magnesia; and it is only when the chloride of calcium is all decomposed that the magnesian chloride is transformed into carbonate of magnesia. This latter reaction can consequently take place only in limited basins, or in portions cut off from the oceanic circulation.

§ 25. It follows from what has been said that the lime-salt may be eliminated from sea-water either as sulphate or as carbonate. In the latter case no concentration is required; while in the former the conditions are two, — a sufficient proportion of sulphates to convert the whole of the lime into gypsum, and such a degree of concentration of the water as to render this insoluble. These conditions meet in the evaporation of modern sea-water; but the evaporated sea-water of earlier periods, with its great predominance of lime-salts, would still contain large amounts of chloride of calcium, — the insolubility of gypsum in this case serving to eliminate all the sulphates from the mother-liquor. Evaporation alone would not suffice to remove the whole of the lime-salts from waters in which the calcium present was more than equivalent to the sulphuric acid; but the intervention of carbonate of soda would be required.

§ 26. In concentrated and evaporating waters freed from lime-salts by either of the reactions just mentioned, but still holding sulphate of magnesia, another process may intervene (*ante*, page 90). The addition of a solution of bicarbonate of lime to such a solution gives rise, by double decomposition, to sulphate of lime and bicarbonate of magnesia. The former, being much the less soluble salt, especially in a strongly saline liquid, is deposited as gypsum; and subsequently the magnesian carbonate is precipitated in a hydrous form. The effect of this reaction is to eliminate from the sea-water both the sulphuric acid and the magnesia, without the permanent addition to it of any foreign element.

§ 27. Gypsum may thus be separated from sea-water by two distinct processes, — the one a reaction between sulphate of magnesia and chloride of calcium, and the other between the same sulphate and carbonate of lime. The latter, involving a separation of bicarbonate of magnesia, can, as we have seen, only take place when the whole of the chloride of calcium has been eliminated; and if we suppose the ancient ocean, unlike the present, to have contained more than an equivalent of lime for each equivalent of sulphuric acid, it is evident that a lake or basin of sea-water free from lime-salts could only have been produced by the intervention of carbonate of soda. The action of this must have eliminated the whole of the lime as carbonate, or at least have so far reduced the amount of this base that the sulphates present would be sufficient to separate the remainder by evaporation in the form of gypsum, and still leave in the mother-liquor a quantity of sulphate of magnesia for reaction with bicarbonate of lime.

The source of the magnesian carbonate, whose union, under certain conditions, with the carbonate of lime, gives rise to dolomite (*ante*, page 90), may thus be due either to the reaction just described between bicarbonate of lime and solutions holding sulphate of magnesia, or to the direct action of carbonate of soda upon waters containing magnesian salts; but, in either case, the previous elimination of the incompatible chloride of calcium must be considered an indispensable preliminary to the production of the magnesian carbonate.

§ 28. To the three principal sources of mineral matters in mineral waters already enumerated, namely, decaying organic matters, decomposing silicates, and the soluble saline matters in rocks, a few other minor ones must be added. One of these is the oxidation of metallic sulphurets, chiefly iron pyrites, giving rise to sulphate of iron, and more rarely to sulphates of copper, zinc, cobalt, and nickel; and by secondary reactions to sulphates of alumina, lime, magnesia, and alkalies. This process of oxidation is necessarily superficial and local, but the soluble sulphates thus formed have probably played a not unimportant part. (§ 9.)

§ 29. Besides the solutions formed by this last process, which contain chiefly neutral and acid salts, there is another class of waters characterized by the presence of free sulphuric or hydrochloric acid, or both together. These acid waters sometimes occur as products of volcanic action; during which both hydrochloric acid and sulphur are often evolved in large quantities. This latter element generally comes to the surface as sulphuretted hydrogen, which by the oxidation of the hydrogen may deposit its sulphur in craters and fissures. In other cases, as shown by Dumas, the sulphur and hydrogen may be slowly and simultaneously oxidized at a low temperature, giving rise directly to sulphuric acid. Not less frequent, however, is probably the direct conversion, by combustion, of the sulphuretted hydrogen into water and sulphurous acid, which, afterwards absorbing oxygen from the air, is converted into sulphuric acid.

§ 30. The source of the hydrochloric acid and the sulphur of volcanoes is probably the decomposition of chlorides and sulphates at high temperatures. It is known that for the decomposition of earthy chlorides, water and an elevated temperature are sufficient; and at a higher temperature, chloride of sodium is readily decomposed in presence of silicious and aluminous minerals, with the intervention of water. Another agency which probably comes into play in volcanic phenomena is that of organic matters, which, reducing the sulphates to sulphurets, enable the sulphur to be subsequently disengaged as sulphuretted hydrogen by the operation of water, either with or without the intervention of carbonic acid or of silicious and argillaceous matters. Even in cases where this reducing action is excluded, the ignition of sulphates in contact with earthy matters must liberate the sulphuric acid as a mixture of sulphurous acid and oxygen; and these uniting in their distillation upward through the strata, may give rise to springs of sulphuric acid.* To reactions similar to those just noticed, involving borates like stassfurthite and hayesine, or boric silicates like tourmaline, etc., are to be ascribed the large amounts of boric acid which are sublimed in some volcanoes, or volatilized with the watery vapor of the Tuscan *suffioni*.

§ 31. The action of subterranean heat upon buried strata containing sulphates and chlorides is then sufficient to explain the appearance of hydrochloric and sulphurous acids and sulphur, even without the intervention of organic matters, which are, however, seldom or never wanting; whether as coal, lignite, bitumen, and pyroschists, or in a more divided condition. The presence of hydrogen and of marsh-gas, as observed by Deville among volcanic products, is an evidence of this. The generation of marsh-gas is, however, in most cases clearly unconnected with volcanic action or subterranean heat.

To the decomposition of carbonates in buried strata by silicious matters, with the aid of heat, is to be ascribed the great amounts of carbonic-acid gas which are in many places evolved from the earth, and, impregnating the infiltrating waters, give rise to acidulous springs. The principal sources of this gas in Europe are in regions adjoining volcanoes, either active or recently extinct; but their occurrence in the palæozoic strata of the United States, far remote from any evidence of volcanic phenomena other than slightly thermal springs, shows that an action too gentle or too deeply seated to manifest itself in igneous eruptions, may evolve carbonic acid abundantly. The sulphuric-acid springs of western New York and Canada, to be described further on, are not less remarkable illustrations of the same fact. [The origin of free carbonic acid in certain cases is, however, doubtless to be found in the reaction pointed out further on in § 66.]

§ 32. The frequent presence of ammoniacal salts in volcanic exhalations is here worthy of notice, especially when considered in connection with the rarity of nitric and ammoniacal

* See the note to § 22, on kieserite.

compounds in natural waters, except in some local conditions, as in the wells of cities, etc., where they are sometimes observed in comparatively large amounts. The explanation of this is evident; for although nitrates themselves are not directly removed from the water, they are, by the reducing action of organic matters, converted into ammonia, which is retained by the soil. In consequence of this affinity the argillaceous strata, whether of the present period or of older formations, hold in a very fixed form a considerable quantity of nitrogen. This, from the slowness with which it is eliminated in the form of ammonia under the influence of alkaline solutions, probably exists as an ammoniacal silicate. (§ 6.) The action of acids, however, as well as alkalies, may be supposed to liberate it from its combination, and thus generate the ammoniacal salts which are such frequent accompaniments of volcanic phenomena. The numerous experiments of Delesse show that ammonia, or at least nitrogen capable of being evolved by heat and alkalies in the form of ammonia, is present in the limestones, marls, argillites, and sandstones of former geological periods, in quantities scarcely inferior to those in similar deposits of modern times, amounting, for most of the ancient sedimentary strata, to from one to five thousandths of nitrogen; * from which it will be seen that the quantity of this element thus retained in the rocky strata of the earth's crust is very great.

§ 33. If we attempt a chemical classification of natural waters in accordance with the principles laid down in the preceding sections, they may be considered under the following heads :—

A. Atmospheric waters.

B. Waters impregnated with the soluble products of vegetable decay.

C. Waters impregnated with the salts from decomposing feldspathic rocks, and holding a portion of carbonate of soda as a characteristic ingredient.

D. Waters holding neutral salts of sodium, calcium, or magnesium from strata where they existed as solid salts or in brines or bitterns.

* Ann. des Mines (5), XVIII. 151–523.

E. Waters holding chiefly sulphates from decomposing pyrites; copperas and alum-waters.

F. Waters holding free sulphuric or hydrochloric acid.

§ 34. The name of mineral waters is popularly applied only to such as contain sufficient foreign matters to give them a decided taste; and hence the waters of the divisions A and B, and many of the feebler ones of C and D, are excluded. Those of E and F have peculiar local sources; but those of C and D are often associated in adjacent geological formations, and their commingling in various proportions gives rise to mineral waters intermediate in composition. In accordance with these considerations, a classification of mineral waters for technical purposes was adopted by me, in 1863, in the Geology of Canada, p. 531, including only those of C, D, and F, which were arranged in six classes.

I. Saline waters containing chloride of sodium, often with large portions of chlorides of calcium and magnesium, with or without sulphates. The carbonates of lime and magnesia are either wanting, or present only in small quantities. These waters are generally bitter to the taste, and may be designated as brines or bitterns.

II. Saline waters which differ from the last in containing, besides the chlorides just mentioned, considerable quantities of carbonates of lime and magnesia. These waters generally contain much smaller proportions of earthy chlorides than the first class, and hence are less bitter to the taste.

III. Saline waters which contain, besides chloride of sodium and the carbonates of lime and magnesia, a portion of carbonate of soda.

IV. Waters which differ from the last in containing but a small proportion of chloride of sodium, and in which the carbonate of soda predominates. The waters of this class generally contain much less solid matter than the three previous classes, and have not a very marked taste until evaporated to a small volume, when they will be found, like the last, to be strongly alkaline.

Of these four classes, I. corresponds to the division D, and IV. to C, while II. and III. are regarded as resulting from the admixture of these in varying proportions. Sulphates are sometimes present in these waters, but never predominate; in their absence, salts of barium and strontium are often met with. The chlorides are generally, if not always, associated with bromides and iodides. Small quantities of potassium-salts are also present, while borates, phosphates, silicates, and small portions of iron, manganese, and alumina, are generally present. These various waters are occasionally sulphurous, and those of the last three classes may be impregnated with carbonic acid.

V. The fifth class includes acid waters remarkable for containing a large proportion of free sulphuric acid, with sulphates of lime, magnesia, portions of iron, and alumina. These waters, which are characterized by their sour and styptic taste, generally contain some sulphuretted hydrogen.

VI. The sixth class includes some neutral saline waters in which the sulphates of lime, magnesia, and the alkalies predominate; chlorides being present only in small quantities. These waters, like the last, are often impregnated with sulphuretted hydrogen.

The above classification, although adopted originally for the convenient description of the mineral waters of Canada, will, it is thought, be found to embrace all known classes of natural waters, with the exception of those included under E, and of some waters from volcanic sources holding hydrochloric acid. These may constitute two additional classes. In the first three of the classes above described, chlorides predominate; in the fourth, carbonates; and in the fifth and sixth, sulphates. The waters of the first, second, and sixth classes are neutral; those of the third and fourth, alkaline; and those of the fifth, acid.

II.

Analyses of Various Natural Waters.

Contents of Sections.—35, 36. Waters of the first class; 37. Their probable origin; the elimination of sulphates; 38. Separation of lime-salts from waters; 39. Earthy chlorides in saliferous formations; brines of New York, Michigan, and England; foot-note on errors in water-analyses; 40. Brines of western Pennsylvania; waters in which chloride of calcium predominates; 41. Origin of such waters; separation of magnesia as an insoluble silicate; 42. Waters of the second class; 43. Waters of the third class; 44. Waters of the fourth class; Chambly; 45. Other waters of the same class; Ottawa River; 46. Waters of Highgate and Alburg; 47. Changes in the Caledonia waters; comparative analyses; 48. Waters of the fifth class; sulphuric-acid springs of New York and Canada; 49. Changes in the composition of these waters; their action on calcareous strata; 50. Waters of the sixth class; their various sources; 51. Examples of neutral sulphated waters; sulphate of magnesia waters.

[§§ 35, 36, in the original paper, contained descriptions and analyses of eight waters of Class I., as defined in § 34. These, with two exceptions, were more concentrated than sea-water, containing from 36 to 50 and even 68 parts of solid matter to 1,000. The composition of three of them is here given: the first is that of a copious spring which issues from the Trenton limestone at Whitby, Ontario; the second is that of a well sunk into the same limestone at Hallowell not far from the last, and one of several wells in the vicinity similar in character, though less concentrated; and the third is from a boring 500 feet deep, sunk through the Medina sandstone into the underlying Hudson River shales at St. Catherine, Ontario.]

Waters of Class I.	Whitby.	Hallowell.	St. Catherine.
Chloride of sodium	18.9158	38.7315	29.8034
" potassium	traces	traces	.3555
" calcium	17.5315	15.9230	14.8544
" magnesium	9.5437	12.9060	3.3977
Bromide of sodium	.2482	.4685	undet.
Iodide "	.0008	.0133	.0042
Sulphate of lime	....	....	2.1923
Carbonate of lime	.0411	....	....
" magnesia	.0227	....	....
" baryta and strontia	undet.	....	....
In 1,000 parts	46.3038	68.0423	50.6075

§ 37. The waters of the first class contain, besides chloride of sodium and a little chloride of potassium, large quantities of the chlorides of calcium and magnesium, amounting together, in several cases, to more than one half the solid contents of the water. Sulphates are either absent, or occur only in small quantities, and the same is true of earthy carbonates. Salts of baryta and strontia are sometimes present, while the proportions of bromides and iodides, though variable, are often considerable.

In the large amount of magnesian chloride which they contain, these waters resemble the bittern or mother-liquor which remains after the greater part of the chloride of sodium has been removed from sea-water by evaporation. The bitterns from modern seas, however, differ in the constant presence of sulphates, and in containing, when sufficiently concentrated, only traces of lime. The reason of this, as already pointed out in § 22, is to be found in the fact that in the waters of the present ocean the sulphates are much more than equivalent to the lime, so that this base separates during evaporation as gypsum.* But as shown in § 23 and § 24, the waters of the ancient seas, which held in the form of chloride of calcium the greater part of the lime since deposited as carbonate, must have yielded by evaporation bitterns containing a large proportion of chloride of calcium. Such is the nature of the brines whose analyses are given above, and such we suppose to have been their origin. The complete absence of sulphates from many of these waters points to the separation of large quantities of earthy sulphates in the Cambrian strata from which these saline springs issue; and the presence in many of the dolomitic beds of the Calciferous sand-rock of small masses of gypsum abundantly disseminated is an evidence of the elimination of the sulphates by evaporation. The frequent occurrence of crystalline masses of sulphate of strontian in the Chazy and Black River limestones of this region is also to be noted as another means by which the sulphates were separated from the waters of the palæozoic seas. From the proportions of chloride

* See further on this point, Bischof, Chem. Geology, I. 413.

of sodium, varying from about one third to more than two thirds of the solid contents of the above waters, it is apparent that in most cases the process of evaporation had gone so far as to separate a part of the common salt; and thus successive strata of this ancient saliferous formation must be impregnated with solid or dissolved salts of unlike composition. The mingling of these in varying proportions affords the only apparent explanation of the differences which appear in the relative amounts of the several chlorides in waters from the same region, and even from adjacent sources.

§ 38. The great solubility of chloride of calcium renders it difficult to suppose its separation from the mother-liquors so as to be deposited in a solid state in the strata.* The same remark applies to chloride of magnesium. It is however to be remarked that the double chloride of potassium and magnesium (carnallite) is decomposed by deliquescence into solid chloride of potassium and a solution of chloride of magnesium; and thus strata like those which at Stassfurth contain large quantities of carnallite (§ 22), might give rise to solutions of magnesian chloride. This, however, would require the presence of a large amount of chloride of potassium in the early seas. It appears from the analyses above referred to that the chloride of magnesium sometimes surpasses in amount the chloride of calcium; and sometimes, on the contrary, is equal to only one half or one fourth of the latter salt. While it is not impossible that the predominance of the magnesian chloride in some waters may be traced to the decomposition of carnallite, it is undoubtedly in most cases connected with the action of solutions of carbonate of soda; the effect of which, as already pointed out, is to first separate the soluble lime-salt as carbonate, leaving to a subsequent stage the magnesian chloride. (§ 18.) As this reaction replaces the calcium-salt by chloride of sodium, it might be expected that there would be an increase in the amount of the latter salt in the water wherever the magnesian chloride predominates, did we not remember that

* [A hydrated double chloride of calcium and magnesium (tachydrite) has since been found at Stassfurth.]

evaporation separates it from the water in the solid form; and that the two processes, one of which replaces the chloride of calcium by chloride of sodium, while the other eliminates the latter salt from the solution, might have been going on simultaneously or alternately. As the nature of the waters now under consideration shows that the process of evaporation had been carried so far as to separate the sulphate in the form of gypsum, and probably also a portion of the chloride of sodium in a solid state, it is evident that we have not yet the data necessary for determining the composition of the water of the ancient Cambrian ocean, as regards the proportions of the sodium, calcium, and magnesium which it held in solution; and we can only conclude from these mother-liquors, that the amount of the earthy bases was relatively very large.

§ 39. As already remarked in § 22, the mother-liquor from modern sea-water contains no chloride of calcium, but, on the contrary, large quantities of sulphate of magnesia; the lime in the modern ocean being less than one half that required to combine with the sulphate present. If, however, we examine the numerous analyses of rock-salt and of brines from various saliferous formations, we shall find that chloride of calcium is very frequently present in both of them; thus supporting the conclusions already announced in § 24 with regard to the composition of the seas of former geological periods. The oldest saliferous formation which has been hitherto investigated is the Onondaga Salt-group of the New York geologists, which belongs to the upper part of the Silurian series, and supplies the strong brines of Syracuse and Salina in New York. These, notwithstanding their great purity, contain small proportions of chlorides of calcium and magnesium, as shown by the analyses of Beck, and the recent and careful examinations of Goessmann. In the brines of this region the solid matters are equal to from 14.3 to 16.7 per cent, and contain on an average, according to the latter chemist, 1.54 of sulphate of lime, 0.93 of chloride of calcium, and 0.88 of chloride of magnesium in 100; the remainder being chloride of sodium.*

* Goessmann, Reports on the Brines of Onondaga: Syracuse, 1862 and 1864; also Report on the Onondaga Salt Co.: Syracuse, 1862.

The nearly saturated brines from the Saginaw valley in Michigan, which have their source at the base of the carboniferous series, contain, according to my calculation from an analysis by Professor Dubois, in 100 parts of solid matters: chloride of calcium 9.81, chloride of magnesium 7.61, sulphate of lime 2.20, the remainder being chiefly chloride of sodium. Another brine in the same vicinity gave to Chilton an amount of chloride of calcium equal to 3.76 per cent.* In a specimen of salt manufactured in this region, Goessmann found 1.09 of chloride of calcium; and in two specimens of salt from the brines of Ohio, from the same geological horizon, 0.61 and 1.43 per cent of the same chloride. The rock-salt from the lias of Cheshire, according to Nicol, contains small cavities, partly filled with air, and partly with a concentrated solution of chloride of magnesium, with some chloride of calcium.†

* Winchell, Amer. Jour. Sci. (2), XXXIV. 311.

† Edin. Neu. Phil. Jour., VII. 111. The results of the analyses by Mr. Northcote of the brines of Droitwich and Stoke in the same region (L. E. & D. Philos. Mag. (4), IX. 32), as calculated by him, show no earthy chlorides whatever, and no carbonate of lime, but carbonates of soda and magnesia, and sulphates of soda and lime. He regarded the whole of the lime present in the water as being in the form of sulphate. If, however, we replace, in calculating these analyses, the carbonate of soda and sulphate of lime by sulphate of soda and carbonate of lime, we shall have for the contents of these brines: — chloride of sodium, with notable quantities of sulphate of soda, some sulphate of lime, and carbonates both of lime and magnesia; a composition which is more in accordance with the admitted laws of chemical combinations. From these results, it would appear that the earthy chlorides, which according to Nicol are present in the rock-salt of this formation, are decomposed by sulphates in the waters which, by dissolving it, give rise to the brines.

It is to be regretted that in many water-analyses by chemists of note, the results are so calculated as to represent the coexistence of incompatible salts. Of the association of carbonates of soda and magnesia with sulphate of lime, as in the analysis just noted, it might be said that I have shown that it may occur in the presence of an excess of carbonic acid (*ante*, page 90). By evaporation, however, such solutions regenerate carbonate of lime and sulphates of soda and magnesia; and by the consent of the best chemists these elements are to be represented as thus combined. But what shall be said when chloride of magnesium, carbonate of soda, and silicate of soda are given as the constituents of a water whose recent analysis may be found in a late number of the Chemical News; or when bicarbonates of soda, magnesia, and lime are represented as coexisting in a water with sulphates and chlorides of magnesium and aluminum? These errors probably arise from determining in

§ 40. The brines from the valley of the Alleghany River, obtained from borings in the coal formation, are remarkable for containing large proportions of chlorides of calcium and magnesium; though the sum of these, according to the analyses of Lenny, is never equal to more than about one fourth of the chloride of sodium. The presence of salts of barium and strontium in these brines, and the consequent absence of sulphates, is, according to Lenny, a constant character in this region over an area of two thousand square miles. (See Bischof, Chem. Geol., I. 377.) A later analysis of another one of these waters from the same region, by Steiner, is cited by Will and Kopp, Jahresbericht, 1861, p. 1112. His results agree closely with those of Lenny. See also the analysis of a bittern from this region by Boyé. (Amer. Jour. Sci. (2), VII. 74.)*

These remarkable waters approach in character to those of Whitby and Hallowell; but in this the chloride of sodium forms only about one half the solid contents, and the proportion of the chloride of magnesium to the chloride of calcium is relatively much greater than in the waters from western Pennsylvania, where the magnesian chloride is equal only to from one third to one fifth of the chloride of calcium; the proportions of the two being subject in both regions to considerable variations.

In this connection may be cited a water from Bras d'Or in the island of Cape Breton, lately analyzed by Professor How, which contains in 1,000 parts, chloride of sodium 4.901, chloride of potassium 0.650, chloride of calcium 4.413, and chloride of magnesium only 0.638, besides sulphate of lime 0.134, carbonates of lime and magnesia 0.085, with traces of iron-oxide and phosphates; = 10.821. (Canadian Naturalist, VIII. 370.)

the recent water, or in water not sufficiently boiled, the lime and magnesia which would by prolonged ebullition be separated as carbonates, together with portions of alumina, silica, etc. In the subsequent calculation of the analyses, these dissolved earthy bases being regarded as sulphates or chlorides, instead of carbonates, there remains an excess of soda, which is wrongly represented as carbonate, instead of chloride or sulphate of sodium.

* [For further examples of waters of this class from western Ontario, see the Supplement to this paper.]

The analyses of European waters furnish comparatively few examples of the predominance of earthy chlorides.*

§ 41. We have already shown in § 38 how the action of carbonate of soda upon sea-water or bittern will destroy the normal proportion between the chlorides of magnesium and calcium by converting the latter into an insoluble carbonate, and leaving at last only salts of sodium and magnesium in solution. A process the reverse of this has evidently intervened for the production of waters like that from Cape Breton, and some others noticed by Lersch, in which chloride of calcium abounds, with little or no sulphate or chloride of magnesium. This process is probably one connected with the formation of a silicate of magnesia. Bischof has already insisted upon the sparing solubility of this silicate, and has asserted that silicates of alumina, both artificial and natural, when digested with a solution of magnesian chloride, exchange a portion of their base for magnesia, thus giving rise to solutions of alumina; which, being decomposed by carbonates, may have been the source of many of the aluminous deposits referred to in § 9. He also observed a similar decomposition between a solution of an artificial silicate of lime and soluble magnesian salts. (Bischof, Chem. Geology, I. 13; also Chap. XXIV.) In repeating and extending his experiments, I have confirmed his observation that a solution of silicate of lime precipitates silicate of magnesia from the sulphate and the chloride of magnesium; and have moreover found that by digestion at ordinary temperatures with an excess of freshly precipitated silicate of lime, chloride of magnesium is completely decomposed; an insoluble silicate of magnesia being formed, while nothing but chloride of calcium remains in solution. It is clear that the greater insolubility of the magnesian silicate, as compared with silicate of lime, determines a result the very reverse of that produced by carbonates with solutions of the two earthy bases. In the one

* Lersch, Hydro-Chemie, Zweite Auflage: Berlin, 1864; *vide* p. 207. This excellent work, which is a treatise on the chemistry of natural waters, in one volume 8vo of 700 pages, was unknown to me when I prepared the first part of this essay.

case the lime is separated as carbonate, the magnesia remaining in solution; while in the other, by the action of silicate of soda (or of lime), the magnesia is removed and the lime remains. Hence carbonate of lime and silicates of magnesia are found abundantly in nature; while carbonate of magnesia and silicates of lime are produced only under local and exceptional conditions. It is evident that the production from the waters of the early seas of beds of sepiolite, talc, serpentine, and other rocks in which a magnesian silicate abounds, must, in closed basins, have given rise to waters in which chloride of calcium would predominate.

[§ 42 of the original paper contains descriptions and analyses of eight waters of Class II., the solid contents of which vary from 9 to 20 parts in 1,000; they rarely contain sulphates. The three given below, which may be taken as examples, rise from the Trenton limestone of the Ottawa and St. Lawrence valleys, the first being that known as the Intermittent Spring of Caledonia.]

Waters of Class II.	Caledonia.	Lanoraie.	St. Léon.
Chloride of sodium	12.2500	11.1400	11.4968
" potassium	.0305	.1460	.1832
" barium	....	.0303	.0019
" strontium	....	.0185	.0019
" calcium	.2870	.2420	.0718
" magnesium	1.0338	.2790	.6636
Bromide of "	.0238	.0283	.0091
Iodide of "	.0021	.0052	.0046
Carbonate of baryta	....	.0106	....
" strontia	....	.0137	....
" lime	.1264	.4520	.3493
" magnesia	.8632	.4622	.9388
" iron	traces	traces	.0145
Silica	.0225	.0552	.0865
Alumina	undet.	undet.	.0145
In 1,000 parts	14.6393	12.8830	13.8365
Specific gravity	1010.9	1009.42	1011.23

[§ 43 gives the description and analysis of eight waters of Class III. which hold from less than 5 to more than 10 parts of solid water in 1,000. Of the three whose analysis is given below, the first rises from the Chazy formation in the Ottawa valley, and the others from the Utica and Hudson River formations in the valley of the St. Lawrence. The alkaline-saline waters of Caledonia, belonging to the same class, which will be mentioned further on in § 47, rise from the Trenton limestone in the former region.]

Waters of Class III.	Fitzroy.	Varennes.	Baie du Febvre.
Chloride of sodium	6.5325	9.4231	4.8234
" potassium	.1160	.1234	.0610
Bromide of sodium	.0217	.0126	undet.
Iodide of "	.0032	.0054	undet.
Phosphate of soda	.0124	....	....
Carbonate of "	.5885	.1705	1.5416
" baryta	traces	.0226	traces
" strontia	"	.0140	"
" lime	.1500	.3540	.2180
" magnesia	.7860	.5433	.4263
" iron	traces	.0048	....
Alumina	.0040	traces	undet.
Silica	.1330	.0465	.2129
In 1,000 parts	8.3473	10.7202	7.2923
Specific gravity	1006.24	1008.15	....

§ 44. Of the waters of Class IV. the first to be noticed is one occurring at Chambly, on the Richelieu River, in the province of Quebec. Here, on a plateau, over an area of about two acres, the clayey soil is destitute of vegetation and impregnated with alkaline waters; which in the dry season give rise to a saline efflorescence on the partially dried up and fissured surface. A well sunk here to the depth of eight or ten feet in the clay, which overlies the Hudson River formation, affords at all times an abundant supply of water, which generally flows in a little stream from the top of the well. Small bubbles of carburetted hydrogen are sometimes seen to escape from the water. The temperature at the bottom of the well was found in October, 1861, to be 53° F., and in August, 1865, to be nearly 54° F. The mean temperature of Chambly can differ but little from that of Montreal, which is 44.6° F., so that this is a thermal water. Another alkaline and saline spring in the same parish has also a temperature of 53° F. The water of the spring here described has a sweetish saline taste, and is much relished by the cattle of the neighborhood. Three analyses have been made of its waters, the results of which are here given side by side. The first was collected in October, 1851; the second in October, 1852; and the third in August, 1864, during a very dry season.

Waters of Chambly, Class IV.	I.	II.	III.
Chloride of potassium	undet.	.0324	.0182
" sodium	.8689	.8387	.8846
Carbonate "	1.0295	1.0604	.9820
" lime	.0540	.0380	.0253
" magnesia	.0908	.0765	.0650
" strontia	undet.	.0045	undet.
" iron	"	.0024	"
Alumina and phosphate	"	.0063	"
Silica	.1220	.0730	.0166
Borates, iodides, and bromides	undet.	undet.	undet.
In 1,000 parts	2.1652	2.1322	1.9917

A portion of barium is included with the strontium salt. The water contains moreover a portion of an organic acid, which causes it to assume a bright brown color when reduced by evaporation. Acetic acid gave no precipitate with the concentrated and filtered water; but the subsequent addition of acetate of copper yielded a brown precipitate of what was regarded as apocrenate of copper. The organic matter of this and of many other mineral springs has probably a superficial origin. The carbonic acid was determined in the third analysis, and was equal in two trials to .903 and .905. The neutral carbonates in this water require .452 parts of carbonic acid.

[§§ 45, 46, give the analyses of six more waters of Class IV., none of which are as highly charged with mineral substances as that of Chambly, though holding from 0.34 to 1.55 parts of solid matter to 1,000. All of these waters are found in the valleys of the St. Lawrence and of Lake Champlain, and are believed to rise from the Utica or Hudson River shales.

The analyses of the three given below may be taken as additional examples of this class. That of St. Ours is remarkable for a large proportion of potassium-salts, about twenty-five per cent of the alkalies, determined as chlorides, being chloride of potassium.]

Waters of Class IV.	St. Ours.	Joly.	Nicolet.
Chloride of sodium	.0207	.0347	.3920
" potassium	.0496	.0076	.0318
Sulphate of potash	.0081	......	......
Carbonate of soda	.1340	.1952	1.1353
" lime	.1740	.0710	undet.
" magnesia	.1287	.0278	"
Iron-oxide, alumina, and phosphates	traces	......	"
Silica	.0161	.0110	"
In 1,000 parts	.5311	.3473	1.5591

To the above may be joined, for comparison, the analysis of the waters of a large river, the Ottawa, which drains a region occupied chiefly by crystalline rocks, covered by extensive forests and marshes. The soluble matters which it contains are therefore derived in part from the superficial decomposition of these rocks, and in part from the decaying vegetation. The water, which was taken at the head of the St. Anne's rapids, on the 9th of March, 1854, before the melting of the winter's snows had begun, had a pale amber-yellow hue, from dissolved organic matter, which gave a dark brown color to the residue after evaporation. The weight of this residue from 10,000 parts, dried at 300° F., was .6975, which after ignition was reduced to .5340 parts. As seen in the table below, one half of the solid matters in this water were earthy carbonates, and more than one third was silica, so that the whole amount of salts of alkaline bases was .088 (of which nearly one half is carbonate of soda); while the St. Ours water, which resembles that of the Ottawa in its alkaline salts, contains in the same quantity 4.248, or more than forty-eight times as much. The alkalies of the Ottawa water equalled as chlorides .0900, of which .0293, or 32.5 per cent, were chloride of potassium. The results of some observations on the silica and the organic matters of this river-water will be given further on (§§ 70, 71). It will be observed that while the contents of all the other waters in this paper are given for 1,000 parts, those of the Ottawa are calculated for 10,000 parts.

Water of the Ottawa River.	
Chloride of potassium	.0169
Sulphate of soda	.0188
" potassium	.0122
Carbonate of soda	.0410
" lime	.2680
" magnesia	.0690
Iron-oxide, alumina, and phosphates	traces
Silica	.2060
In 10,000 parts	.6116

§ 47. It was an interesting question to determine whether the composition of these various waters remains constant. Having collected and analyzed, in September, 1847, the waters of three springs in Caledonia, Ontario, belonging to Class III., and not far from the spring of Class II. in the same town, noticed in § 42, I again visited and collected for examination the waters of the same springs in January, 1865, after a lapse of more than seventeen years. The results, when compared as below, show that considerable changes have occurred in the composition of each of these springs, and tend to confirm in an unexpected manner the theory which I had long before put forward, — that the waters of the second and third classes owe their origin to the mingling of saline waters of the first class with alkaline waters of the fourth class. It will be observed that the three Caledonia waters in 1847 were all alkaline, although the proportions of carbonate of soda were unlike. Sulphates were then present in all of them, but most abundant in the Sulphur Spring, which, although holding the smallest amount of solid matters, was the most alkaline. In January, 1865, however, the first and second of these waters had ceased to be alkaline, and contained, instead of carbonate of soda, small quantities of earthy chloride, causing them to enter into the second class. They no longer contained any sulphates, but, on the contrary, portions of baryta and strontia. Only the Sulphur Spring, which in 1847 contained the largest proportion of carbonate of soda and of sulphates, still retained these elements, though in diminished amounts, and was feebly impregnated with sulphuretted hydrogen. If we suppose these waters to arise from the commingling of saline waters of the first or second class, like those of Whitby and Lanoraie, containing earthy chlorides and salts of baryta and strontia, with a water of the fourth class holding carbonate and sulphate of soda, it is evident that a sufficient quantity of the latter water would decompose the earthy chlorides and precipitate the salts of baryta and strontia present, while an excess would give use to alkaline-saline waters containing sulphate and carbonate of soda, such as were the three springs of Caledonia in 1847. A falling off in the supply of the sulphated alkaline water may be supposed to have taken place, and the result is seen in the appearance of chloride of magnesium and of baryta and strontia in two of the springs, and in a diminished proportion of carbonate of soda in the Sulphur Spring.*

These later analyses being directed chiefly to the determination of these changes, no attempt was made to determine potassium, iodine, or bromine. For the purposes of comparison, the two series of analyses† are here put in juxtaposition; the element just mentioned being included with the chloride of sodium, and the figures reduced to three places of decimals. The precipitate by a solution of gypsum from the concentrated and acidulated water was regarded as sulphate of strontia, and calculated as such, but was in part sulphate of baryta.

* [The Harrowgate springs, in England, have undergone changes not unlike those of Caledonia. Several of the Harrowgate waters, all of which were found by Dr. Hofman, in 1854, to contain sulphate of lime, were examined by Mr. Davis, in 1866, and found, with one exception, to be free from sulphate, and to contain instead salts of baryta, even in the sulphuretted waters. Great differences are there, as elsewhere, observed between closely adjacent springs; and in one of them, a strong saline holding chloride of barium, Dr. Muspratt detected a small amount of protochloride of iron. (Chemical News, Vol XIII., *passim.*)]

† [The complete earlier analyses are given in the original paper.]

Table showing the Changes in the Caledonia Springs.

	1. Gas Spring.		2. Saline Spring.		3. Sulphur Spring.	
	1847.	1865.	1847.	1865.	1847.	1865.
Chlor. sodium .	7.014	6.570	6.488	6.930	3.876	3.685
" magnesium .	......	.024	......	.026	......	......
Sulph. potash .	.005	......	.005	......	.018	.021
Carb. soda . .	.048	......	.176	......	.456	.091
" lime .	.148	.096	.117	.095	.210	.077
" magnesia .	.526	.455	.517	.469	.294	.228
" strontia .	......	.009	......	.012	......	......
Silica . . .	.021	.020	.042	.015	.084	.021
In 1,000 parts .	7.772	7.174	7.345	7.547	4.938	4.123

In the later analyses of these waters, the carbonic acid in the Gas Spring was found to equal, for 1,000 parts, .671; of which .278 were required for the neutral carbonates. The Saline Spring contained .664 of carbonic acid; of which .290 go to make up the neutral carbonates. The Sulphur Spring, in like manner, gave of carbonic acid .573; while the neutral carbonates of the water require only .191. All of these waters, in January, 1865, thus contained an excess of carbonic acid above that required to form bicarbonates with the carbonated bases present; while the analyses of the same springs in 1847 showed a quantity of carbonic acid insufficient for the formation of bicarbonates with these bases. The questions of the cause of this deficiency, and of the variation in the amount of carbonic acid in these and other waters, will be considered in the third part of this paper.

§ 48. The waters of our fifth and sixth classes, as defined in § 34, are distinguished by the presence of sulphates; the former being acid, and the latter being neutral waters. In the fifth class the principal element is sulphuric acid, associated with variable and accidental amounts of sulphates of alkalies, lime, magnesia, alumina, and iron. Apart from the springs of this kind which occur in regions where volcanic agencies are evidently active, the only ones hitherto studied are those of New York and western Canada, which issue from almost horizontal Silurian rocks (§ 31). The first account of these remarkable waters was given in the Amer. Jour. Sci. in 1829 (Vol. XV. p. 238), by the late Professor Eaton, who described two acid springs in Byron, Genesee County, N. Y.; one yielding a stream of distinctly acid water sufficient to turn a mill-wheel, and the other affording in smaller quantities a much more acid water. The latter was afterwards examined by Dr. Lewis Beck (Mineralogy of New York, p. 150). He found it to be colorless, transparent, and intensely acid, with a specific gravity of 1.113; which corresponds to a solution holding seventeen per cent of oil of vitriol. No chlorides, and only traces of lime and iron, were found in this water, which was nearly pure dilute sulphuric acid. Professor Hall (Geology of New York, 4th District, p. 134) has noticed, in addition to these, several other springs and wells of acid water in the adjacent town of Bergen. Farther westward, in the town of Alabama, is a similar water, whose analysis by Erni and Craw will be found in the Amer. Jour. Sci. (2), IX. 450. It contained in 1,000 parts about 2.5 of sulphuric acid, and 4.6 parts of sulphates, chiefly of lime, magnesia, iron, and alumina. In this, as in the succeeding analyses, hydrated sulphuric acid, SO_3,HO, is meant.

The earliest quantitative analyses of any of these waters were those by Croft and myself of a spring at Tuscarora, in 1845 and 1847, of which the detailed results appear in the Amer. Jour. Sci. (2), VIII. 364. This, at the time of my analysis in September, 1847, contained, in 1,000 parts, 4.29 of sulphuric acid, and only 1.87 of sulphates; while the previous analysis by Professor Croft gave approximatively 3.00 of neutral sulphates, and only about 1.37 of sulphuric acid. Similar acid waters occur on Grand Island above Niagara Falls and at Chippewa.

All of these springs, along a line of more than 100 miles from east to west, rise from the outcrop of the Onondaga salt-group; but in the township of Niagara, not far from Queenston, are two similar waters which issue from the Medina sandstone. One of these is in the southwest part of the township, and fills a small basin in yellow clay, which, at a depth of three or four feet, is underlaid by red and green sandstones. The water, which, like those of Tuscarora and Chippewa, is slightly impregnated with sulphuretted hydrogen, is kept in constant agitation from the escape of inflammable gas. It contained in 1,000 parts about two parts of free sulphuric acid, and less than one part of neutral sulphates. This water was collected in October, 1849, and at that time another half-dried-up pool in the vicinity contained a still more acid water. Another similar spring occurs near St. David, in the same township. In connection with the suggestion made in § 31 as to their probable origin at great depths, it would be very desirable to have careful observations as to the temperature of these acid springs. When, on the 19th October, 1847, I visited the Tuscarora spring, the water in two of the small pools had a temperature of 56° F.; but on plunging the thermometer in the mud at the bottom of one of these it rose to 60.5°.

§ 49. It appears from a comparison of the analysis of Croft with my own that the waters of the Tuscarora spring underwent a considerable change in composition in the space of two years; the proportion of the bases to the acid at the time of the second analysis being little more than one third of that in the analysis of Croft. This change was indeed to be expected, since waters of this kind must soon remove the soluble constituents from the rocks through which they flow, and eventually become, like the water from Byron, little more than a solution of sulphuric acid. The observations of Eaton at Byron, and my own at Tuscarora, show that half-decayed trees are still standing on the soil which is now so impregnated with acid waters as to be unfit to support vegetation. Reasoning from the changes in composition, it may be supposed that these waters were at first neutral, the whole of the acid being saturated by the calcareous rocks through which they must rise. It was from this consideration that I was formerly led to ascribe to the action of these waters the formation of some of the masses of gypsum which appear along the outcrop of the Onondaga salt-group (Amer. Jour. Sci. (2), VII. 175). That waters like those just mentioned must give rise to sulphate of lime by their action on calcareous rocks is evident; and some of the deposits of gypsum in this region, as described by good observers, would appear to be thus formed. So far, however, as my personal observations of the gypsums of western Canada have extended, these appear to be in all cases contemporaneous with the shales and dolomites with which they are interstratified, and to have no connection with the sulphuric-acid springs which are so common throughout that region. (Ibid. (2), XXVIII. 365; and Geology of Canada, 352.)

§ 50. We have included in a sixth class the various neutral saline waters in which sulphates predominate, sometimes to the exclusion of chlorides. The bases of these waters are soda, potash, lime, and magnesia; which are usually found together, though in varying proportions. For the better understanding of the relations of these sulphated waters, it may be well to recapitulate what has been said about their origin; and to consider them, from this point of view, under two heads.

First, those formed from the solution of neutral sulphates previously existing in a solid form in the earth. Strata enclosing natural deposits of sulphates of soda and magnesia, sometimes with sulphate of potash (§§ 17, 19), afford the most obvious source of these waters. The frequent occurrence of gypsum, however, points to this salt as a more abundant source of sulphated waters. Solutions of gypsum may in some case exchange their lime for the soda of insoluble silicates, or this salt may be decomposed by solutions of carbonate of soda (§§ 7, 19). The decomposition of the sulphate of lime by hydrous carbonate of magnesia, as explained in § 21, is doubtless in many cases the source of sulphate of magnesia, which, more frequently than sulphate of soda, is a predominant element in mineral waters. In connection with a suggestion made in the section last cited, it may be remarked that I have since found that predazzite, in virtue of the hydrate of magnesia which it contains, readily decomposes solutions of gypsum holding dissolved carbonic acid, and gives rise to sulphate of magnesia.

In the second place, sulphuric-acid waters, like those described in § 47, by their action upon calcareous and magnesian rocks, or by the intervention of carbonate of soda, may, as already suggested, give rise to neutral sulphated waters of the sixth class. It is evident also that waters impregnated with sulphates of alumina and iron from oxidizing sulphates, as mentioned in § 28, may be decomposed in a similar manner, and with like results.

Neutral sulphated waters generated by any of the above processes are evidently subject to admixtures of saline matters from other sources, and may thus become impregnated with chlorides and carbonates. Indeed, it is rare to find waters of the sixth class without some portion of chlorides; and a transition is thus presented to the waters of the first four classes, in which also portions of sulphates are of frequent occurrence. The presence of sulphates being one of the conditions required for the generation of sulphuretted hydrogen (§ 10), we find that the waters of the sixth class are very often sulphurous.

§ 51. Waters of the sixth class are very frequently met with in the palæozoic rocks of New York and western Canada, and are probably derived from the gypsum which is found in greater or less abundance at various horizons, from the Calciferous sand-rock to the Onondaga salt-group. It is, however, not improbable that the sulphuric-acid waters which abound in this region (§ 48) may, by their neutralization, give rise to similar springs. In the waters of the district under consideration, the sulphate of lime generally predominates over the sulphates of the other bases, and chlorides are frequently present in considerable quantities. For numerous analyses of these waters, see Beck, Mineralogy of New York. The results of an examination by me of the Charlotteville spring, remarkable for the amount of sulphuretted hydrogen which it contains, will be found in the Amer. Jour. Sci. (2), VIII. 369. A copious sulphur spring which issues from a mound of calcareous tufa in Brant, in Ontario, overlying the Corniferous limestone, is distinguished by the absence of any trace of chlorides; in which respect it resembles the acid waters of the fifth class from the adjacent region. A partial analysis of a portion of it collected in 1861 gave, for 1,000 parts, sulphate of lime 1.240, sulphate of magnesia .207, and carbonate of lime .198. From a slight excess in the amount of sulphuric acid, it is probable that a little sulphate of soda was also present.

Of waters of this class, in which sulphate of magnesia predominates, but few have yet been observed in this country. A remarkable example of this kind, from Hamilton, Ontario, was examined by Professor Croft, of Toronto, and described by him in the Canadian Journal for 1853 (page 153). It had a specific gravity of 1006.4, and gave, for 1,000 parts, —

Chloride of sodium	.5098
Sulphate of soda	1.6985
" lime	1.1246
" magnesia	4.7799
	8.1128

The rocks exposed at Hamilton include the Medina sandstone and the Niagara limestone, with the intermediate Clinton group. Along the outcrop of the latter, crystalline crusts of nearly pure sulphate of magnesia are observed to form in many localities, during the dry season of the year. (Geology of Canada, 460.)

III.

Chemical and Geological Considerations.

Contents of Sections. — 52. Salts of alkaline metals; proportion and sources of potash; 53. Potassium and sodium in the primitive sea; 54. Salts of lime and magnesia; relations of chlorides and carbonates; 55. Solubility of earthy carbonates; 56. Supersaturated solutions of carbonates of lime and magnesia; 57. Salts of barium and strontium; solution of their sulphates; 58. Iron, manganese, alumina, and phosphates; 59. Bromides and iodides; the small portion of bromine and the excess of iodine in saline springs as compared with the modern ocean; 60. Probable relation of iodides to sediments; 61. Sulphates, their elimination from waters; 62. Water holding a soluble sulphuret; 63. Borates, their detection; 64. Analysis of a borax-water from California; 65. Carbonates, their amount in the Caledonia waters; 66. Intervention of neutral carbonate of soda; 67. Deficiency of carbonic acid in waters; 68. Reactions of various waters; 69. Silica, its source and its proportion; 70. Its conditions; formation of silicates; 71. Organic matters; 72. Geological position of the waters here described; 73. Succession of palæozoic strata; lithological relations of successive formations; 74. Quebec group, its waters; 75. Sources of various classes of waters; 76. Their relation to the formations; 77. Associations of unlike waters; changes in constitution; 78. Temperature of springs; thermal waters; 79. Geological interest of the above analyses; possible results of the evaporation of these springs.

§ 52. Salts of the Alkaline Metals. — These salts abound in most saline waters, and, except in the few cases in which sulphate of magnesia prevails, form a large part of the soluble matters present. The salts of sodium are by far the most abundant, and the proportion of potassium-salt is generally small. The chloride of potassium in modern sea-water constitutes three or four hundredths of the alkaline chlorides, while in the brines from old rocks, and in saline waters of the first two classes alike from Germany, England, the United States, and Canada, its proportion is much less, sometimes amounting to traces only. In the waters of Classes III. and IV., where alkaline carbonates appear, and even predominate, the proportion of potassium-salt becomes greater. Thus of the waters of the latter class (§ 45), the alkalies of the Nicolet spring calculated as chlorides contain 1.89 per cent of chloride of potassium, and those of the Jacques-Cartier 2.95; while for the St. Ours spring the chloride of potassium is equal to not less than 25.0 per cent. There does not, however, appear to be any relation between the proportion of alkaline carbonate and that of potassium, since the salts from the waters first named are more alkaline than those of St. Ours; while those of the alkaline water of Joly contain less than one per cent of potassic chloride.

The amount of this salt obtained from the water of the Ottawa River is worthy of notice, being equal to not less than 32.0 per cent of the alkaline chlorides, while in the waters of the St. Lawrence it amounts to 16.0 per cent.* A large proportion of potassium relatively to the sodium has already been observed in the case of many ordinary river and spring waters, and this is readily explained when we consider the extent to which potash is set free by the decomposition of both vegetal and mineral matters at the earth's surface. The process by which this base is eliminated in filtering through soils has already been explained in § 5. The occasional presence of considerable amounts of potash in sulphated mineral waters (Lersch, Hydro-chemie, page 346) is explained by the power of solutions of gypsum to set free this alkali from soils (§ 7), and also probably in some cases by the dissolution of double potassic salts like polyhallite. Strata holding glauconite, which occurs alike in palæozoic and more recent formations,† may also be conceived to yield potash-salts to infiltrating waters.

§ 53. It will be seen that the waters above noticed, in which the proportion of the potash to the soda is large, are but feebly saline, so that the real amount of potassium is in no case great. The fact of especial importance as regards the alkaline metals in the waters whose analyses we have given

* See London, Edinburgh and Dublin Phil. Mag. (4), XIII. 239, and Geology of Canada, page 565, where analyses of both of these waters may be found.

† For a notice, with analyses by the author, of a green hydrated silicate of alumina, iron and potash, allied to glauconite, from the palæozoic rocks of Canada and of the Mississippi valley, see the Geology of Canada, pages 487, 488; where also will be found an analysis by the author of the glauconite from the cretaceous formation of New Jersey. See also Amer. Jour. Science (2), XXX. 277.

in this paper is the very small amount of potassium in the strongly saline muriated waters of the first three classes, which we conceive to be more or less directly derived from the waters of the ancient ocean. To this primeval sea, almost destitute of potassium, the process of mineral decay has been for ages adding potash-salts, and, despite the partial elimination of these by vegetation (§ 5), and by the formation of glauconite, we find a notable proportion of potash in the waters of the modern ocean.

In the analyses of the saline waters here given lithia was sought for in a few instances, and was detected in the waters of Varennes. Most of these analyses were made before the discovery of the new metals cæsium and rubidium.

§ 54. Salts of Calcium and Magnesium. — We have to consider under this head the relations both of the chlorides and the carbonates of these bases. The bitter saline waters of the first class, although containing large quantities of chlorides of calcium and magnesium, are, as we have seen, generally destitute of earthy carbonates. These latter, however, are found in small quantities in the alkaline waters of the fourth class, and in somewhat larger amounts in those intermediate waters which form Classes II. and III., and are apparently formed by admixtures of the two classes previously mentioned. Besides the carbonates of lime and magnesia which the waters of the fourth class hold in solution, the carbonate of soda which they contain gives rise, by its reaction with the chlorides of calcium and magnesium, to additional quantities of the carbonates of these bases. In the bitter saline waters of Kingston (described in the original paper) a large amount of chloride of calcium is associated with earthy carbonates, and these waters thus offer a passage from the first to the second class.

In most of the waters of the second class, as will be seen from the table in § 42, there appears but a small amount of chloride of calcium; and even this depends upon the manner in which the analysis has been conducted. We may suppose in the recent water such a partition of bases between the chlorine and the carbonic acid that chloride of calcium, chloride of magnesium, bicarbonate of lime, and bicarbonate of magnesia coexist. When such a solution is submitted to evaporation at ordinary temperatures, provided there is present a sufficient amount of chloride of calcium, carbonate of lime alone is deposited, and chloride of magnesium remains in solution. In case the chloride of calcium is insufficient, the lime is still first deposited as carbonate, and the more soluble magnesian carbonate is precipitated by further evaporation. When, however, such a water is boiled, a reverse process takes place, — the carbonate of lime slowly decomposes the magnesian chloride, and carbonate of magnesia is deposited, while chloride of calcium remains in solution. Hence if the amount of chloride of magnesium be great enough, and the ebullition sufficiently prolonged, the precipitate will at length contain only carbonate of magnesia; while an equivalent of chloride of calcium, now found in the solution, represents the carbonate of lime which the analysis of the precipitate at an earlier stage of the ebullition would have furnished.

As an example of this may be cited the analysis of a water of Class II. from Ste. Genevieve, where the precipitate, after a few minutes' boiling, contained carbonates of lime and magnesia in the proportion 12 : 750. When, however, another portion was boiled down to one sixth, the precipitate was found to be pure carbonate of magnesia. The water of another spring of the same class, that of Plantagenet, [described in the original paper,] gave as the result of ebullition a precipitate of .8904 of carbonate of magnesia and .0330 of carbonate of lime; while the liquid retained a portion of lime equal to .1364 of chloride of calcium, besides .2452 of chloride of magnesium, in 1,000 parts. When, however, this water is left to spontaneous evaporation, the whole of the lime separates as carbonate, and the liquid remains for a time charged with carbonate of magnesia, probably as sesqui-carbonate. This solution is, however, after a time spontaneously decomposed even in closed vessels, with deposition of a portion of crystalline hydrated carbonate of magnesia; another portion remains in solution, together with chloride of magnesium, but is precipitated by ebullition. (Amer. Jour. Science (2), XXVII. 173.)

§ 55. Bicarbonate of magnesia and chloride of calcium, when brought together in solution, undergo mutual decomposition with separation of carbonate of lime, if the solutions are not too dilute. At the ordinary temperature and pressure, water saturated with carbonic acid will not hold more than about one gramme of carbonate of lime to the litre (1 : 1,000), equal to only 0.88 grammes of carbonate of magnesia. (The solubility of carbonate of lime in pure water is well known to be much less, and is, according to Bineau, equal to 1 : 30,000 or 1 : 50,000.) We should not, therefore, expect to find that water holding chloride of calcium in solution would yield, by boiling, more than the latter amount of magnesian carbonate; so much might evidently be formed by the action of dissolved carbonate of lime which the water might hold as bicarbonate. I have elsewhere described a series of experiments on the solubility of bicarbonate of lime both in pure water and in saline solutions, and have shown that the presence of salts of soda, lime, and magnesia does not increase the amount of bicarbonate of lime which water is capable of holding *permanently* in solution. . . . . Recent experiments have, however, shown me that supersaturated solutions of a certain stability may be obtained, in which comparatively large quantities of neutral carbonates of lime and magnesia exist in the presence of sulphates and chlorides of calcium and magnesium.

§ 56. In a memoir on the salts of lime and magnesia published in 1859 (Amer. Jour. Science (2), XXVIII. 171), it was shown that by the addition of bicarbonate of soda to a solution holding chlorides of sodium, calcium and magnesium, with or without sulphate of soda, and saturated with carbonic acid, it is possible to obtain transparent solutions holding from 3.40 to 4.16 grammes of carbonate of lime to the litre. Of this, however, the greater part was deposited after twenty-four hours, when the solutions were found to contain somewhat less than 1.0 gramme, in the form of bicarbonate. Boutron and Boudet had previously shown that by saturating lime-water with carbonic acid, solutions were obtained holding in a litre 2.3 grammes of carbonate of lime; of which one half was soon deposited, even when the solution was kept under a pressure of several atmospheres. It would thus seem that saline liquids favor this temporary solubility of the carbonate of lime as bicarbonates.

In all of the above experiments an excess of carbonic acid was present, but this I have since found is not essential, since supersaturated solutions may be obtained holding as much as 1.2 grammes of carbonate of lime, together with sulphate of magnesium and chloride of calcium, in a litre of water, without any excess of carbonic acid. The power of alkaline chlorides and of chloride of calcium to prevent the precipitation of chloride of calcium by carbonate of soda has already been observed by Storer (Dictionary of Solubilities, page 110). I have found that the precipitate produced by the admixture of solutions of these two salts is readily dissolved, when recent, by a solution of chloride of calcium or of sulphate of magnesia; and thus liquids may be prepared holding at the same time from 1.0 to 1.2 grammes of neutral carbonate of lime and 1.0 of neutral carbonate of magnesia, in presence of sulphate of magnesia. These solutions of carbonate of lime, which are strongly alkaline, may be kept for twelve hours or more without perceptible change at ordinary temperatures, but after a time deposit crystals of hydrated carbonate of lime. The addition of alcohol immediately throws down the whole of the carbonate of lime in an amorphous condition.

The carbonate of magnesia is still more soluble than the carbonate of lime under similar conditions, and it is possible to obtain 5.0 grammes of neutral carbonate of magnesia dissolved in a litre of water holding seven per cent of hydrated sulphate of magnesia, without any excess of carbonic acid. These solutions, which are strongly alkaline to test-papers, yield a precipitate by heat, which redissolves on cooling.

It is evident that the mingling of saline and alkaline waters may give rise to solutions like those just described, and thus explain apparent anomalies in the composition of certain saline waters. See also in this connection the observations of Bineau, and my own on the properties of solutions of sesqui-carbonate of magnesia. (Amer. Jour. Science (2), XXVII. 173.)

§ 57. Salts of Barium and Strontium. — The salts of these two bases are found in very many of the saline and alkaline waters of Canada. Their carbonates probably sustain to the magnesian chloride a similar relation with that of calcium, and hence these bases appear in some of the analyses partly as carbonates, and partly as chlorides of barium and strontium. The precipitate formed in the concentrated and acidulated water by dilute sulphuric acid was, whenever submitted to analysis, found to contain both barium and strontium. For the separation of these, the mixed sulphates were first converted into chlorides; the barium was then thrown down as silico-fluoride, and the strontium subsequently precipitated by a solution of gypsum.

The insolubility of its sulphate must have excluded baryta from the waters of the primeval sea, and when set free, as we may suppose, by the decomposition of its silicated compounds existing in the primitive crust (§ 12), its soluble bicarbonate carried down to the sea would there be precipitated by the sulphates present. A similar process must still go on with all the dissolved barytic salts which find their way to the ocean.

The sulphate of baryta thus accumulated in sedimentary strata may be partially decomposed by infiltrating solutions of alkaline carbonates, and thus be rendered capable of being subsequently dissolved as carbonate; but the most probable mode of its solution is, we conceive, through its previous reduction by organic matters to the form of a soluble sulphuret (§ 10), ready to be converted into carbonate or chloride of barium. In this way we may explain the frequent occurrence of baryta-salts in the saline waters of the first three classes, and the consequent absence of sulphates, which will be further considered in § 61. From the similarity of its chemical reactions, the preceding remarks apply to strontia as well as baryta.

§ 58. Iron, Manganese, Alumina, and Phosphates.—None of the waters of the four classes here described contain any notable quantity of iron, yet this element is never wanting in those waters which contain earthy carbonates. Whenever a portion of one of these waters, or better the earthy precipitate separated from it by boiling, is evaporated to dryness with an excess of hydrochloric acid, the residue treated with acidulated water yields a portion of silica, and the solution will then be found to yield with ammonia a precipitate. This, which is partially soluble in caustic alkalies, is often colorless, and will be found to consist of alumina and peroxide of iron, with phosphoric acid and a trace of manganese, which latter metal is seldom or never absent. The small quantity of alumina which these waters contain appears not to be derived from suspended argillaceous matters, but to be held in a state of solution. The phosphates are generally present only in very small quantities in these waters, for the reason pointed out in § 5. The largest amount which I have met with was in an alkaline water of Class III. from Fitzroy (§ 43), where it is equal to .0124 of tribasic phosphate of soda in 1,000 parts of water.

§ 59. Bromides and Iodides.—The chlorides in these ancient mineral waters are always accompanied by bromides and iodides, but the proportion of the bromides to the chlorides appears to be much less than in the waters of the modern seas. According to Usiglio, 100 parts of the salts from the Mediterranean contain 1.48 of bromide of sodium; while ten analyses by Von Bibra of the waters of different oceans give from 0.86 to 1.46, affording for 100 parts of salts a mean of 1.16 of bromide of sodium, equal to 1.04 parts of bromide of magnesium. The waters of Whitby and Hallowell, on the contrary, which are the richest in bromides of those described in this paper, contain only 0.54 and 0.69 parts of bromide of sodium in 100 parts of solid matters; while few of the saline springs of the second class contain more than one half of this proportion, and some of them very much less.

With regard to the iodides in many of these waters, however, the case is very different. The waters of the modern ocean, as is well known, contain but traces of iodine,* and in some strongly saline springs of the first class, like that of Whitby, it is only in the alcoholic extract of the salts from this water that iodine can be detected. The Hallowell water (§ 36), which closely resembles this in its general composition, and in the proportion of bromides, is, however, so rich in iodine that its presence can readily be discovered without previous evaporation. It is sufficient to add to the recent water acidulated by hydrochloric acid a little solution of starch and a few drops of nitrite of potash, to produce an intense blue color. The iodide of sodium in the first-named water was found equal to 0.0017 parts of the solid matters, and that of the second to 0.019, or nearly twelve times as much. The unconcentrated saline waters from the two springs of Ste. Genevieve, which belong to the second class, also give a strong reaction for iodine, and when acidulated with hydrochloric acid, without previous evaporation, yield with a salt of palladium a precipitate of iodide of palladium after a few hours. The salts from these two springs of Ste. Genevieve, though poorer in bromides, are much richer in iodides, than the waters of Hallowell; one of the former containing in 100 parts of salts no less than 0.138 of iodine, so that there appears to be no constant proportion between the chlorides, bromides and iodides of these saline waters.

§ 60. The relations of bromides and iodides to argillaceous sediments have yet to be determined. It would appear from the facts just cited that bromine has in the course of ages been slowly eliminated from insoluble combinations, and, like potassium, has accumulated in the waters of the ocean; while the facts in the history of iodine seem to point to a process the reverse of this, — in other words, to a gradual elimination of iodine from the sea-waters, and its fixation in the earth's crust. The observations of numerous chemists unite to show the frequent occurrence of small portions of iodine in some unknown combination, in sedimentary rocks of various kinds; from which

* [See in this connection the late researches of Sonstadt, noticed at the end of Essay XII.]

we may conjecture that it was in former times abstracted from the sea, either directly or through the intervention of organic bodies (as in the case of potash, which is separated and fixed by means of algæ, § 5). Experiments after the manner of those of Way and Voelcker may throw light upon this interesting question. We are aware that insoluble combinations of soluble chlorides with silicates of alumina are found under certain conditions, as appears in sodalite, eudialyte, and the chloriferous micas, and it is not improbable that the soluble iodides may give rise to similar compounds. By such a process might be explained the rarity of this element in modern seas, while the occasional re-solution of the iodine from these insoluble compounds by infiltrating waters would help to explain the variable and often large proportions in which this element is met with in some of the waters noticed above.

§ 61. Sulphates. — In the preceding sections we have already discussed the principal facts in the history of those neutral waters in which sulphates predominate, or prevail to the exclusion of chlorides (§§ 50, 51). The history and the probable origin of those curious springs which contain free sulphuric acid has also been considered (§§ 31, 48, 49); and it now remains to notice the relation of sulphates to the muriated waters. The first fact that excites our attention is that of the total absence of sulphates from numerous springs of the first, second, and third classes, as shown in the preceding analyses, and also in the observations of Lenny and others on the saline waters over a great area in western Pennsylvania (§ 40).

The elimination of sulphate in the form of gypsum from evaporating waters containing an excess of chloride of calcium has already been discussed in § 37; but the bitterns resulting from such a process still retain small portions of sulphates; while it is to be remarked that the saline waters under consideration contain no traces of sulphates, and in many instances hold portions of baryta and strontia, bases incompatible with the presence of sulphates. The modes in which this complete elimination of sulphates may be effected are two in number. The first has already been suggested in § 10, and depends upon the deoxidizing power of organic matters, which reduce the sulphates to sulphurets. These in their turn may be converted into carbonates, the sulphur being separated either as sulphuretted hydrogen (giving rise by oxidation to free sulphur), or as insoluble metallic sulphurets. This reducing action not only decomposes the soluble sulphates of soda, lime, and magnesia, but also, as has been pointed out in § 57, may extend to sulphate of baryta, and thus sulphuret or carbonate of baryta be formed. It is the action of these soluble baryta-salts which constitutes the second mode of desulphatizing waters ; and this, if we may judge from the frequence with which baryta-salts occur in the saline waters in question, appears to have been the most general process.

It is a fact worthy of notice, that a saline spring at Sabrevois, in the province of Quebec, near Lake Champlain, which holds both baryta and strontia in solution, is at the same time slightly impregnated with sulphuretted hydrogen. Another saline and sulphurous spring, which rises within ten feet of this, contains, however, a portion of sulphates. (Geology of Canada, page 542.)

§ 62. I am indebted to Professor Croft, of Toronto, for some notes of a recent examination by himself of a saline of the first class, which contains at the same time a soluble sulphuret. This water, from a boring in Chatham, Ontario, at a depth of 600 feet, and about 236 feet below the summit of the Corniferous limestone, had a specific gravity of 1039.3, and yielded for 1,000 parts about 51 of solid matters. It contained large portions of chlorides of calcium and magnesium, with very little sulphate, traces of carbonate, and no free carbonic acid. The water, which gave an alkaline reaction with turmeric, was greenish in color, very sulphurous to the taste, and yielded a purple color with nitroprusside of sodium, and a black precipitate of sulphuret with a solution of sulphate of iron. A current of carbonic acid rendered the recent water opalescent, and by exposure to the air it deposited sulphur.*

* [For further studies of waters of this class, see the Supplement to this paper.]

§ 63. Borates. — The reddening of the yellow color of turmeric-paper in presence of free hydrochloric acid affords, with certain precautions, the ordinary means for detecting small portions of boric acid. Most of the waters of the third and fourth classes, and some of those of the second, have been tested in this way, and have never failed when reduced to a small volume, and acidulated with hydrochloric acid, to give this reaction; which was, however, most marked with the waters of the fourth class. . . . .

§ 64. I have recently had an opportunity of examining from California the waters of a borax-lake, which contains, beside borate and carbonate of soda, a portion of chloride, and a little silicate, traces only of phosphate, and no sulphate. It held in solution very small quantities of earthy carbonate, and was remarkable for the large proportion of potassium-salt which it contains. The evaporated and fused saline residue was treated by the ordinary methods for the determination of the chlorine, carbonic acid, and silica; while the bases were obtained in the form of sulphates by the aid of sulphuric and hydrofluoric acids, and afterwards separated as chlorides by the aid of chloride of platinum. From the data thus obtained the following ingredients were found by calculation for 1,000 parts of the water: —

Carbonate of soda	9.476
Biborate of soda	4.395
Chloride of sodium	1.702
Carbonate of potash	1.818
Silica	0.129
	17.520

The potassium, as above determined, equals 11.46 per cent of the bases weighed as chlorides; another trial gave 11.41. Although for convenience we have represented the potassium as carbonate, it will be seen that the amount of chlorine is such that it might, for the greater part, have been represented as chloride of potassium, with an equivalent portion additional of carbonate of soda.

§ 65. Carbonates. — In examining in 1847 the alkaline-saline waters of Caledonia, it was found that these contained a quantity of carbonic acid insufficient to form bicarbonates with the carbonated bases present. It was partly with this fact in view that, after an interval of more than seventeen years, I undertook the new analyses of these waters, which in § 47 are given side by side with the earlier results. In these later analyses, as there remarked, a slight excess of carbonic acid was met with. In the interval the springs had undergone changes in composition, and while the third one still retained in a slight degree its alkaline character, the other two had become waters of the second class, holding, instead of carbonate and sulphate of soda, chloride of magnesium, and baryta salts. The amount of carbonic acid had, however, undergone but little change; and as will be seen by comparing the figures below with those in the table in § 47, the slight diminution in the first and third corresponds very closely with the falling off in the amount of solid matters between 1847 and 1865; while, on the contrary, the augmentation in the amount of carbonic acid in the second is accompanied with a corresponding increase in the amount of fixed matters present.

CARBONIC ACID IN ONE LITRE OF THE CALEDONIA WATERS.

	1847.	1865.
Gas Spring	.705 grammes.	.671 grammes.
Saline Spring	.648 "	.664 "
Sulphur Spring	.590 "	.573 "

While the amounts of fixed matters and of carbonic acid in the several waters have undergone but little change, we find, however, that there has been a great diminution in the proportion of carbonated bases. Thus in the Gas Spring in 1847 the carbonic acid required for the neutral carbonates found in the analysis was .356, while for the same water in 1865 only .278 of carbonic acid was required. In the Sulphur Spring, in like manner, the neutral carbonates required .449, or more than three fourths of the carbonic acid present; while the falling off in the amount of carbonates in 1865 is such that only .191 of carbonic acid, or just about one third of the carbonic acid present, is required for the neutral carbonate. Nor is this change due entirely to a less amount of carbonate of soda; the carbonates of lime and magnesia in 1847 required .246, and in 1865 only .153, of carbonic acid. The changed conditions which we here meet with may be explained by supposing that the carbonated bases are due to the mingling in different proportions of neutral carbonate of soda (generated by the reaction indicated in § 13) with an earthy-saline water holding a constant amount of free carbonic acid; which, in some cases, is more than is required to form bicarbonates, but in others, as we have seen above, shows a deficiency.

§ 66. If we admit, as I have already assumed, that the waters of the second and third classes have been generated by the mingling of solutions of carbonate of soda with waters of the first class, it can readily be shown that these solutions contained chiefly or exclusively the neutral carbonate. If we add a solution of bicarbonate of soda to earthy-saline waters of the first class, it is easy to obtain solutions holding twenty grammes or more of bicarbonate of magnesia to the litre; while in none of the natural waters of the second class do our analyses show the existence of much over one gramme to the litre. Again, if we suppose any considerable amount of chloride of calcium to be decomposed by bicarbonate of soda, the separation of the lime in the form of neutral carbonate, and the liberation of the second equivalent of carbonic acid, would yield waters holding an excess of carbonic acid above that required to form the bicarbonates of the solution. From the absence of such an excess, as appears in the case of the waters of Caledonia, Varennes, and St. Leon, and from the small amount of bicarbonate of magnesia in these waters, it may be concluded that the alkaline salt whose addition has changed their character was the neutral carbonate of soda.

§ 67. Examples are not wanting of waters in which, as in those of Caledonia in 1847, the carbonic acid is insufficient to form bicarbonates (or even neutral carbonates) with the bases uncombined with sulphuric acid or chlorine. Thus, according to Pagenstecher and Müller, the spring and well waters of Berne do not contain sufficient carbonic acid for the lime present, a part of which they suppose to be held in solution as a silicate. See Bischof, Chem. Geology, I. 5; who remarks that Löwig seems to have observed the same fact in the thermal spring of Pfaffers. For further examples of this kind see Lersch, Hydro-Chemie, page 333. The carbonic acid in the water of Töplitz is, according to him, not sufficient to form bicarbonates unless the silica present be supposed to be combined with a portion of bases; while in the alkaline thermal spring of Bertrich, according to the analysis of Mohr, a similar deficiency of carbonic acid exists; leading to the conclusion that a part of the earthy bases present is in combination with silica and organic matters. The existence of solutions holding comparatively large amounts of neutral carbonates of lime and magnesia, as described in § 56, is not without interest in this connection; since it at once affords an explanation of the nature and origin of all such alkaline waters, and waters deficient in carbonic acid, as contain earthy sulphates and chlorides.

§ 68. It was found that the waters of Chambly in 1864, and of the Sulphur Spring of Caledonia in 1865, gave with lime-water a precipitate which was soluble in an excess of these mineral waters, but to a much less extent than in the acidulous saline water from the High-Rock Spring of Saratoga. The latter, which contains bicarbonate of soda, and is highly charged with carbonic acid, turns to a wine-red the blue color of litmus-tincture, which is not changed by the Chambly or the Caledonia water. The Saratoga water, after some time, gives a feeble alkaline reaction with dahlia-paper; this is more distinctly but slowly changed by the Caledonia water, and almost immediately turned to green by that of Chambly. This latter water readily changes to brown yellow tumeric-paper, which is scarcely affected by the water of Caledonia.

§ 69. Silica. — The silica which exists in solution in cold saline springs is generally very small in amount, as might be expected from the insolubility of earthy silicates, which is such that superficial drainage waters in filtering through the soil lose the silica which they held in solution (§ 5). We have further shown that as a result of this tendency to the formation of insoluble silicates, the silicate of soda liberated in the sediments by the decomposition of feldspar generally appears at the surface as carbonate of soda, having been decomposed by earthy carbonates (§ 13).

In two cases, however, considerable quantities of silica are found dissolved in natural waters. The first is met with where the rapid solvent and decomposing action of heated waters is exerted upon alkaliferous silicious minerals (§ 14), as seen in springs like the Geysers. The second case is that of those rivers and streams which drain surfaces covered with decaying vegetation and decomposing silicates, from both of which they derive dissolved silica. Such waters contain but small amounts of solid matters, but the proportion of silica is relatively considerable, amounting, as we have seen in the water of the Ottawa River, which contains, in 10,000 parts, 0.6116 of solid matters, to 0.2060, or thirty-two per cent; while in the St. Lawrence, which contains, for the same amount of water, 1.6056, the silica equals .3700, or twenty-four per cent, of the solid ingredients. The analysis by H. Ste-Claire Deville of the river-waters of France show, in like manner, large amounts of silica, which seem to have been hitherto overlooked in the analyses of most chemists. (Ann. de Chim. et Phys. (3), XXIII. 32.)

It will be seen by a reference to the tables of analyses given in the second part of this paper, that in the waters of the second class the amount of silica is equal to from 0.15 to 0.60 parts for 100 of solid matter. In the alkaline waters of the third and fourth classes its proportion is greater, and up to a certain point appears to increase with that of the carbonate of soda. . . . . The amount of silica which these waters contain does not in any case exceed one or two ten-thousandths.

§ 70. Inasmuch as carbonic acid, according to Bischof (Chem. Geol., I. 2), decomposes not only the silicates of soda, but those of lime and magnesia, when they are in solution, it might be supposed that the silica in the above waters exists either in a free state or as a soluble silicate with a great excess of acid. The latter view, especially in the case of magnesia, is rendered probable by numerous experiments which form a part of the series already mentioned in § 41. From these it appears that free soluble silica, when mingled with a solution of bicarbonate of magnesia, or with the neutral carbonate dissolved in sulphate of magnesia in the manner described in § 56, whether separating immediately or by a slower process of gelatinization, always carries down with it, in combination, a few hundredths of magnesia.

In these experiments, besides the carbonate of magnesia, sulphate or chloride of magnesium was present; but the silicated natural waters now under discussion are alkaline from the presence of carbonate of soda, and whatever partition of bases between carbonic and silicic acids may exist in the recent waters, we may suppose that when they are boiled a silicate of soda is formed, with the expulsion of carbonic acid. The silicate thus produced reacts on the earthy bases present, with the production of silicates of lime and magnesia, which are in part precipitated with the earthy carbonates. Berzelius and Kersten long since observed the separation of such silicates during the evaporation of the waters of Carlsbad and of Marienbad (Bischof, I. 5); while a silicate of lime is said to be deposited from the waters of Wiesbaden. But the silicates thus formed are but partially precipitated, — a portion remaining in solution till a late period of the evaporation. Dr. J. Lawrence Smith long since remarked the existence of a dissolved silicate of lime, apparently combined with soda, in the concentrated alkaline waters of Broosa, in Asia Minor. (Amer. Jour. Science (2), XII. 377.)

Many facts in accordance with the above were observed in the analyses of the waters described in this paper. Thus a water of Class III. from Belœil, Quebec, which held in 1,000. parts .114 of silica, besides .608 of carbonate of soda, and carbonates of lime, magnesia, and baryta, when evaporated to one tenth deposited with the carbonates .050 of silica, and the hydrochloric solution of this precipitate became gelatinous during evaporation. The water thus evaporated still retained in solution, besides a portion of lime, .064 of silica; which was completely separated when the alkaline liquid was evaporated to dryness in contact with the earthy carbonates previously precipitated. When, however, these were removed by filtration, it was found that during the evaporation to dryness a reaction took place by which the precipitated silicate of lime was partially decomposed, the separated silica being redissolved by the alkaline carbonate. In the case of the Chambly water in 1852, which contained in 1,000 parts .073 of silica, .042 parts still remained in solution in the water evaporated to one twentieth; and in that of the Ottawa River when reduced to one fortieth there still remained in solution from 10,000 parts of water, .075 of silica and .028 of lime. Similar results were observed with the alkaline-saline waters of Varennes and Fitzroy, and all of these yielded, by further evaporation, precipitates containing silica and lime, and in one instance magnesia.

It is not, however, probably from alkaline waters like these, but from neutral sea-water, that the silicates of magnesia (and of lime), which abound in stratified rocks, have been for the most part formed. See further on this point, § 41.

§ 71. Organic Matters. — In § 44 we have described some of the reactions of the organic matter found in the Chambly water, and it is to be remarked that small portions of a similar substance were found in all alkaline waters of the third and fourth classes, and caused them to become brownish when evaporated to a small volume. This, it has been already suggested, may have a superficial origin, the organic matters carried down by surface-waters being kept in solution by the alkaline salts; it is not, however, impossible that this same menstruum may remove the organic matters which abound in the pyroschists and other materials of organic origin in the ancient rocks. Thus, for example, the coprolites of the lower palæozoic limestones contain so much animal matter as to evolve an odor like burning horn when exposed to heat. (Geology of Canada, 462.)

The Ottawa water (§ 46), when boiled to one tenth, deposits a precipitate in small bright brown iridescent scales. This was found to contain silica, carbonate of lime, and a small portion of an organic substance which was dissolved in dilute potash ley. The brown solution thus obtained was not disturbed by acetic acid and acetate of copper, but by the subsequent addition of carbonate of ammonia yielded a white precipitate. The concentrated water retained a large proportion of organic matter, and when reduced to a small bulk was dark brown, alkaline to turmeric-paper, and continued by evaporation to deposit opaque films of silicate of lime. The finally dried residue was dark brown in color, and carbonized by heat, burning like tinder and diffusing an agreeable odor. The residue of 10,000 parts dried at 300° F. weighed .6974, and lost by gentle ignition .1635, consisting partly of organic matter. No chemical examination was made of this matter held in solution by the concentrated water. From the late researches of Peligot, however, it appears that the organic matter precipitated by nitrate of lead from the water of the Seine has nearly the composition of the apocrenic acid of Berzelius. It gave, on analysis, carbon 53.1, hydrogen 2.7, nitrogen 2.4, oxygen 41.8, and is evidently related to the soluble form of vegetable humus. (Comptes Rendus, April 25, 1864.) When exposed to heat this substance evolved ammonia, with the odor of burning wool, while the organic matter from the Ottawa water, on the contrary, gave an odor like burning turf.

GEOLOGICAL POSITIONS OF THE PRECEDING WATERS.

§ 72. The palæozoic area from which the above-described waters are derived includes the basin of the St. Lawrence from Lake Erie to near Quebec, with its extensions in the valleys of the Lower Ottawa and Lake Champlain. Over the greater part of this champaign region the strata are nearly horizontal, but towards its eastern part there are various minor folds and undulations. It is in this disturbed region that by far the greater number of the mineral springs already described occur; and although it is often difficult to establish the presence or to trace the extent of faults in the strata, on account of the alluvial deposits which generally cover the palæozoic strata of the region, it is apparent that in a great number of cases the mineral springs occur along the lines of disturbance, and it is probable that a constant relation of this kind exists. The great western portion of the basin, which is less disturbed than its eastern part, presents but few mineral springs; yet the wells of strongly saline water which have been obtained by boring at Kingston, Hallowell, St. Catherine's, Chatham, and elsewhere in Ontario, show that the undisturbed rocky strata are charged with saline matters. For a better understanding of the relations of these waters, a list of the palæozoic formations in which the mineral springs here described occur is given below, numbered in ascending order. [Of these the first six correspond to the first and second palæozoic faunas, the Cambrians of Sedgwick and the Lower Silurian of Murchison, while 7 – 12 include the third fauna, or true Silurian, and the remaining three the lower part of the Devonian series.]

Palæozoic Formations of the St. Lawrence Basin.

15. Hamilton, — shales.
14. Corniferous, — limestone.
13. Oriskany, — sandstone.
12. Lower Helderberg, — limestone.
11. Onondaga, or Salina, — dolomite and shales.
10. Guelph, — dolomite.
9. Niagara, — dolomite.
8. Clinton, — dolomite and shales.
7. Medina, — sandstone.
6. Hudson River, — shales.
5. Utica, — shales.
4. Trenton, — limestone.
3. Chazy, — limestone.
2. Calciferous, — dolomite.
1. Potsdam, — sandstone.

§ 73. Of the above series the Trenton group includes the Birds-eye and Black River limestones, as well as the Trenton limestone of the New York geologists, and is non-magnesian, enclosing beds of chert, silicified fossils, and petroleum; in all of which characters it resembles the Corniferous limestone above. In like manner the Potsdam is represented by the Hudson River and Medina formations, while the gypsiferous dolomite of the so-called Calciferous sand-rock corresponds to the great mass of dolomite which constitutes Nos. 9, 10, and 11, and includes the gypsum and the salt-bearing strata of the Onondaga formation. These repetitions of similar strata mark successive recurrences of similar geological and geographical conditions, which form great cycles in the history of the continent.

§ 74. [Within the eastern border of this basin, stretching along the western base of the Green Mountains, and thence northeast to Quebec, and beyond it on the southeast shore of the St. Lawrence, is spread a great series including about 7,000 feet of limestones, dolomites, shales, and sandstones. These rocks, which are more or less disturbed, constitute what Logan has called the Quebec group, and are the Taconic of Emmons, or the Primal and Auroral of Rogers, containing organic remains of the first palæozoic fauna, and corresponding to the Lower and Middle Cambrian of Sedgwick, of which the first three formations in the above table are but incomplete and littoral, or shallow-water deposits. (See further, paper XV., part 3.)]

None of the waters described in the present paper belong to this Quebec group, which, nevertheless, presents several mineral springs, some of which are described in the Geology of Canada. Of these, the salines of Cacouna, Green Island, Rivière Ouelle, and Ste. Anne de la Pocatière are bitter waters belonging to the first class; while a sulphurous spring at the latter place, and another at Quebec, are alkaline waters of the fourth class.

§ 75. Of the waters of the region which is considered in this paper, many have been qualitatively analyzed which are not here described. Including two from Vermont, twenty-one alkaline waters of the third and fourth classes have been examined. Of these, as already stated, the waters of Caledonia rise from the Trenton group, and that of Fitzroy from the Chazy or

Calciferous, while two others, at Ste. Martine and Rawdon, appear to have their source in the Potsdam. All the other waters of these two classes issue from the Hudson River shales, with the exception of those of Varennes and Jacques Cartier, which seem to rise from the Utica formation.

Of the waters of the second class, of which about thirty have been examined, some five or six issue from the shale formations Nos. 5 and 6, but all the others are from the underlying limestones. The bitter salines of the first class flow from the limestones of the Trenton group, with the exception of one at Ancaster, which is from a well sunk in the Niagara formation, and that of St. Catherine's, from a boring carried through the Medina down into the Hudson River shales. The source of both of these is probably, like that of the other very similar waters, the underlying limestones.

§ 76. From this distribution of the waters of the first four classes it would appear that the source of the neutral salts, which consist of alkaline and earthy chlorides, is in the limestones and other strata from the Potsdam to the Trenton inclusive, while the alkaline carbonates are derived from the argillaceous sediments which make up the Utica and Hudson River formations. The sediments are never deficient in alkaline silicates, whose slow decomposition yields to infiltrating waters (§ 13) the alkaline carbonates which characterize the mineral springs of the fourth class. These, mingling in various proportions with the brines which rise from the limestones beneath, produce the waters of the second and third classes in the manner already explained. The appearance of several springs of the third class, as those of Caledonia and Fitzroy, from these lower limestones, is not surprising, when it is considered that the Chazy formation in the Ottawa Valley includes a considerable thickness of shales, sandstones, and argillaceous limestones, approaching in composition to the sediments of the Hudson River formation.

§ 77. As an evidence that the different classes of waters have their origin in different strata, may be cited the fact that springs very unlike in composition are often found in close proximity, and apparently rising from a common fissure or dislocation. Thus in the seigniories of Nicolet and La Baie du Febvre, I have examined six springs, all of which rise through the Utica formation along a line, in a distance of about eight miles. Of these springs two belong to the second, two to the third, and two to the fourth class; these last being probably derived entirely from the shales, while the others have their source in the underlying limestones, and are more or less modified in their ascent. Again, at Sabrevois, within a few feet of each other, are two springs of the second class, of which one contains salts of baryta and strontia, and the other soluble sulphates. In like manner at Ste. Anne de la Pocatière a spring of the second class and one of the fourth are found not far apart. The springs of Caledonia offer another and not less remarkable example. In 1847 there were to be seen, not far from a spring of the second class, three others of the third class very near together, one of them sulphurous, but all sulphated, and differing in the proportions of carbonate of soda present. In 1865, while one of these still retained its character of a sulphurous sulphated water of the third class, the others were changed to waters of the second class, and held salts of baryta in solution. These relations, which we have already pointed out (§ 47), not only show waters holding incompatible salts issuing from different strata along the same fissure, but mingling in such varying proportions as to produce from time to time changes in the constitution of the resulting springs.

§ 78. The temperature of none of the springs which we have here described exceeds 53°, which has been observed for two springs at Chambly, about twelve miles from Montreal (§ 44). No other springs in Canada are known to present so high temperature, unless possibly the acid waters of the fifth class (§ 48). St. Léon spring was found to be 46°, while that of Caxton, near the last, and like it of Class II., was 49° F.

§ 79. The extended series of analyses which we have given in the preceding pages presents many points of interest. Nowhere else, it is believed, has such a complete systematic examination of the waters of a region, and of a great geological series, been made. Additional importance is given to these results by the fact that the waters are all derived from palæozoic strata. We are thus enabled to compare these saline materials of an ancient period with those which issue from, and in many cases owe their saline impregnation to, strata of comparatively modern origin (§ 39).

It is a consideration not without interest, that the valley of the St. Lawrence might, under different meteorological conditions, become a region abounding with saline lakes affording sea-salt, natron, and borax, the results of the evaporation of the numerous saline and alkaline springs which have here been described.

SUPPLEMENT.

[From the Report of the Geological Survey of Canada for 1863-66, pages 272-277.]

As further examples of saline waters of the first class, such as are described in §§ 35-40 of the preceding paper, I here give the results of the analyses of two from western Ontario, both which were met with in boring for petroleum. The first of these is from a well on Manitoulin Island in Lake Huron, and was found at a depth of 192 feet from the surface, after passing through the black slates of the Utica formation, and for sixty feet in the underlying Trenton limestone. The water was intensely bitter and saline to the taste; it contained no trace of sulphates, nor yet of barium nor strontium. It was not examined for bromides or iodides, which, however, were probably present. The analysis of this water gave, for 1,000 parts, as follows: —

Chloride of sodium	4.800
Chloride of potassium	.792
Chloride of calcium	12.420
Chloride of magnesium	3.650
	21.662

This water is remarkable for the amount of chloride of calcium which it contains, equal to more than one half of the solid contents, a much larger proportion than in any of the bitter saline waters hitherto examined in Canada, or elsewhere. In most waters of this class, the proportion of chloride of potassium (as shown in § 52) is small, rarely attaining to one hundredth of the alkaline chlorides; but in the Manitoulin water it amounts to not less than 16.6 per cent of these or more than 3.7 per cent of the entire solid matters, a proportion as great as in modern sea-water. This peculiarity, not less than the absence of sulphates, would lead us to suspect that this water may be derived, by dilution, from an ancient bittern, from which, owing to the excess of lime in the primitive seas, the sulphates have been eliminated in the form of gypsum, in the process of evaporation. Further analyses of waters from this region are needed to complete their history.

The second water to be noticed is from a well sunk at Bothwell, Ontario, for petroleum, in 1865. At a depth of 475 feet from the surface, and probably at or near the base of the Corniferous limestone, a copious spring was met with, which rose to the surface, and on the 16th of September, 1865, was yielding at the rate of about 700 gallons per hour of bitter, very sulphurous water, with a little petroleum. The temperature of this water was 54° F., or about 7° above the mean temperature of the region, which is traversed by the isothermal line of 47° F. The water was placed in carefully filled and well-corked bottles, which were laid on their sides, and thus transported to my laboratory at Montreal. Its specific gravity was 1020.9, and that of another portion collected five days later, on the 21st September, was 1021.1. The water, which at the well was transparent and colorless, was found on opening the bottles to have become slightly yellowish. By further exposure to the air it turned greenish-yellow from the formation of a persulphide, and soon became coated with a film of sulphur, the liquid after a while again growing colorless. The color was at once destroyed by a little hydrochloric acid, the water becoming opalescent from the separation of sulphur. The recent water was feebly alkaline to litmus, but did not affect the color of curcuma-paper.

These characters showed the recent water to contain a soluble monosulphide, whose presence was further indicated by the addition of a solution of green vitriol, which gave an abundant precipitate of sulphide of iron. Nitroprusside of sodium gave a fine purple color with the water, which was rendered more intense by the previous addition of a little caustic soda.

When boiled, the recent water evolves an abundance of sulphuretted hydrogen, and after twenty minutes of ebullition the reaction of sulphur disappears from the water; which becomes turbid, from the separation of a hydrate of magnesia, readily soluble in a cold solution of sal-ammoniac. Crystals of gypsum are also deposited during the boiling. This volatilization of the sulphur is evidently due to the well-known decomposition of sulphide of magnesium, by boiling, into hydrated oxide of magnesium and sulphuretted hydrogen gas. It was, however, a question whether the whole of the sulphur in the recent water existed as a sulphide of sodium or magnesium, or whether a portion was present as sulphide of hydrogen, giving with the former a double sulphide MgS,HS. This problem, of considerable delicacy, can only be solved by indirect means. For the determination of the whole amount of sulphide in the recent water, having at the well no other suitable reagent, I added to two bottles of the water a few grammes each of sulphate of copper; the sulphide thus precipitated was afterwards collected and analyzed. In that from one bottle the amount of sulphur in the precipitate was directly determined, while in the other it was deduced from that of the copper. These two results gave, respectively, .460 and .464 grammes of sulphur to the litre of water, the mean of which, .462, is equal to .491 grammes of sulphide of hydrogen. In addition to these, a determination was made with the water brought to the laboratory. This, when mingled with an acid solution of terchloride of arsenic, gave a quantity of tersulphide of arsenic equal to .460 grammes of sulphuretted hydrogen, indicating a slight loss of sulphur.

When a double sulphide of sodium and hydrogen exists in an alkaline water, it is possible, by boiling, to destroy the compound, and by expelling the sulphide of hydrogen, to determine the amount of sulphur which is present as a fixed monosulphide. But when the double sulphide has a base of magnesium, or exists in a water containing an excess of a soluble magnesian salt, the ready decomposition of sulphide of magnesium will cause the whole of the sulphur to be carried off by boiling, in the form of sulphuretted hydrogen, with separation of hydrate of magnesia, as is the case of the Bothwell water. The following experiment was, however, devised, which shows the existence of a double sulphide in this water, and at the same time enables us to suggest a method which may probably be used for the complete analysis of this and of similar waters.

It is well known that solutions of alkaline and earthy sulphides dissolve tersulphide of arsenic, yielding double sulphides or sulpharsenites, whose formula, for the alkaline bases, is, according to Berzelius, AsS_3,3MS, and for the earthy bases, AsS_3,2MS. If these protosulphides are combined with sulphide of hydrogen, forming double salts, MS,HS, the latter will be displaced by the arsenious sulphide. The presence of such a compound in the Bothwell water was shown by adding to it freshly precipitated and carefully washed tersulphide of arsenic, which was rapidly dissolved, with an abundant disengagement of sulphuretted hydrogen gas. The solution, after digestion for a few minutes at 36° Centigrade, was filtered from the excess of undissolved sulphide, and supersaturated with acetic acid, which threw down a quantity of sulphide of arsenic equal to .925 grammes to the litre. Another portion of the same bottle of water, treated with an acid solution of terchloride of arsenic, gave an amount of sulphide of arsenic equal to 1.110 grammes to the litre.

If now we suppose the dissolved sulphide of arsenic in the experiment above described to have been in the state of a sulpharsenite of magnesium, AsS_3,2MgS, in which the amount of sulphur in the two terms is as 3 : 2, we should have (3 : 2 : : .925 : .617) .617 grammes of the sulphide of arsenic in the last determination derived from the magnesian sulphide, leaving 1.110 — .617 = .493 grammes due to the sulphide of hydrogen in the water. If, however, the arsenious sulphide was dissolved as sulpharsenite of sodium, $AsS_3,3NaS$, in which the sulphur ratio is 3 : 3, we have evidently .925 of sulphide of arsenic derived from the sulphide of sodium in the water, leaving only .185 to be formed by the sulphide of hydrogen. Since, however, the water contains large proportions alike of the chlorides of sodium, calcium, and magnesium, we may suppose that there is a partition of bases, so that portions both of alkaline and earthy sulphides may be present. The excess of magnesian chloride would in any case produce the complete decomposition, observed in boiling, into magnesia and sulphuretted hydrogen.

Two questions then suggest themselves in the analysis of this water; the first as to the relative proportions of sulphide of hydrogen and the monosulphides of fixed bases, and the second as to the base or bases of these fixed sulphides. To resolve the first question, the following method suggests itself: add to one measured portion of the water, at the spring, an acid solution of terchloride of arsenic, by which the whole amount of sulphide in the water may be determined. To another portion add a neutral solution of chloride of zinc or protochloride of iron, which will precipitate the sulphur of the fixed sulphides only, liberating the sulphide of hydrogen. Having removed this by boiling, or by filtration, the insoluble metallic sulphide might be treated with a mixture of a solution of terchloride of arsenic and hydrochloric acid, by which means its sulphur would be obtained as sulphide of arsenic, whose weight, as compared with that from the former determination, would show the quantities both of fixed and volatile sulphide in the water. In connection with this, a determination of the solvent power of the recent water for tersulphide of arsenic would afford the means of solving the second question.

For the analysis of the Bothwell water, the sulphate of lime being determined by the amount of sulphuric acid, the chlorides were calculated from the quantities of bases present, the sulphur corresponding to the dissolved sulphide of arsenic being provisionally estimated as sulphide of sodium. We have thus for 1,000 parts of the water, as follows : —

Chloride of sodium	14.4460	
Chloride of potassium	.3350	
Chloride of calcium	3.1830	
Chloride of magnesium	5.7950	
Sulphate of lime	3.0580	
Sulphide of sodium	.8797	} =.4600 HS.
Sulphide of hydrogen	.0767	
	27.7734	

Waters like this of Bothwell are not unfrequently met with in the borings in the adjacent region, especially in those in Enniskillen, where in a well at Petrolia, at a depth of 471 feet from the surface, and 171 feet from the summit of the Corniferous limestone, a copious spring of this kind was struck, which filled the bore of the well, and flowed in a copious stream, bearing with it a little petroleum. It was a very bitter saline, which dissolved sulphide of arsenic, and gave a purple color with nitroprusside of sodium, but was less strongly sulphurous than that of Bothwell. Waters apparently similar are pumped from several of the oil-wells in the vicinity.

From the facts observed in these wells of Bothwell, Petrolia, and that of Chatham mentioned already in § 62, it would appear that these waters occur beneath the Corniferous limestone, and in the upper part of the Onondaga or saliferous formation of the region. The great density of that of Chatham, which much surpasses that of sea-water, shows it to be derived from a bittern, the result of the evaporation of the waters of an ancient sea. The sulphurous impregnation is doubtless to be ascribed to the reducing action of hydrocarbonaceous matters upon the sulphates which these waters contain. It may therefore happen that the proportion of sulphides in them will be found subject to considerable variations.

APPENDIX.

ON THE POROSITY OF ROCKS.

[From the Report of the Geological Survey of Canada for 1863-66, pages 281-283.]

ALL rocks are more or less porous, and most uncrystalline sedimentary ones possess this character to a very considerable degree. Such rocks when taken from the quarries are more or less completely saturated with water, from which, indeed, they have never been free since the time of their formation. This water they gradually lose when exposed to the air, and, as is well known in the case of many building-stones, become much harder than before. The porosity of rocks is of considerable importance in relation to their value as building materials. The open spaces between the particles diminish the cohesion of the mass, and, in addition to this, the water held in the pores of a rock, when exposed to cold, tends, by its expansion in freezing, to disintegrate the mass, and cause it to crumble, a consideration of much importance in a cold climate. Other things being equal, it may probably be said that the value of a stone for building purposes is inversely as its porosity or absorbing power.

The study of the porosity of rocks is, moreover, of much interest from a geological point of view. As I have elsewhere endeavored to show (*ante*, pages 103 and 163), the origin of most of the muriated saline springs is to be sought in old sea-waters and bitterns imprisoned in ancient sedimentary strata, which must now hold in their pores an amount of water bearing a considerable proportion to the entire volume of the present ocean. The observations here given were made in 1864, with reference to both of the above considerations.

The method of investigation was as follows: Small broken fragments of the rocks — generally from twenty to forty grammes in weight — were selected, and freed from scales or loose grains, which might, by falling off during the experiment, vitiate the results. These specimens were carefully dried at about 200° F., till they ceased to lose weight; most of them had, however, been long preserved in a dry room, and were found to be nearly free from moisture. The weight of these having been determined, they were placed with their lower portions in water, and allowed to remain for some hours, after which they were covered with water and placed under the exhausted receiver of an air-pump, by which process a large portion of air was removed. The exhaustion of the receiver was several times repeated, at intervals, until the portions of rock were as nearly as possible saturated, and bubbles ceased to escape on further exhaustion. They were then removed, carefully wiped with blotting-paper, and again weighed, — first in air, and then in water. These three weighings furnish the data necessary for determining, —

I. The specific gravity of the mass, or the apparent specific gravity, compared with water as unity.
II. The specific gravity of the particles, or real specific gravity.
III. The volume of water absorbed by 100 volumes of the rock.
IV. The weight of water absorbed by 100 parts by weight of the rock.

The loss in weight of the saturated rock when weighed in water, being equal to that of the volume of water displaced by the mass, enables us to determine the specific gravity of the latter; while this loss in weight, less the weight of the water absorbed by the mass, gives the true volume of water displaced by its particles, and hence the means of determining their specific gravity. The division, by the volume of water displaced, of the amount of water absorbed, gives the absorption by volume; and the division of the weight of the water absorbed by that of the dry mass, the absorption by weight: —

a = the weight of the dry rock.
b = the weight of water which the rock can absorb.
c = the loss of weight, in water, of the saturated rock.

We have then the following equations: —

I. $c : a :: 1.000 : x$ = specific gravity of the mass, or apparent specific gravity, water being 1.000.
II. $c - b : a :: 1.000 : x$ = specific gravity of the particles, or real specific gravity, water being 1.000.
III. $c : b :: 100 : x$ = volume of water absorbed by 100 volumes of the rock.
IV. $a : b :: 100 : x$ = weight of water absorbed by 100 parts by weight of the rock.

From these the following table has been calculated, the results given under the last four columns corresponding to the four equations above: — *

* A similar series of results will be found in a report to the British House of Commons, in 1839, by Messrs. Barry, Delabeche, and Smith, made with

TABLE OF THE DENSITY AND POROSITY OF VARIOUS ROCKS.

		I.	II.	III.	IV.
1	Sandstone, Potsdam, — hard and white	2.607	2.644	1.39	0.50
2	Sandstone, Potsdam, — hard and white	2.560	2.638	2.72	1.06
3	Sandstone, Potsdam, — hard and white	2.563	2.633	2.26	0.88
4	Sandstone, Potsdam, — hard and white	2.557	2.618	2.47	0.96
5	Sandstone, Potsdam, with Scolithus .	2.453	2.636	6.94	2.83
6	Sandstone, Potsdam	2.432	2.641	7.90	3.25
7	Sandstone, Potsdam, with Lingula .	2.366	2.611	9.35	3.96
8	Sandstone, Sillery,—green, argillaceous	2.719	2.795	2.73	1.00
9	Sandstone, Sillery,—green, argillaceous	2.642	2.719	2.85	1.08
10	Sandstone, Medina, — red, argillaceous	2.529	2.767	8.37	3.31
11	Sandstone, Medina, — red, argillaceous	2.481	2.776	10.06	4.04
12	Sandstone, Devonian, — fine, gray .	2.110	2.646	20.24	9.59
13	Sandstone, Devonian, — fine, gray .	2.099	2.645	20.62	9.85
14	Sandstone, Devonian, — fine, gray .	2.086	2.649	21.27	10.22
15	Shale, Sillery, — red, argillaceous .	2.674	2.784	3.96	1.49
16	Shale, Hudson River, — black, argil'ous	2.529	2.747	7.94	3.14
17	Shale, Utica, — pyroschist	2.317	2.334	0.75	0.32
18	Shale, Utica, — pyroschist	2.373	2.396	0.93	0.39
19	Shale, Utica, — pyroschist	2.370	2.421	2.10	0.88
20	Limestone, Trenton, — black, compact	2.706	2.714	0.30	0.11
21	Limestone, Trenton, — gray, compact	2.707	2.715	0.32	0.11
22	Limestone, Trenton,—gray, crystalline	2.643	2.673	1.16	0.44
23	Limestone, Trenton,—gray, crystalline	2.671	2.708	1.34	0.50
24	Limestone, Trenton,—gray, crystalline	2.638	2.684	1.70	0.65
25	Dolomite, Niagara, — gray, crystalline	2.537	2.679	5.27	2.08
26	Dolomite, Calciferous	2.772	2.833	2.15	0.78
27	Dolomite, Calciferous	2.737	2.838	3.53	1.28
28	Dolomite, Calciferous	2.635	2.822	6.61	2.51
29	Dolomite, Calciferous	2.601	2.832	7.22	2.77
30	Dolomite, Guelph.	2.527	2.829	10.60	4.19
31	Dolomite, Guelph.	2.528	2.810	10.04	3.97
32	Dolomite, Onondaga	2.517	2.825	10.92	4.33
33	Dolomite, Chazy, argillaceous . . .	2.442	2.824	13.55	5.55
34	Dolomite, Chazy, argillaceous . . .	2.717	2.823	3.75	1.39
35	Dolomite, Chazy, argillaceous . . .	2.693	2.825	4.69	1.73
36	Dolomite, Chazy, argillaceous . . .	2.598	2.891	10.12	3.89
37	Limestone, Tertiary (Caen, France) .	1.859	2.637	29.49	15.85
38	Limestone, Tertiary (Caen, France) .	1.860	2.644	26.93	14.48
39	Limestone, Tertiary (Caen, France) .	1.839	2.611	29.54	16.05

reference to the choice of building-stones for the Houses of Parliament. They made use of blocks of an inch cube, which were first soaked in water and then placed under the vacuum of an air-pump, as in my own experiments. The following examples are taken from a table in the above report, giving the results for thirty-six specimens of building-stones. The value of x in III., or the absorption of water for 100 volumes of rock, as determined by them, is as follows : For three silicious limestones, 5.3, 8.5,

The rocks in the preceding table, with the exception of six, are from the palæozoic formations of Canada, including, as will be seen, pure limestones of the Trenton formation, dolomites of the Calciferous sand-rock, the Chazy, the Onondaga (or Salina), the Niagara, and the Guelph, a local formation resting upon the Niagara. The sandstones are from the Potsdam, the Medina, and the Sillery, a member of the Quebec group, which is associated with the argillaceous shale No. 15, with which are compared the argillaceous shale of the Hudson River group and the compact pyroschists of the Utica formation. I have given in Nos. 12, 13, and 14 determinations with three specimens of a fine gray and very porous sandstone from Ohio, of Devonian or Lower Carboniferous age, and much used for building. Nos. 37, 38, and 39 are three specimens of the well-known soft limestone of Caen, in France, so much employed in that country for architectural purposes.

10.9; for four nearly pure limestones from the oolite, 18.0, 20.6, 24.4, 31.0; for four magnesian limestones, 18.2, 23.9, 24.9, 26.7; and for six sandstones, 10.7, 11.2, 14.3, 15.6, 17.4, and 22.1. These numbers represent the absorption obtained by the aid of the air-pump, without which it is impossible to remove all the air from the pores of the previously dried rock. Thus a cube of two inches of a sandstone which takes up in this way 14.3 of water only absorbed 8.0 by prolonged immersion in water; an oolitic limestone, capable of holding 20.6, in like manner absorbed only 13.5; and a magnesian limestone only 9.1, instead of 24.4. (See, also, On the Porosity of Rocks, Delesse, Bull. Soc. Geol. de France (2), XIX. 64.)

X.

ON PETROLEUM, ASPHALT, PYROSCHISTS, AND COAL.

In the following paper on the Oil-bearing Limestone of Chicago, read before the American Association for the Advancement of Science, in 1870, and published in the American Journal of Science for June, 1871, will be found a summary of my conclusions on the geological history of petroleum. To it are appended extracts from an earlier paper in the same Journal for March, 1863, On Bitumens and Pyroschists, and some later observations by Dawson and myself on the vegetable tissues forming coal. The reader is also referred in connection with petroleum to my paper on the Geology of Southwestern Ontario, in the same Journal for November, 1868, and to Notes on the Oil-Wells of Terre Haute, Indiana, in that for November, 1871.

When, in 1861,* I first published my views on the petroleum of the great American palæozoic basin, I expressed the opinion that the true source of it was to be looked for in certain limestone formations which had long been known to be oleiferous. I referred to the early observations of Eaton and Hall on the petroleum of the Niagara limestone, to numerous instances of the occurrence of this substance in the Trenton and Corniferous formations, and, in Gaspé, in limestones of Lower Helderberg age. Subsequently, in this Journal for March, 1863, and in the Geology of Canada, I insisted still further upon the oleiferous character of the Corniferous limestone in southwestern Ontario, which appears to be the source of the petroleum found in that region. I may here be permitted to recapitulate some of my reasons for concluding that petroleum is indigenous to these limestones, and for rejecting the contrary opinion, held by some geologists, that its occurrence in them is due to infiltration, and that its origin is to be sought in an unexplained process of distillation from pyroschists or so-called bituminous shales. These occur at three

* See the Appendix to this paper.

distinct horizons in the New York system, and are known as the Utica slate, immediately above the Trenton limestone, and the Marcellus and Genesee slates which lie above and below the Hamilton shales; the latter being separated from the underlying Corniferous limestone by the Marcellus state.

First, these various pyroschists do not, except in rare instances, contain any petroleum or other form of bitumen. Their capability of yielding volatile liquid hydrocarbons or pyrogenous oils, allied in composition to petroleum, by what is known to chemists as destructive distillation, at elevated temperatures, is a property which they possess in common with wood, peat, lignite, coal, and most substances of organic origin, and has led to their being called bituminous, although they are not in any proper sense bituminiferous. The distinction is one which will at once be obvious to all those who are familiar with chemistry, and who know that pyroschists are argillaceous rocks containing in a state of admixture a brownish insoluble and infusible hydrocarbonaceous matter, allied to lignite or to coal.

Second, the pyroschists of these different formations do not, so far as known, in any part of their geological distribution, whether exposed at the surface or brought up by borings from depths of many hundred feet, present any evidence of having been submitted to the temperature required for the generation of volatile hydrocarbons. On the contrary, they still retain the property of yielding such products when exposed to a sufficient heat, at the same time undergoing a charring process by which their brown color is changed to black. In other words, these pyroschists have not yet undergone the process of destructive distillation.

Third, the conditions in which the oil occurs in the limestones are inconsistent with the notion that it has been introduced into these rocks by distillation. The only probable or conceivable source of heat, in the circumstances, being from beneath, the process of distillation would naturally be one of ascension, the more so as the pores of the underlying strata would be filled with water. Such being the case, the petroleum of the Silurian and Lower Devonian limestones must have been derived from the Utica slate beneath. This rock, however, is unaltered, and moreover, the intermediate sandstones and shales of the Loraine, Medina, and Clinton formations are destitute of petroleum, which must, on this hypothesis, have passed through all these strata to condense in the Niagara and Corniferous limestones. More than this, the Trenton limestone, which, on Lake Huron and elsewhere, has yielded considerable quantities of petroleum, has no pyroschists beneath it, but on Lake Huron rests on ancient crystalline rocks, with the intervention only of a sandstone devoid of organic or carbonaceous matter. The rock-formations holding petroleum are not only separated from each other by great thicknesses of porous strata destitute of it, but the distribution of this substance is still further localized, as I many years since pointed out. The petroleum is, in fact, in many cases, confined to certain bands or layers in the limestone, in which it fills the pores and the cavities of fossil shells and corals, while other portions of the limestone, above, below, and in the prolongation of the same stratum, although equally porous, contain no petroleum. From all these facts the only reasonable conclusion seems to me to be that the petroleum, or rather the materials from which it has been formed, existed in these limestone rocks from the time of their first deposition. The view which I put forward in 1861, that petroleum and similar bitumens have resulted from a peculiar "transformation of vegetable matters, or in some cases of animal tissues analogous to these in composition," has received additional support from the observations of Lesley* in West Virginia and Kentucky, and from the more recent ones of Peckham.†

The objections to this view of the origin and geological relations of petroleum have been for the most part founded on incorrect notions of the geological structure of southwestern Ontario, which has afforded me peculiar facilities for studying

* Rep. Geol. Canada, 1866, 240; and Proc. Amer. Philos. Soc., X. 33, 187.

‡ Ibid., X. 445.

the question. In this region, it has been maintained by Winchell that the source of the petroleum is to be sought in the Devonian pyroschists. I however showed in 1866, as the result of careful studies of the various borings : first, that none of the oil-wells were sunk in the Genesee slates, but along denuded anticlinals, where these rocks have disappeared, and where, except the thin layer of Marcellus slate sometimes met with at the base of the Hamilton shales, no pyroschists are found above the Trenton limestone. Second, that the reservoirs of petroleum in the wells sunk into the Hamilton shales are sometimes met with in this formation, and sometimes, in adjacent borings, only in the underlying Corniferous. Examples of this have been cited by me in wells in Enniskillen, Bothwell, Chatham, and Thamesville, where petroleum was only found at depths of from thirty to one hundred and twenty feet in the Corniferous limestone, in all of these places overlaid by the Hamilton shales. It was also shown, that in two localities in this region, namely, at Tilsonburg and in Maidstone, where the Corniferous is covered only by post-pliocene clays, petroleum in considerable quantities has been obtained by sinking into the limestone.* That the supplies of petroleum in such localities are less abundant than in parts where a mass of shales and sandstones overlies the oil-bearing limestone, is explained by the fact that both the pores and the fissures in the superior strata serve to retain the oil, in a manner analogous to the post-pliocene gravels in some parts of this region, which are the sources of the so-called surface oil-wells. It is, therefore, not surprising that examples of pyroschists impregnated with oil should sometimes occur, but the evidence of the existence of indigenous petroleum, which is so clear in the various limestones, is wanting in the case of the pyroschists; although concretions holding petroleum, have been observed in the Marcellus and the Genesee slates of New York. There is, however, reason to believe, as I have elsewhere pointed out, that much of the petroleum of Pennsylvania, Ohio, and the

* American Journal of Science (2), XLVI. 360 ; and Report Geol. Canada, 1866, pp. 241 - 250.

adjacent regions is indigenous to certain sandstone strata in the Devonian and Carboniferous rocks.*

At the meeting of the American Association for the Advancement of Science at Chicago, in August, 1868, in a discussion which followed the reading of a paper by myself on the Geology of Ontario,† it was contended that, although the various limestones which have been mentioned are truly oleiferous, the quantity of petroleum which they contain is too inconsiderable to account for the great supplies furnished by oil-producing districts, like that of Ontario, for example. This opinion being contrary to that which I had always entertained, I resolved to submit to examination the well-known oil-bearing limestone of Chicago.

This limestone, the quarries of which are in the immediate vicinity of the city, is filled with petroleum, so that blocks of it which have been used in buildings are discolored by the exudation of this substance, which, mingled with dust, forms a tarry coating upon the exposed surfaces. The thickness of the oil-bearing beds, which are massive and horizontal, is, according to Professor Worthen, from thirty-five to forty feet, and they occupy a position about midway in the Niagara formation, which has in this region a thickness of from 200 to 250 feet. As exposed in the quarry, the whole rock seems pretty uniformly saturated with petroleum, which exudes from the natural joints and the fractured surfaces, and covers small pools of water in the depressions of the quarry. I selected numerous specimens of the rocks from different points and at various levels, with a view of getting an average sample, although it was evident that they had already lost a portion of their original content of petroleum. After lying for more than a year in my laboratory they were submitted to chemical examination. The rock, though porous and discolored by petroleum, is, when freed from this substance, a nearly white, granular, crystalline, and very pure dolomite, yielding 54.6 per cent of carbonate of lime.

Two separate portions, each made up of fragments obtained

* Report Geol. Canada, 1866, p. 240.

† American Journal of Science (2), XLVI. 355.

by breaking up some pounds of the specimens above mentioned, and supposed to represent an average of the rock exposed in the quarry, were reduced to coarse powder in an iron mortar. Of these two portions, respectively, 100 and 138 grammes were dissolved in warm dilute hydrochloric acid. The tarry residue which remained in each case was carefully collected and treated with ether, in which it was readily soluble with the exception of a small residue. This, in one of the samples, was found equal to .40 per cent, of which .13 was volatilized by heat with the production of a combustible vapor having a fatty odor; the remainder was silicious. The brown ethereal solutions were evaporated, and the residuum, freed from water and dried at 100° C., weighed, in the two experiments, equal to 1.570 and 1.505 per cent of the rock, or a mean of 1.537. It was a viscid reddish-brown oil, which, though deprived of its more volatile portions, still retained somewhat of the odor of petroleum which is so marked in the rock. Its specific gravity, as determined by that of a mixture of alcohol and water in which the globules of the petroleum remained suspended, was .935 at 16° C. Estimating the density of the somewhat porous dolomite at 2.6, we have the proportion .935:2.600::1.537:4.260; so that the volume of the petroleum obtained equalled 4.26 per cent of the rock. This result is evidently too low, for two reasons: first, because the rock had already lost a part of its oil, while in the quarry and subsequently, before its examination; and secondly, because the more volatile portions had been dissipated in the process of extraction just described.

In assuming 100.00 parts of the rock to hold 4.25 parts by volume of petroleum, we are thus below the truth in the following calculations. A layer of this oleiferous dolomite one mile (5,280 feet) square and one foot in thickness will contain 1,184,832 cubic feet of petroleum, equal to 8,850,069 gallons of 231 cubic inches, and to 221,247 barrels of forty gallons each. Taking the minimum thickness of thirty-five feet, assigned by Mr. Worthen to the oil-bearing rock at Chicago, we shall have in each square mile of it 7,743,745 barrels, or in round numbers seven and three quarter millions of barrels of petroleum. The total produce of the great Pennsylvania oil-region for the ten years from 1860 to 1870 is estimated at twenty-eight millions of barrels of petroleum, or less than would be contained in four square miles of the oil-bearing limestone formation of Chicago.

It is not here the place to insist upon the geological conditions which favor the liberation of a portion of the oil from such rocks, and its accumulation in fissures along certain anticlinal lines in the broken and uplifted strata. These points in the geological history of petroleum were shown by me in my first publications on the subject in March and July, 1861, referred to on the next page, and independently, about the same time, by Professor E. B. Andrews in this Journal for July, 1861.*

The proportion of petroleum in the rock of Chicago may be exceptionally large, but the oleiferous character of great thickness of rock in other regions is well established, and it will be seen from the above calculations that a very small proportion of the oil thus distributed would, when accumulated along lines of uplift in the strata, be more than adequate to the supply of all the petroleum wells known in the regions where these oil-bearing rocks are found. With such sources existing ready formed in the earth's crust, it seems to me, to say the least, unphilosophical to search elsewhere for the origin of petroleum, and to imagine it to be derived by some unexplained process from rocks which are destitute of the substance.

* American Journal of Science (2), XXXII. 85. See also papers on the subject by Andrews and by Professor Evans, Ibid. (2), XL. 33, 334; and one by the author (2), XXXV. 170; also Report Geological Survey of Canada, 1866, pp. 256, 257.

APPENDIX.

ON BITUMENS AND PYROSCHISTS.

(1861-1863.)

This paper is reprinted from the American Journal of Science for March, 1863, but many of the facts and deductions which it contains appeared in an earlier paper, entitled Notes on the History of Petroleum, in the Canadian Naturalist for July, 1861, reprinted in the Chemical News, and also in the Report of the Smithonian Institution for 1862. I had for some time previously maintained that the source of the petroleum of the West was not, as was generally thought, to be found in the Devonian pyroschists, but in the underlying fossiliferous limestones, and had shown the relation of the oil-springs to anticlinals.— See a report of my lecture before the Board of Arts of Lower Canada, in the Montreal Gazette of March 1, 1861.

It is proposed in the following pages to bring together some facts and theoretical considerations bearing upon the nature, origin, and distribution of bitumens, together with a few remarks on the rocks commonly called bituminous shales. Under the general name of bitumen, as is well known, are included both the liquid forms, petroleum and naphtha, and the solid varieties known as asphalt or mineral pitch. The related substances guayaquillite and berengelite, and the substance known as idrialine, seem from the modes of their occurrence to have a similar origin to asphalt, and thus to be distinct from fossil resins. The characters of fusibility and solubility in liquids like benzole and sulphuret of carbon, serve to distinguish the solid bitumens from coal and some other matters about to be noticed. It is to be remarked that the chemical composition of these bodies varies considerably; the earlier analyses of petroleum and naphtha give a composition which approaches C_nH_n; but the later investigations of De la Rue and Muller on the products distilled from the petroleum of Rangoon, and those of Uelsmann on that from Sehnde, show a slight excess of hydrogen, the various hydrocarbons having, for the most part, the formula C_nH_{n+2}. The first formula C_nH_n may however be adopted, as expressing approximatively the composition of the liquid bitumens. The different analyses of asphalt show a diminished quantity of hydrogen, and small quantities of oxygen. Thus the elastic bitumen from Derbyshire gave to Johnston results which may be represented by $C_{24}H_{22}O_{0\cdot3}$; * of two varieties of asphalt analyzed by Ebelmann, the

* In these formulas, which have been calculated for twenty-four equivalents of carbon, to compare with cellulose, $C_{24}H_{20}O_{20}$, I have designed to represent

one from Bastennes gave $C_{24}H_{16}O_{0.7}$, while that from near Naples may be represented by $C_{24}H_{14.6}O_2$, and an asphalt from Mexico gave to Regnault $C_{24}H_{17}O_2$. The analyses of Johnston shows that guayaquillite and berengelite do not differ greatly from these in the proportions of carbon and hydrogen. Passing from the asphalts to idrialine, the results of whose analysis are represented by $C_{24}H_8$, we have a hydrocarbon with a minimum of hydrogen. It is well in this place to compare the above results with the formula $C_{24}H_{15.9}O_{1.6}$, which is deduced from Wetherell's analysis of the so-called albertite or Albert coal. A "lignite passing into mineral resin" gave to Regnault $C_{24}H_{15}O_{3.3}$, and five analyses of bituminous coal by the same chemist yield from $C_{24}H_8O_{0.9}$ to $C_{24}H_{10}O_{3.3}$, while the mean composition deduced by Johnston from several analyses of coal was $C_{24}H_9$, with from O_2 to O_4. From these results it will be seen that some asphalts approach bituminous coals in composition. That of Naples, which is completely fusible at 140° C., contains less hydrogen and more oxygen than the albertite, while the idrialine is near in composition to certain bituminous coals, which are thus almost isomeric with some fusible bitumens; so that it is easy to conceive the same organic matters giving rise either to coal or to asphalt, even without losing their structure. Such appears to be the case in the tertiary strata of Trinidad and Venezuela, the bitumen of which, from Mr. Wall's researches, seems to have arisen from "a special mineralization of vegetable remains in certain strata, which has resulted in the production of bitumen, instead of coal or lignite." This conversion, according to him, "is not attributable to heat, nor of the nature of a distillation, but is due to chemical reactions at the ordinary temperature, and under the normal conditions of climate." Mr. Wall also describes portions of wood from these deposits, which have been partially converted into bitumen, and

simply the results of analysis, without attempting to fix the constitution of the matters in question.

In the notation employed, H = 1, C = 6, and O = 8. As it is not generally used in the American Journal of Science, I have not thought necessary to adopt, in this paper, the double equivalent of the latter elements, now employed by so many chemists. I may, however, call attention to the fact that I was, I believe, the first to propose such a change, when, in 1853, I asserted that the even coefficients of oxygen, sulphur, and carbon in ordinary formulas seem to furnish a conclusive reason for doubling their equivalents, or for dividing those of hydrogen, chlorine, nitrogen, and the metals, according as four volumes or two volumes are taken as the equivalent. (Theory of Chemical Changes, Am. Jour. of Science (2), XV. p. 230. [Reprinted as Essay XVI. of the present volume.])

leave, when this is removed by solvents, a residue of woody tissue. (Proc. Geol. Soc. London, May, 1860.) These observations have been confirmed by an eminent microscopist and chemist, whose results, lately communicated to me by himself, are not yet published.

The chemical changes by which the conversion of woody tissue into peat, lignite and bituminous coal is effected, are too well known to be repeated here. The abstraction of variable proportions of water, carbonic acid, and marsh-gas may give rise either to hydrocarbons like $C_{24}H_8$, which represents idrialine and the basis of most bituminous coals, to $C_{24}H_{16}$, which is the approximate formula of the hydrocarbons of many asphalts, or to $C_{24}H_{24}$, which represents petroleum. The removal of further amounts of marsh-gas, C_2H_4, may even convert bituminous coal into anthracite, as Bischof has pointed out; and we conceive that although heat has in many cases given rise to this conversion, by a subterranean coking, the change may often have been the result of decompositions going on at ordinary temperatures. Anthracite or nearly pure carbon, on the one hand, and petroleum or carbon with a maximum of hydrogen, on the other, represent the two extremes of a series of which bituminous coals and asphalts are intermediate terms.

Petroleum, as is well known, impregnates certain rocks, from which it flows spontaneously, and the solid forms of bitumen are often disseminated throughout limestones or sandstones, from which they may be in part removed by heat, and more completely by solvents such as benzole. To such rocks the term "bituminous" may be correctly applied, but it is often inappropriately given to substances like coal and certain combustible schists, which contain little or no bitumen, but yield, by destructive distillation, volatile hydrocarbons, more or less resembling those obtained from asphalt or petroleum. Analogous products are, however, obtained by the distillation of lignite, peat, and even of wood, so that the epithet "bituminous," applied to hydrogenous coals and combustible schists, raises a false distinction, and perpetuates an error. I therefore proposed some time since to distinguish these so-called bituminous schists, the *brandschiefer* of the Germans, by the name of *pyroschists*. This is the equivalent of the German term, and has a precedent in the name of pyrorthite, given by Berzelius to a substance which appears to be a mixture of orthite with a combustible hydrocarbonaceous matter. Pyroschists are well known to occur in almost every geological group from the Cambrian to the tertiary, and are often, like coal, employed as valuable sources of volatile hydrocarbons, although like it they contain little or no bitumen. They may be regarded as clays or marls, holding, in a state of intimate admixture, a variable proportion of a matter approaching to coal in its chemical characters. Although frequently dark brown or black in color, they are sometimes light brown or even yellowish-gray, as is the case with the Jurassic pyroschists of the department of the Doubs, and those of tertiary age near Clermont, both in France. Remarkable examples of this are also given by Professor J. D. Whitney in the pyroschists from the Utica formation in Iowa, which were yellowish-brown, weathering to a bluish-ash color. They, however, blackened when exposed to heat, burning with a bright flame, and contained from eleven to twenty per cent of combustible matter.* . . . . A pyroschist of the Utica formation, from Collingwood on Lake Huron, examined by me, gave to dilute hydrochloric acid from fifty-three to fifty-eight per cent of carbonate of lime, besides a little magnesia and oxide of iron. The insoluble residue was snuff-brown in color, and, when heated, gave off a bituminous odor. When ignited in a close vessel, it lost 12.6 per cent of volatile and combustible matters, and left a coal-black residue, which, by calcination in the open air, lost 8.4 per cent additional, making in all 21.0 per cent of volatile and carbonaceous matters, and left an ash-gray argillaceous residue. This schist, however, contained but a very small amount of bitumen ; for, on treating the residue from a dilute acid with boiling benzole, there was dissolved about 1.0 per cent of a brown bituminous matter. The residue, when heated, no longer evolved the odor of bitumen, but rather one like burning lignite, and still gave, by ignition in a close vessel, 11.8 per cent of volatile and inflammable matters. When boiled with a solution of caustic soda, this was scarcely discolored. In its insolubility, therefore, the organic matter of this rock resembles true coal rather than lignite. Attempts have been made, on a large scale, to distil this calcareous schist of Collingwood, which was found to yield from 3.0 to 5.0 per cent of oily and tarry matter, besides combustible gases and water.

Overlying the Hamilton formation in Ontario are found black pyroschists, which are supposed to be the equivalent of the Genesee slates of New York. A specimen of these from Bosanquet on Lake

* For numerous analyses of pyroschists from this geological horizon, see a note appended to this paper in the American Journal of Science (2), XXXV. 160.

Huron lost, by ignition in a closed vessel, 12.4 per cent, and left a black residue, which was not calcareous. A portion in fine powder was digested for several hours with heated benzole, which took up 0.8 per cent of brown combustible matter. The residue, carefully dried at 200° F., then lost, by ignition in a close vessel, 11.3 per cent, and by subsequent calcination 11.6 additional, equal to 23.7 per cent of combustible and volatile elements. The calcined residue was gray in color. By distillation in an iron retort there were obtained from this shale 4.2 per cent of oily hydrocarbons, besides a large quantity of inflammable gas, and a portion of ammoniacal water.

The pyroschists of Bosanquet belong to the Devonian series, and contain the remains of land-plants, so that a partially decayed vegetation may be supposed to have been the source of the organic matter which is intimately mingled with the earthy base of the rock. Such was probably the case in the abundant pyroschists of the coal period; but in the pyroschists of the Utica formation (which are Upper Cambrian) the chief organic remains to be detected are graptolites, with a few brachiopods and crustaceans. No traces of terrestrial vegetation are known to have existed at that time, nor do the schists contain the evidences of any marine plants. The pyroschists of mesozoic age, in several parts of Europe, contain, on the contrary, numerous fossil fishes, from the soft parts of which, or other animal matters, the combustible substance of these rocks is generally supposed to be derived. (Dufrénoy, Mineralogie, IV. p. 603.) Similar questions arise with regard to the origin of the bitumens of the various geological formations already noticed; for while in some cases, as in the tertiary rocks of Trinidad, they are clearly traced to a vegetable source, bitumens are also met with in Cambrian, Silurian, and Devonian limestones of marine origin, which abound in shells and corals, but afford no traces of vegetable remains. When, however, it is considered that the lower forms of animals contain considerable portions of a non-azotized tissue analogous in its composition to that of plants, and that even muscular tissue, *plus* the elements of water, contains the elements of cellulose and ammonia, it is easy to understand that vegetable and animal remains may, by their slow decomposition, give rise to similar hydrocarbonaceous bodies.* The various fermentations of which sugar is susceptible

* This relation was first pointed out by me in 1849. (American Journal of Science (2), VII. p. 109.) I then endeavored to show that the albuminoid bodies might be regarded as a nitryl of cellulose, or some isomeric hydrate of carbon, and represented by the formula $C_{24}H_{17}N_3O_8$. I had already pro-

suggest analogies to the different transformations of organic tissues which have resulted in the formation of anthracite, coal, lignite, asphalt, and petroleum, together with carbonic acid and gaseous hydrocarbons as accessory products. (See note on page 182.)

[The conclusions of the remaining nine pages of the above paper are briefly summed up in the preceding one on The Oil-bearing Limestones of Chicago. As a supplement to the remarks on the origin of coal I may here make some extracts from a paper on Spore-Cases in Coal, by Dr. J. W. Dawson, in the American Journal of Science for April, 1871, including also a note by myself. Dawson has there shown that while some exceptional beds of coal are to a large extent made up of spores and spore-cases, probably of lepidodendron, it is by no means true that these are, as some have conjectured, the principal source of coal. On the other hand, it is clear

posed to regard bone-gelatine as an analogous nitryl, $C_{24}H_{20}N_4O_8$; which corresponds to one equivalent of glucose and four of ammonia, less 8 HO. These nitryls, it was conceived, might, under certain conditions, regenerate ammonia and a hydrate of carbon. I also adduced evidence that in a case of diabetes, sugar was generated at the expense of ingested gelatine. (American Journal of Science (2), V. p. 75; VI. p. 259; and Silliman's Elements of Chemistry, p. 517.) The analyses of cartilage-gelatine, or chondrine, in like manner correspond very nearly to a nitryl formed from $C_{24}H_{22}O_{22}$ (cane-sugar) and three equivalents of ammonia. The formula thus deduced, $C_{24}H_{19}N_3O_{10}$, requires 14.7 of nitrogen.

In 1856, Dusart, starting, as he tells us, from my theoretical views, endeavored to produce the albuminoid bodies by the action of a solution of ammonia on starch, lactose, or glucose at temperatures of 150° and 200° C. In this way he obtained, after several days, an azotized body, which resembled gelatine. It was precipitated by alcohol in elastic filaments, formed an imputrescible compound with tannin, and, when heated, gave off the odor of burning horn. Its proportion of nitrogen was 14.0 per cent, which is near that of chondrine. (Comptes Rendus de l'Académie, May, 1861, p. 974.) Schoonbroodt has since asserted the possibility of converting sugar into an albuminoid substance, and reiterated my suggestion that the albuminoids are veritable nitryls of the amyloids; under which convenient term he includes those hydrates of carbon which are susceptible of conversion into glucose. (Ibid., May, 1860, p. 856.)

In 1861, Messrs. Fischer and Boedeker announced the production of fermentescible sugar by the action of dilute acids on cartilage, and showed that the ingestion of gelatine increases the amount of sugar in normal human urine. These authors seem, by the abstract before me (Repertoire de Chimie Pure, July, 1861, from Ann. der Chem. und Pharm., CXVII. p. 111), to ignore alike my own observations and those of Gerhardt, who twenty years since showed that, by long boiling with dilute sulphuric acid, there is formed from gelatine a sweet fermentescible sugar, together with a large amount of sulphate of ammonia. (Précis de Chimie Organique, II. p. 521.)

from the microscopical studies of Dawson and others that, although it is doubtless true that cellulose may yield bodies having the chemical composition of bituminous coal, and even bitumens, by a process of alteration such as I have described above, the chief source of such coal in the older coal measures has been epidermal tissues, which differ from cellulose in being much richer in carbon and hydrogen. These tissues, as remarked by Dawson, "are very little liable to decay, and resist more than most other vegetable matters aqueous infiltration, properties which have caused them to remain unchanged and resist the penetration of mineral substances more than other vegetable tissues. These qualities are well seen in the bark of our American white birch (*Betula alba*). It is no wonder that materials of this kind should constitute considerable portions of such vegetable accumulations as the beds of coal, and that when present in large proportion they should afford richly bituminous beds. All this agrees with the fact apparent on examination of common coal, that the greater number of its purest layers consist of the flattened bark of sigillariæ and similar trees, just as any single flattened trunk imbedded in shale becomes a layer of pure coal. It also agrees with the fact that other layers of coal, and also the cannels and earthy coals, appear under the microscope to consist of finely comminuted particles, principally of epidermal tissues, not only of the fruits and spore-cases of plants, but also of their leaves and stems."

In this connection I noticed in the same paper the chemical composition of the epidermal or cortical tissue of plants, to which the name of suberin has been given, and compared it with that of the spores of lycopodium, and at the same time with cellulose and with forms of coal and related bodies. The nitrogen which the first two mentioned bodies contain probably represents a portion of albuminoid matter, which in lycopodium is considerable in amount. For the purpose of comparison empirical formulas corresponding to twenty-four equivalents of carbon have been calculated for these bodies, as already done on page 176. We have then as follows: —

Cellulose	$C_{24}H_{20}O_{20}$
Cork	$C_{24}H_{18\cdot2}O_{6\cdot7}$
Lycopodium	$C_{24}H_{19\cdot4}NO_{5\cdot6}$
Peat (Vaux)	$C_{24}H_{14\cdot4}O_{10}$
Brown coal (Schrötter)	$C_{24}H_{14\cdot3}O_{10\cdot6}$
Lignite (Vaux)	$C_{24}H_{11\cdot3}O_{6\cdot4}$
Bituminous coal (Regnault)	$C_{24}H_{10}O_{3\cdot3}$

I further said, "It will be seen from this comparison that in ultimate composition cork and lycopodium are nearer to lignite than to woody fibre (cellulose), and may be converted into coal with far less loss of carbon and hydrogen than the latter. They, in fact, approach closer in composition to resins and fats than to wood; and, moreover, like these substances, repel water, with which they are not easily moistened, and are thus able to resist those atmospheric influences which effect the decay of woody tissue."

The nitrogen present in the lycopodium spores, as remarked by Dawson, "no doubt belongs to the protoplasm in them, which would soon perish by decay; and, subtracting this, the cell-walls of the spores and the walls of the spore-cases would be most suitable material for the production of bituminous coal. But this suitableness they share with the epidermal tissue of the scales of strobiles and of the stems and leaves of ferns and lycopods, and above all with the thick corky envelope of the stems of sigillariæ and similar trees .... which, from its condition in the prostrate and in the erect trunks contained in the beds associated with coal, must have been highly carbonaceous and extremely enduring, and impermeable to water." The substance known as mineral charcoal is, according to Dawson, derived from woody tissue and the fibres of bark. (See in this connection his paper on the Conditions of the Accumulation of Coal, Quarterly Geological Journal, XXII. 95.)

[NOTE to page 180. The petroleum of Pennsylvania, according to Pelouze and Cahours, yields by fractional distillation various liquids having the common formula C_nH_{2n+2} ($C = 12$), the value of n ranging from 4 to 15, (corresponding to $C_8H_{10} \ldots C_{30}H_{32}$ in the notation adopted in the preceding pages), and the boiling-point from 0° to 160° C. Of this series, which also includes the paraffines, the first term is marsh-gas or formene, and the second and third belong to the ethylic and propylic groups, being C_2H_4, C_4H_6 and C_4H_8 in the above notation. The latter two, according to Ronalds, are found in solution in the crude petroleum. The researches of Foucou and Fouqué (Comptes Rendus, November 23, 1868) show that while the inflammable gases from the so-called Burning Spring near Niagara Falls, and from an oil-well in Wirtz County, West Virginia, are marsh-gas with small admixtures of carbonic acid, the gases from an oil-well in Petrolia, Ontario, and from Fredonia, Chatauque County, New York, are mixtures, in about equal parts, of the second and third hydrocarbons of the above series. The gas at the latter locality is from a well sunk into the Genessee slates, at the summit of the Hamilton formation, which gives no petroleum, but has for many years furnished the supply of gas for lighting a small town. The gas from an oil-well in Venango County, Pennsylvania, contained besides the first three bodies of the series a portion of the fourth, C_8H_{10}. Neither acetylene, free hydrogen, carbonic oxide, nor olefiant gas or its homologues were detected.]

XI.

ON GRANITES AND GRANITIC VEIN-STONES.

(1871 - 1872.)

This paper appeared in three parts in the American Journal of Science for February and March, 1871, and for February, 1872. The license by which the title is made to include a description of certain calcareous vein-stones is explained to the reader under §§ 35 - 37. Part I., as originally printed, included §§ 1 - 15; part II., §§ 16 - 31; and part III., §§ 32 - 49.

CONTENTS OF SECTIONS. — 1, 2. Definitions of granite and syenite; 3. Structure of granitic and gneissic rocks; 4, 5. Felsites and felsite-porphyries; 6. Gneisses and granites of New England; 7. Granitic dikes and granitic vein-stones; 8. Scheerer's theory of granitic veins; 9 - 10. Elie de Beaumont on granites and granitic emanations; 11. Granitic distinguished from concretionary veins; 12. Von Cotta on granitic veins; 13, 14. The author's views on the concretionary origin of granitic veins; 15. The banded structure of granitic veins; 16. Granitic veins of Maine, Brunswick; 17. Topsham, Paris; 18. Westbrook, Lewiston; crystalline limestones; 19. Danville, Ketchum; 20. Denuded granitic masses; 21. Banded veins; Biddeford, Sherbrooke; 22. Veins at various New England localities; 23. Mineral species of these veins; 24. Veins in erupted granites; 25. Geodes in granites; 26. Veins distinguished from dikes; 27. Volger and Fournet on the origin of veins; 28, 29. Certain fissures and geodes distinguished from veins opening to the surface; 30, 31. Temperatures of crystallization of granitic minerals; 32. Laurentian gneisses; 33. Pyroxenites and limestones; 34. Absence of mica-schists; 35. Classes of veins; 36. Granitic vein-stones; 37. Similar veins in Norway; 38. Minerals of granitic veins; 39. Evidences of concretionary origin; banded structure; 40. Incrustations of crystals; 41. Skeleton-crystals; 42. Rounded crystals; 43. Quartz crystals in metalliferous veins; 44. Types of vein-stones; feldspathic; 45. Calcareous vein-stones; 46. Order of succession of minerals; 47. Attitude of the veins; 48. Calcareous vein-stones in higher rocks; 49. Supposed eruptive limestones.

§ 1. THE name of granite is employed to designate a supposed eruptive or exotic unstratified composite rock, granular, crystalline in texture, and consisting essentially of orthoclase-feldspar and quartz, with an admixture of mica, and frequently of a triclinic feldspar, either oligoclase or albite. This is the definition of granite given by most writers on lithology, and applies to a great portion of what are commonly called granitic rocks; there are, however, crystalline granite-like aggregates in which the mica is replaced by a dark colored hornblende or amphibole, and to such a compound rock many authors have given the name of syenite, while to those in which mica and hornblende coexist the name of syenitic granite is applied. It is observed that in certain of these hornblendic granites the quartz becomes less in amount than in ordinary granites, and finally disappears altogether, giving rise to a rock composed of orthoclase and hornblende only. To such a binary aggregate Von Cotta and Zirkel would restrict the term "syenite," which was already defined by D'Omalius d'Halloy to be a crystalline aggregate of hornblende and feldspar; by which orthoclase-feldspar may be understood, since he describes varieties of syenite as passing into diorite, — a name by most modern lithologists restricted to a compound of albite, or some more basic triclinic feldspar, with hornblende. It is apparently by failing to appreciate the distinction between orthoclase and triclinic feldspar, in this connection, that Haughton has lately described, under the name of syenite, rocks which are composed of crystalline labradorite and hornblende.

§ 2. Naumann, regarding orthoclase and quartz as the essential constituents of granite, designates those aggregates which contain mica as mica-granites, and thus distinguishes them from hornblende-granites, in which the mica is replaced by hornblende. These definitions seem the more desirable, as the name of granite is popularly applied both to the hornblendic and the micaceous aggregates of orthoclase and quartz. There are not wanting examples of well-defined rocks of this kind in which both mica and hornblende are almost or altogether wanting. Such rocks have been designated binary granites, a term which it will be well to retain. Chloritic and talcose granites, into the composition of which chlorite and talc enter, need only be mentioned in this connection. The name of syenite, so often given to hornblendic granites, will, in accordance with the views already expressed, be restricted to rocks destitute of quartz. While the disappearance of this mineral from hornblendic granites is held to give rise to a true syenite, the same process with micaceous granites affords a quartzless rock consisting of orthoclase and mica, for which we have no name. Great masses of an eruptive rock, granite-like in structure, and consisting of crystalline orthoclase or sanidin, without any quartz, occur in the province of Quebec. This rock contains in some cases a small admixture of black mica, and in others an equally small proportion of black hornblende. The latter variety might be described as syenite, but for the former we have no distinctive name; and I have described both of these by the name of granitoid trachytes, a term which I adopted the more willingly on account of the peculiar composition of the feldspar, and also because compact and finely granular rocks in the same region, having a similar chemical composition, present all the characters of typical trachytes, and apparently graduate into the granitoid rocks just noticed.* In all attempts to define and classify compound rocks, it should be borne in mind that they are not definite lithological species, but admixtures of two or more mineralogical species, and can only be arbitrarily defined and limited.

§ 3. Having thus defined the mineral composition of granitic rocks, we proceed to notice their structure. Gneiss has the same mineral elements as granite, but is distinguished by the more or less stratified and parallel arrangement of its constituents; and lithologists are aware that in certain varieties of gneiss this structure is scarcely evident, except on a large scale; so that the distinction between gneiss and granite rests rather on geognostical than on lithological grounds. To the lithologist, in fact, the granitoid gneisses are simply more or less stratiform granites, while it belongs to the geologist to consider whether this structure has resulted from a sedimentary

* American Journal of Science (2), XXXVIII. 95. See also Zirkel, Petrographie, II. 179.

deposition, or from the flowing of a semi-fluid heterogeneous mass giving rise to a stratiform arrangement.*

§ 4. The rocks having the mineralogical composition of granites present a gradual passage from the coarse structure of

[* This process has been particularly described in my Contributions to Lithology, where also the principles of lithological classification are discussed at length. (American Journal of Science for March and July, 1864.) A stratiform structure in eruptive rocks is there said to be due to "the arrangement of crystals during the movement of the half-liquid crystalline mass, but it may in some instances arise from the subsequent formation of crystals arranged in parallel planes." In the same paper, in describing the dolerite of Montarville, the alternations of a coarse variety, porphyritic from the presence of large crystals of augite, with a finer grained and whiter variety is noticed; the two being "arranged in bands, whose varying thickness and curving lines suggest the notion that they have been produced by the flow and the partial commingling of two fluid masses." At Mount Royal also, as there described, "mixtures of augite with feldspar are met with, constituting a granitoid dolerite, in parts of which the feldspar predominates, giving rise to a light grayish rock. Portions of this are sometimes found limited on either side by bands of nearly pure black pyroxenite, giving at first sight an aspect of stratification. The bands of these two varieties are found curiously contorted and interrupted, and, as at Montarville, seem to have resulted from movements in a heterogeneous pasty mass, which have effected a partial blending of an augitic magma with another more feldspathic in its nature."

Further illustrations of this are given by the author in a communication to the Boston Society of Natural History, January 7, 1874. Among these was a specimen from Groton, Connecticut, in which a large angular fragment of strongly banded micaceous gneiss is enclosed in a fine-grained eruptive granite, the mica plates in which are so arranged as to show a beautiful and even stratification in contact with the broken edges of the gneiss, but at right angles to the strata of the latter. Another example is afforded by the eruptive diorite from the mesozoic sandstone of Lambertville, New Jersey, which is conspicuously marked by light and dark bands due to the alternate predominance of one or the other of the constituent minerals; and still another in a fine-grained dark micaceous dolerite dike from the Trenton limestone at Montreal, in which the abundant laminæ of mica (probably biotite) are arranged parallel to the walls of the dike. A similar banded structure is seen in glacier-ice and in furnace-slags. Some geologists have from facts of this kind been led to suppose that the banded structure of great areas of gneiss was caused by movements of flow in a solidifying mass, and not by successive deposits of dissolved or suspended material from a watery medium. While admitting the frequent occurrence of this structure in eruptive rocks, and the necessity in many cases of a careful geognostical study to determine to which class a stratiform rock should be referred, it was maintained that the great areas of gneissic rocks are of aqueous origin, and were deposited in successive horizontal layers with their associated limestones, quartzites, and iron-oxides.]

ordinary micaceous, hornblendic, and binary granites to finely granular and even impalpable mixtures of the constituent minerals, constituting the rocks known as felsite, eurite, and petrosilex. These rocks are often porphyritic from the presence of crystals of orthoclase, and sometimes of crystals or grains of quartz imbedded in the finely granular or impalpable paste. These felsites and felsite-porphyries (orthophyres) are, in very many cases at least, stratified or indigenous rocks, and they are sometimes found associated with granular aggregates of different degrees of coarseness, which show a transition from true felsites into granitic gneisses. The resemblances in ultimate composition between felsites, granites, and granitic gneisses are so close that it cannot be doubted that their differences are only structural.

§ 5. Felsites and felsite-porphyries or orthophyres are well known in eastern Massachusetts, at Lynn, Saugus, Marblehead, and Newburyport, and may be traced from Machias and Eastport in Maine, along the southern coast of New Brunswick to the head of the Bay of Fundy, with great uniformity of type, though in every place subject to considerable variations, from a compact jasper-like rock to more or less coarsely granular varieties, all of which are often porphyritic from feldspar crystals, and sometimes include grains or crystals of quartz. The colors of these rocks are generally some shade of red, varying from flesh-red to purple ; pale yellow, gray, greenish, and even black varieties are however occasionally met with. These rocks are, throughout this region, distinctly stratified, and are closely associated with dioritic, chloritic, and epidotic strata. They apparently belong, like these, to the great Huronian system.

[Stratiform rocks, seemingly identical with these quartziferous feldspar-porphyries, abound in Missouri, where they are associated with the iron-ores of Iron Mountain and Shepard Mountain. I have also found them over a considerable area along the north shore of Lake Superior, on an island south of St. Ignace, and for some distance along the coast to the southwest. The breccia and conglomerate in which is found the native copper of the Calumet and Hecla and the Boston and Albany mines of the Keweenaw peninsula, on the south shore of the same lake, is made up in large part of the ruins of similar orthophyres.]

§ 6. Many of the so-called granites of New England are true gneisses; as, for example, those quarried in Augusta, Hallowell, Brunswick, and many other places in Maine, which are indigenous rocks interstratified with the micaceous and hornblendic schists of the great White Mountain series. To this class also, judging from lithological characters, belong the so-called granites of Concord and Fitzwilliam, New Hampshire. These indigenous rocks are tenderer, less coherent, and generally finer grained than the eruptive granites, of which we have examples in the micaceous granite of Biddeford, Maine, and the hornblendic granites of Marblehead and Stoneham, Massachusetts, and Newport, Rhode Island, in all of which localities the contact of the eruptive mass with the enclosing rock is plainly seen, as is also the case farther eastward, on the St. Croix and St. John's Rivers in New Brunswick, and in the Cobequid Hills and elsewhere in Nova Scotia. The hornblendic granites of Gloucester, Salem, and Quincy, Massachusetts, seem also, from their lithological characters, to belong to the class of exotic or true eruptive granites.* The further discussion of the nature and origin of these gneisses and granites is reserved for another occasion, and we now proceed to notice the history of granitic veins.

§ 7. The eruptive granitic masses just noticed not only include fragments of the adjacent rocks, especially near the line of contact, but very often send off dikes or veins into the surrounding strata. The relation of these with the parent mass is however generally obvious, and it may be seen that they do not differ from it except in being often finer grained. These injected or intruded veins are not to be confounded with a third class of granitic aggregates, which I have elsewhere described as granitic vein-stones, or, to express their supposed

* T. S Hunt on the Geology of Eastern New England, American Journal of Science for July, 1870, p. 88; also Notes on the Geology of the Vicinity of Boston, Proc. Boston Nat. Hist. Soc., Oct. 19, 1870.

mode of formation, endogenous granites. They are to the gneisses and mica-schists, in which they are generally enclosed, what calcite veins are to stratified limestones, and although long known, and objects of interest from their mineral contents, have generally been confounded with intrusive granites.

§ 8. Scheerer, in his famous essay on granitic rocks, which appeared in the Bulletin of the Geological Society of France in 1847 (Vol. IV. p. 468), conceives the congealing granitic rocks to have been impregnated with "a juice," which was nothing else than a highly heated aqueous solution of certain mineral matters. This, under great pressure, oozed out, penetrating even the stratified rocks in contact with the granite, filling cavities and fissures in the latter, and depositing therein crystals of quartz and of hornblende, the arrangement of which shows them to have been of successive growth. Neither Scheerer nor Virlet d'Aout, who supported his views, however (Ibid., IV. p. 493), extended them to feldspathic veins, though Daubrée, at an earlier date, had described certain granitic veins in Scandinavia as having been formed by secretion, rather than by igneous injection, as maintained by Durocher.

§ 9. Elie de Beaumont, starting from the hypothesis of a cooling liquid globe, imagined "a bath of molten matter on the surface of which the first granites crystallized." From the ruins of these were formed the first sedimentary deposits, but directly beneath were other granitic masses, which became fixed immediately afterward. "Some parts of these masses, coagulated from the commencement of the cooling process, but not completely solidified, were then erupted through the sedimentary deposits" just mentioned. "In these jets of pasty matter" were contained many of the rarer elements of the granitic magma, which were thus concentrated in the outermost portions of the granitic crust, and in the ramifications formed by these portions in the masses through which they were forced by the eruptive agents. Those portions of the granitic masses, and their ramifications, in which these rarer elements are concentrated, are distinguished from the rest of the masses alike by their exterior position and their peculiar structure. They are often coarse-grained, and include the pegmatites, tourmaline-granites, and veins carrying cassiterite and columbite, frequently abounding in quartz. These mineral products are to be regarded as emanations from the granite, and are described as a *granitic aura*, constituting what Humboldt has called the *penumbra* of the granite. (Bull. Soc. Geol. de France (2), IV. 1249. See particularly pages 1295, 1321, and 1323.)

§ 10. While Fournet, Durocher, and Rivière conceived the granitic magma to have been purely anhydrous, and in a state of simple igneous fusion, Elie de Beaumont maintained, with Poulett Scrope and Scheerer, that water had in all cases intervened, and that a few hundredths of water might, at a low red heat, have given rise to the condition of imperfect liquidity which he imagined for the material of the injected granites. The coarsely crystalline granitic veins were, according to him, veins of injection, and he speaks of them as examples in which "the phenomena essential to the formation of granite had been manifested with the greatest intensity." The granitic emanations, which are supposed to have furnished the material of these veins, appear to be regarded by him as the result of a process of eliquation from the congealing granitic mass. De Beaumont is careful to distinguish between them and those emanations which are dissolved in mineral waters, or are exhaled as volcanic vapors (page 1324). To the agency of such waters he ascribes the formation of concretionary veins, which are generally characterized by their symmetrically banded structure. He further adds that granites, as to their mode of formation, offer a character intermediate between ordinary veins and volcanic and basic rocks. This is conceivable as regards granitic veins, since these, according to him, although formed by injection, and not by concretion, result from a process of emanation from the parent granitic mass, which may be described as a kind of segregation.

I have thus endeavored to give, for the most part in his own words, the views on the origin of granites enunciated by the great French geologist in his classic essay on Volcanic and Metalliferous Emanations, published in 1847. They belong to the history of our subject, and are remarkable as a clear and complete expression of those modified plutonic views which are probably held by a great number of enlightened geologists at the present time. My reasons for dissenting from them, and the theories which I offer in their stead, will be shown in the sequel.

§ 11. Elie de Beaumont, while regarding the formation of granitic veins as a process in which water intervened to give fluidity to the magma, was careful to distinguish the process from that of the production of concretionary veins from aqueous solution, and supposed the fissures to have been filled by the injection of a jet of pasty matter derived from a consolidating granitic mass. Daubrée and Scheerer, in describing the granitic veins of Scandinavia, conceive the material filling them to have been derived from the enclosing crystalline strata, instead of from an unstratified granitic nucleus, but do not, so far as I am aware, compare their formation to that of concretionary veins. Their publications on this subject, it should be said, are both anterior to the essay of De Beaumont.

§ 12. The notion that all granitic veins are the result of some process of injection, and not to be confounded with concretionary veins, seems indeed to have been general up to the present time. Even Von Cotta, while strongly maintaining the aqueous and concretionary origin of metalliferous veins in general, when describing those consisting of quartz, mica, feldspar, tourmaline, garnet, and apatite, with cassiterite, wolfram, etc., which occur at Zinnwald and at Johanngeorgenstadt, is at a loss whether to regard these veins, from their granitic character, as igneous-fluid injections or as concretionary lodes. In support of the latter view he refers to their more or less regular and symmetrically banded structure, and while recalling the fact that mica and feldspar may both be formed in the humid way, considers the nature of these veins to be very problematical, and the question of their origin a difficult one. (Ore Deposits, Prime's translation, 1870, pages 110 – 124.)

§ 13. I have for several years taught that granitic veins of the kind just referred to are concretionary and of aqueous origin. In 1863 I described certain veins in the crystalline schists of the Appalachian region of Canada, "where flesh-red orthoclase occurs so intermingled with chlorite and white quartz as to show the contemporaneous formation of the three species. The orthoclase generally predominates, often reposing upon or surrounded by chlorite; at other times it is imbedded in quartz, which covers the latter. Drusy cavities are also lined with small crystals of the feldspar, and have been subsequently filled with cleavable bitter-spar, sometimes associated with specular iron, rutile, and sulphuretted copper ores." A study of these veins shows a transition from those "containing quartz and bitter-spar, with a little chlorite or talc, through others in which feldspar gradually predominates, until we arrive at veins made up of orthoclase and quartz, sometimes including mica, and having the character of a coarse granite; the occasional presence of sulphurets of copper and specular iron characterizing all of them alike. It is probable that these, and indeed a great proportion of quartzo-feldspathic veins, are of aqueous origin, and have been deposited from solutions in fissures of the strata, precisely like metalliferous lodes. This remark applies especially to those granitic veins which include minerals containing the rarer elements. Among these are boron, phosphorus, fluorine, lithium, cæsium, rubidium, glucinum, zirconium, tin, and columbium; which characterize the mineral species apatite, tourmaline, lepidolite, spodumene, beryl, zircon, allanite, cassiterite, columbite, and many others." (Geology of Canada, pp. 476, 644; and *ante*, p. 33.)

In this connection I referred to the occurrence of orthoclase with quartz, calcite, zeolites, epidote, and native copper in certain mineral veins of Lake Superior, so well described by Professor J. D. Whitney. (American Journal of Science (2), XXVIII. 16.) The associations, according to him, show the contemporaneous crystallization of the copper, natrolite, calcite, and feldspar, which last was found by analysis to be a pure potash-orthoclase.

§ 14. In 1864 this view was still further insisted upon in the Journal just cited ((2), XXXVII. 252), where, in speaking of mineral vein-stones "which doubtless have been deposited from aqueous solution," it is added, "while their peculiar arrangement, with the predominance of quartz and non-silicated species, generally serves to distinguish the contents of these veins from those of injected plutonic rocks, there are not wanting cases in which the predominance of feldspar and mica gives rise to aggregates which have a certain resemblance to dikes of intrusive granite. From these, however, true veins are generally distinguished by the presence of minerals containing boron, fluorine, phosphorus, cæsium, rubidium, lithium, glucinum, zirconium, tin, columbium, etc.; elements which are rare, or found only in minute quantities in the great mass of sediments, but are here accumulated by deposition from waters which have removed these elements from the sedimentary rocks and deposited them subsequently in fissures."

In the Report of the Geological Survey of Canada for 1866 (p. 192), I have, in describing the veins of the Laurentian rocks, insisted still further on the distinction just drawn between granitic dikes and granitic vein-stones, which latter I have proposed to call endogenous rocks, to indicate the mode of their formation, and to distinguish them from intrusive or exotic rocks, and sedimentary or indigenous rocks.

§ 15. The peculiar banded arrangement, which is so characteristic in concretionary veins not granitic in composition, is probably not less marked in granitic vein-stones, and often appears in these in a remarkable manner, showing that they have been formed by successive depositions of mineral matter, and generally in open fissures. This structure, and various peculiarities to be observed in granitic vein-stones, will be best illustrated by descriptions of various localities, most of which I have personally examined. It is proposed to notice, first, the veins of the gneiss and mica-schist series of New England; and, secondly, those of the Laurentian rocks of New York and Canada. In the latter class will be noticed the more or less calcareous vein-stones into which the Laurentian granitic veins are found to graduate.

§ 16. It is in the series of micaceous schists with interstratified gneisses (§ 6) which I have elsewhere provisionally designated the Terranovan series * [since called Montalban], that I have seen concretionary granitic veins in the greatest abundance and on the grandest scale. This stratified system, which is well seen in the White Mountains, appears to extend southward along the Blue Ridge as far as Georgia, and northeastward beyond the limits of Maine. It is in this State that I have particularly studied the granitic vein-stones of this system, whose history may be illustrated by a few examples from notes taken on the spot. In Brunswick the strata near the town are fine grained, friable, dark colored, micaceous, and hornblendic, passing into mica-schist on the one hand, and into well-marked gneiss on the other, and dipping to the southeast at angles of from 15° to 40°. Very similar beds are found in the adjoining town of Topsham, and in both places they include numerous endogenous granitic veins. The course of these veins is generally northwest, or at right angles to the strike, though occasionally for short distances with the strike, and intercalated between the beds; the veins vary in breadth from a few inches to sixty feet, and even more. They generally consist in great part of orthoclase and quartz, with some mica and tourmaline, and offer in the associations and grouping of these minerals many peculiarities, which are met with not only in different veins, but in different parts of the same vein. In some cases, colorless vitreous quartz greatly predominates, and encloses crystals of milk-white orthoclase, often modified, and from one to several inches in diameter. At other times pure vitreous quartz forms one or both walls, or the centre of the vein, or else is arranged in bands parallel with the sides of the vein, and sometimes a foot or more in thickness, alternating with similar bands consisting wholly or in great part of orthoclase,

* American Journal of Science for July, 1870, page 83, and Can. Naturalist, V. p. 198. — The rocks of this White Mountain series are, in the present state of our knowledge, supposed to be newer than the Huronian system noticed in § 5, to which, with Macfarlane and Credner, I refer the crystalline schists, with associated serpentines and diorites, of the Green Mountains. [See further in this connection Paper XIII. and its Appendix; also the third part of Paper XVI. and the Introduction to III.]

or of an admixture of this mineral with quartz, having the peculiar structure of what is called graphic granite, or else presenting a finely granitoid mixture of the two minerals, with little or no mica, and with small crystals of deep red garnet. Prisms of black tourmaline are also met with in these veins, and more rarely beryl and even chrysoberyl. In the rock-cutting on the Lewiston Railroad, just below Topsham bridge over the Androscoggin, there is a fine exhibition of these veins, which present alternate coarser and finer grained layers, traversed by long spear-shaped crystals of dark mica passing from one layer to another.

§ 17. A remarkable example of a vein of considerable dimensions is seen in the feldspar-quarry in Topsham, which occurs in a dark fine-grained friable micaceous schist. At the time of my visit, in 1869, the limits of the vein were not seen, though large quantities of white orthoclase and of vitreous quartz had already been extracted. These were each nearly pure, and in alternate bands, the quartz presenting drusy cavities lined with remarkable tabular crystals. One band was made up in great part of large crystals of mica, and portions of the vein consisted of a granular saccharoidal feldspar. The famous locality of red, green, and blue tourmalines, with beryl, lepidolite, amblygonite, cassiterite, etc., at Mount Mica in Paris, Maine, is a huge granitic vein, which, with many others, is included in a dark-colored very micaceous gneiss.

§ 18. In Westbrook numerous small veins of this kind, holding coarsely lamellar orthoclase with black tourmaline and red garnet, intersect strata of fine-grained whitish granitoid gneiss. In Windham the dark-colored staurolite-bearing mica-schist of this series is traversed by a granitic vein holding crystals of beryl. In Lewiston a large vein of coarse graphic granite, holding black tourmaline, and showing fine-grained bands, cuts a great mass of bluish gneissoid limestone, which forms an escarpment near the railroad, about half a mile below the town. This limestone, which dips eastward about 15°, is interlaminated with thin quartzite beds, which are seen on weathered surfaces to be much contorted. The bluish crystalline limestone is mixed with grains of greenish pyroxene, and includes nodular granitic masses of white crystalline orthoclase with quartz, enclosing large plates of graphite, crystals of hornblende, and more rarely of apatite. These associations of minerals are met with in the granitic veins of the Laurentian limestones, to be noticed elsewhere. The limestone of Lewiston, however, appears to be included in the great mica-schist series of the region; where similar beds, though less in extent, are met with in various places, sometimes associated with pyroxene, garnet, idocrase, and sphene. A thin band of impure pyroxenic limestone, like that of Lewiston, occurs with the mica-schists on the Maine Central Railroad, near Danville Junction; and beds of a purer crystalline limestone were formerly quarried in the southeast part of Brunswick, where they are interstratified with thin-bedded dark hornblendic and micaceous gneiss, dipping southeast at a high angle.

§ 19. At Danville Junction strata of hornblendic and micaceous gneiss, passing into mica-schists, dip southeast at moderate angles, and include huge veins of endogenous granite. Two of these appear in the hill just south of the railroad-station, apparently running with the strike of the beds. They are seen to rest upon the mica-schist, and in one of them a mass of this rock, three feet in width, is enclosed like a tongue in the granite, which has a transverse breadth of about seventy-five feet. Notwithstanding the apparent intercalation of these granitic masses, the proof of their foreign origin is evident in a transverse fracture and slight vertical dislocation of the mica-schist, around the broken edges of which the granite is seen to wrap. The endogenous character of this granite is well shown by its banded structure; belts of white quartz some inches wide alternate with others of coarsely cleavable orthoclase, while other portions hold black tourmalines and garnets of considerable size.

The evidence of disturbance of the strata in connection with these endogenous granites is seen on a large scale at the falls of the Sunday River in Ketchum. There, mica-schists and gneisses, similar to those already noticed, enclose great masses of endogenous granite, which are seen to be transverse to the strata. On one side of such a mass more than sixty feet wide, the schistose strata are twisted from their regular northeast strike to the northwest, and so enclosed in the granite as to appear interstratified with it for short distances. The banded structure of the transverse granite veins is here very marked. Some portions present cleavage-planes of orthoclase six inches in diameter; other parts, which are less coarse, abound in mica. Similar banded granite veins abound in the adjoining towns of Newry and North Bethel, and sometimes present layers of quartz six inches or more in thickness, beside large crystals of mica, and more rarely apatite.* These veins are often irregular in shape and bulging at intervals, and they sometimes run partially across the beds, which seem to have been distended and disturbed; a fact which was also observed in the thin-bedded schists in contact with some of the veins in Brunswick, and is apparently due to the expansive force of crystallization, as noticed in § 27.

§ 20. The locality already described at Danville offers an instructive example of a phenomenon often met with in the region now under consideration, where granitic masses, resisting the actions which have degraded the soft enclosing schists, stand out in relief on the surface, and seem to constitute the rock of the country. A careful search will however show that they are simply veins or endogenous masses of very limited dimensions, rising from out of the mica-schists, which are often concealed by the soil. This is well seen about the lower falls of the Presumpscott, near Portland, where the mica-schists, with some fine-grained gneisses, dipping southeast at angles of from 30° to 40°, enclose large numbers of granitic veins, which, though sometimes but a few inches in breadth, often measure twenty or even fifty feet, and are usually very coarse grained, with white mica, black tourmaline, and more rarely beryl.

* A good example of a large vein of this kind of intersecting rocks of the White Mountain series may be seen in the Ramble in the Central Park in the city of New York. Its place is marked by a great erratic block perched directly over the vein.

They are sometimes transverse to the stratification, but more often parallel, and, standing above the soil, are very conspicuous.

§ 21. We have already noticed the exotic granites of Biddeford, which are intruded among fine-grained bluish or grayish silicious strata. These latter are traversed by numerous veins of endogenous granite, which are very unlike in aspect to the intrusive rock. One of these veins, near Saco Pool, has a diameter of about an inch and a half, and presents on either wall a layer of yellowish crystalline feldspar about one fourth of an inch in thickness, which includes long plates of dark brown mica. These penetrate the central portion of the vein, which is a broadly crystalline bluish orthoclase, enclosing small portions of quartz, after the manner of a graphic granite. The yellowish and less coarsely crystalline feldspar, with its accompanying mica, had evidently lined the walls of the vein while the centre yet remained open, and had moreover entirely filled a small lateral branch. The same conditions are seen in the filling of other veins in this vicinity, which are often much larger, and present upon their walls bands of an inch or two of the yellowish feldspar, with mica.

The successive filling of a granitic vein is still more clearly shown in a specimen from Sherbrooke, Nova Scotia, which I owe to the kindness of Professor H. Y. Hind. The vein, which is seen to be transverse to the adherent fine-grained mica-schist, has a breadth of nearly four inches, about two thirds of which is symmetrical, and is included between two layers, perpendicular to the walls, consisting of a fine-grained mixture of white feldspar and quartz, each about one fourth of an inch thick, and marked by subordinate zones, more or less quartzose. Within these two bands is a coarser aggregate, consisting of two feldspars, with some quartz and muscovite, plates of which, and crystals of pink orthoclase, penetrate an irregular layer of smoky quartz varying from one eighth to one half an inch in diameter. This fills the centre of the symmetrical portion of the vein, on one side of which is the mica-schist, while the other is bounded by a band of more than half an inch of fine-grained granite with yellowish-green mica, presenting large crystals of feldspar near the outer margin, where it is succeeded by a layer of pure smoky vitreous quartz of about the same thickness, whose outer surface, against the wall, shows irregular bosses or nodular masses, the depressions between which are occupied by a finely granular micaceous aggregate unlike any other part of the vein in texture.* This description may be read in connection with the remarks in § 27.

Dana has described and figured a similar granitic vein, banded with quartz, observed by him at Valparaiso in Chili (Manual of Geology, 1862, p. 713),† and has moreover maintained that such granitic veins, like ordinary metalliferous lodes, are clearly concretionary in their origin, and have been filled by slow and successive deposits from aqueous solutions. His testimony to the views which I have advocated in this paper had been overlooked by me, or it would have been noticed in § 12.

§ 22. The numerous granitic veins so well known to mineralogists in the mica-schists and gneisses of New Hampshire, Massachusetts, and Connecticut, including, among other familiar localities, Grafton, Acworth, Royalston, Norwich, Goshen, Chesterfield, Middletown, and Haddam, seem, from descriptions and from their mineral constituents, to be similar to those of Maine, already mentioned. With the exception of Royalston and Haddam, however, these localities are as yet only known to me from specimens and descriptions. It is noteworthy that at the former the finely crystallized beryls are directly imbedded in vitreous quartz, and the same is the case with the beryls of Acworth and the blue and green tourmalines of Goshen. A remarkable example of a vein of this character occurs in Buckfield, Maine, described to me by Professor Brush,

* The banded structure is well shown in a granitic vein which I owe to Professor Haughton of Trinity College, Dublin, got from Three Rock Mountain, near that city. It consists of white orthoclase, with quartz and some mica and garnet, and exhibits near the middle two bands of prisms of black tourmaline pointing towards the centre, which is filled with a coarsely crystalline orthoclase.

† From U. S. Exploring Expedition, Report on the Geology, 1849, p. 570.

where large isolated crystals of white orthoclase, nearly colorless muscovite, and brown tourmaline occur in a vein of vitreous quartz. At Paris and at Hebron, Maine, tourmalines are found penetrating crystals of quartz. The flattened tourmalines and garnets found in muscovite at several localities in New England are well known to collectors, and a curious example of enclosure has been observed by Professor Brush at Hebron, where crystals of muscovite are encased in lepidolite.

§ 23. The following list includes the principal mineral species found in these granitic veins in New England : apatite, amblygonite, triphylline, autunite, yttrocerite, orthoclase, albite, oligoclase, spodumene, iolite, muscovite, biotite, lepidolite, cookeite, chlorite, chlorophyllite, garnet, epidote, tourmaline, beryl, zircon, quartz, chrysoberyl, automolite, cassiterite, rutile, brookite, uraninite, columbite, pyrochlore, scheelite, and bismuthine. As I am not aware that chlorite has hitherto been mentioned as a constituent of these veins, it may be said that it occurs in one at Albany, Maine. To the above should be added the rare species nepheline, cancrinite, and sodalite, which have long been known in bowlders of a granite-like rock in Maine. According to information given me by Professor Brush, green elæolite with white orthoclase and black biotite occurs in a granitic vein twenty feet in breadth, lately observed in the northwest part of Litchfield, Maine.

§ 24. We have seen that these endogenous veins are found alike in the gneisses, mica-schists, limestones, and quartzose strata of this region. They are also met with in the eruptive granites, small fissures in which are sometimes filled with coarsely crystalline orthoclase, smoky quartz, various micas, and zircon. Examples of this are seen in the granites of Hampstead, New Brunswick, and Mount Uniacke, Nova Scotia. The fine green feldspar of Cape Ann, Massachusetts, and the micas, cryophyllite and lepidomelane, with zircon, described by Professor Cooke, from the same region, occur in veins in the hornblendic granites of that locality. Small veins cutting a somewhat similar rock at Marblehead contain crystallized green epidote with white quartz and red orthoclase.

§ 25. The veins which we have described are frequently of very limited extent, and seem to occupy short and irregular fissures, while in other cases the mineral aggregates which characterize them occur in nests or geodes. This is seen near Fall Brook, in the Nerepis valley, in New Brunswick, where the red micaceous granite is in one part very friable, and presents irregular geode-like cavities, sometimes several inches in diameter, which are partially filled by radiating prisms of black tourmaline, accompanied with quartz and albite crystals, and more rarely small octahedrons of purple fluorite. The enclosing granite is composed of deep red orthoclase, with small portions of a white triclinic feldspar, smoky quartz, and black mica. The conditions seen at this place recall the description of the famous locality of feldspars, etc., at Fariolo, near Baveno, in northern Italy. The rock of that place, described as a granite, resembles, in a specimen before me, some of the intrusive granites of New Brunswick, and contains a pink and a white feldspar, with a little black mica. It includes veins of graphic granite, and also spheroidal masses, which differ in texture from the mass of the rock, and present geodes of considerable size, lined with fine large red and white crystals of orthoclase, accompanied by albite, epidote, quartz, fluorite, and a greenish mica (or chlorite), all of which, according to Fournet, are so mingled and interlocked as to show that they are of contemporaneous origin. To these are to be added, as occurring in the geodes, prehnite, calcite, hyalite, and specular iron. The orthoclase crystals often have adhering to their opposite faces crystalline plates of albite, which are larger than the planes to which they are attached. The crystals of orthoclase, moreover, frequently present hollowed-out or hopper-shaped faces, which Fournet happily describes as resulting from the forming of the framework or skeleton of the crystals, when the material was not sufficient for their completion. A process analogous to this is often seen in crystallization, whether from fusion, solution, or vaporous condensation, giving rise in some cases to external depressions, and in others to internal cavities in the resulting crystals. Fournet ascribes the formation of the geodes in the granite of Fariolo to a process of shrinking, and a subsequent segregation filling the resulting cavities, in which he is forced to recognize the intervention of water, though by no means admitting the aqueous origin of veins, since he holds even those of quartz to have been formed by igneous injection. (Géologie Lyonnaise, *278.)

§ 26. When we consider the cause which has produced the fissures in the mica-schists and gneisses of New England, which hold the granitic veins already described, it is to be remarked that their comparative abundance, their shortness and their irregularity, distinguish them from the fissures which are filled with eruptive rocks. Examples of the latter may be seen near Danville, Maine, where dikes of fine-grained dolerite are posterior to the endogenous granitic veins here occurring in the mica-schist. These dikes may be supposed to be dependent upon movements in the earth's crust opening deep fissures which connected with some softened rock far below. Through such openings were extravasated the exotic rocks, whether granites or dolerites, — more or less homogeneous mixtures, often widely different in composition from the encasing rocks. The endogenous veins, on the contrary, are distinguished not only by their more or less heterogeneous and often banded structure, but by the fact that their principal constituents are generally the mineral species common in the adjacent strata.

§ 27. Volger has attributed the formation of the openings containing concretionary veins to the force of crystallization, which is shown to be very great in the congelation of water and the crystallizing of salts in cavities and fissures. Such a process once commenced in an opening in a rock would, he conceived, be sufficient to make still wider the fissure, which might be fed by fresh solutions passing by capillarity through the pores of the rock. If this process were to become concentrated around several points, the intermediate spaces might be so opened that free crystallization could go on, resulting in the production of goedes in veins thus formed.

Fournet, on the other hand, suggests that contraction in the cooling of erupted granites gave origin to the fissures and geodes now filled or partially filled with crystalline minerals at Fariolo ; we may readily suppose that a process of contraction attendant upon the crystalline aggregation of the materials of sedimentary strata would give rise to rifts or fissures therein. The lesions thus produced in the solid rocks become more or less completely repaired, if we may so speak, by an effusion of mineral matter from the walls, and thus are generated geodes, irregular masses, and many veins. That the process imagined by Volger may in some cases intervene, and may act subsequently to the one just imagined, is highly probable, though we are disposed to assign it but a secondary place in the production of vein-fissures. It offers, however, the most plausible explanation of the distortion of the thin-bedded strata already noticed in connection with some of the concretionary granitic veins of Maine, which seem, by a process of growth, to have bent outward the adjacent beds. The vertical transverse veins are, in many cases at least, unsymmetrical, as if they had grown from one side, while the distortion of the beds, sometimes attended by irregular concretions in the banded veinstone, appears at the opposite wall. The notion that the vein-fissures opened as crystallization advanced has been defended by Grüner.

§ 28. It is not here the place to discuss how far the greater and deeper fissures of the earth are dependent upon the contraction of sediments, as just explained, or upon the wider spread movements of the earth's crust, though even of these it may be said that they are more or less directly the results of a process of contraction. It should, however, be noted that while some fissures of this kind are filled with dikes of erupted rocks (§ 26), others hold concretionary veins, which are to be distinguished from the class of veins just described, inasmuch as the openings in which they were deposited evidently communicated with the surface of the earth. Examples of these are seen in the lead and zinc-bearing veins with calcite and barytine, which traverse vertically the carboniferous limestone in England, and enclose in their central portions material of liassic age, abounding in the remains of a marine and a fresh-water fauna, which show these veins to have been deposited in fissures communicating with the surface-waters of the liassic period. For a description of these veins by Mr. Charles Moore, see the Report of the British Association for 1869, and Amer. Jour. of Science (2), L. 365. Similar evidence is afforded by the existence of rounded pebbles imbedded in veins, as observed in Bohemia and also in Cornwall, where numerous pebbles both of slate and quartz were found at a depth of six hundred feet in a lode, cemented by cassiterite and sulphuret of copper. (Lyell, Student's Elements of Geology, p. 593.) Not less instructive in this connection are the observations of Mr. J. Arthur Phillips, on the silicious vein-stones now in process of formation in open fissures in Nevada. (L. E. and D. Phil. Mag. (4), XXXVI. 321, 422; Amer. Jour. of Science (2), XLVII. 138.) We cannot doubt that the ancient, like these modern veins have been channels for the discharge of subterranean mineral waters; and it would seem that while the deposition of the incrusting materials on the walls of the fissure is in part due to cooling, and in part perhaps to infiltration, in some cases, of precipitants from lateral sources, it is chiefly to be ascribed to the reduction of solvent power consequent upon the diminution of pressure as the waters rise nearer to the surface.* This conclusion, deducible from the researches of Sorby on the relation of pressure to solubility (*ante*, page 65), I have pointed out in the Geological Magazine for February, 1868, p. 57. See also Amer. Jour. of Science (2), L. 27.

§ 29. There is evidently a distinction to be drawn between veins which have been open channels and the segregated

* Of this a remarkable example was afforded in 1866 at Goderich, in Ontario, where, in a boring at a depth of 1,000 feet, a bed of rock-salt was met, from which for a time a saturated or rather supersaturated brine was obtained. As an evidence of this, I saw a cube of pure salt, one fourth of an inch in diameter, which had formed upon and around a projecting point of an iron valve in the pump, above the surface of the ground. The liquid beneath a pressure of 1,000 feet of brine, equal to about 1,200 feet of water, or thirty-six atmospheres, having taken up more salt than it could hold at the ordinary pressure, deposited a portion of it as it reached the surface, and actually obstructed thereby the action of the pump. After a few months of pumping, however, the well ceased to afford a fully saturated brine.

masses and geodes formed in cavities which appear to have been everywhere limited by the enclosing rock. In the former case a free circulation of the mineral solution would prevail, while in the latter there could be no renewal of it except by percolation or diffusion through the rock. A comparison between the contents of geodes and fissure-veins, whether in granitic rocks or in fossiliferous limestones, will however show that these differences do not sensibly affect the mineral constitution of the deposits.

§ 30. The range of conditions under which the same mineral species may be formed is apparently very great. Sorby, from his investigations of the fluid-cavities of crystals, concludes that the quartz which occurs with cassiterite, mica, and feldspar in the granitic veins of Cornwall must have crystallized at temperatures from 200° to 340° Centigrade, and under great pressure; conditions which we can hardly suppose to have presided over the production of the crystallized quartz found in the unaltered tertiaries of the Paris basin, or the auriferous conglomerates of California. In like manner beryl, though a common mineral of the tin-bearing granite veins, like those studied by Sorby, occurs at the famous emerald-mine of Muso, in New Grenada, in veins in a black bituminous limestone, holding ammonites, and of neocomian age, its accompaniments being calcite, quartz, and carbonate of lanthanum (parisite). Small crystals of emerald are disseminated through this argillaceous, somewhat magnesian limestone, which contains, moreover, a small amount of glucina in a condition soluble in acids. (Léwy, Annales de Chimie et de Physique, LIII. 1 – 26; and Fournet, Géol. Lyonnaise, 455.)

§ 31. To these we may add the production of various hydrated crystallized silicates, including apophyllite, harmotome and chabazite, during the historic period in the masonry of the old Roman baths at Plombières and Luxeuil, and by the action of waters at temperatures of from 46° to 70° Centigrade (*ante*, page 25); the presence of apophyllite, natrolite, and stilbite in the lacustrine tertiary limestones of Auvergne; apophyllite incrusting fossil wood, and chabazite crystals lining shells in a recent deposit in Iceland. The association of such hydrated silicates with orthoclase, as already noticed (§ 13), and as described by Scheerer, where natrolite and orthoclase envelop each other, showing their contemporaneous formation, with many other facts of a similar kind, lead to the conjecture that orthoclase, like beryl and quartz, and perhaps some other constituents of granitic veins, may have crystallized in many cases at temperatures much lower than those determined by Sorby, and that the conditions of their production include a considerable range of temperature; a conclusion which is, however, probably true to some extent of zeolites also.

§ 32. It is now proposed to consider the granitic vein-stones found in Laurentian rocks. The stratified rocks of this ancient gneissic series, as I have elsewhere pointed out, differ considerably from those of the White Mountain series, which, with their vein-stones, have been treated of in §§ 16 – 23.

The Laurentian series, the Lower Laurentian of Sir William Logan, as studied by him in a region to the north of the Ottawa, the only area in which it has yet been examined in detail, appears to consist of an alternation of conformable gneissic and calcareous formations. The latter are three in number, each from 1,000 to 2,000 feet or more in thickness, and separated by still more considerable formations of gneiss and quartzite, a mass of gneiss of great but unknown thickness forming the base. (Geology of Canada, page 45.) The gneissic rocks of the series are very firm and coherent, reddish or grayish in color, often very coarse grained and granitoid, sometimes with but obscure marks of stratification; and frequently porphyritic from the presence of large cleavable masses of reddish orthoclase, occasionally with a white triclinic feldspar. They are often hornblendic, and sometimes contain small quantities of dark colored mica. A white granitoid gneiss, composed chiefly of orthoclase and quartz, sometimes contains an abundance of red iron-garnet. The latter mineral is often disseminated, or forms subordinate beds in the quartzites of the series.

§ 33. With the crystalline limestones of the calcareous parts of the series are often found strata made up of other minerals, to the entire exclusion of carbonate of lime, by an admixture of which, however, they graduate into the adjacent limestones. These beds generally consist of pyroxene, sometimes nearly pure, and at other times mingled with a magnesian mica, or with quartz and orthoclase, often associated with hornblende, serpentine, magnetite, sphene, and graphite. These pyroxenite rocks are generally gneissoid or granitoid in structure, and sometimes very coarse grained. They occasionally assume a great thickness, and are then often interstratified with beds of granitoid orthoclase-gneiss, into which the quartzo-feldspathic pyroxenites pass by a gradual disappearance of the pyroxene. The limestones often include serpentine, pyroxene, hornblende, phlogopite, quartz, orthoclase, magnetite, and graphite; so that the same minerals are common to them and to the pyroxenic strata, which may be looked upon as marking the transition between the gneissic and the calcareous parts of the series. These strata, marked by the predominance of calcareous and magnesian silicates, appear, so far as known, to accompany each of the limestone formations of the Laurentian, sometimes, however, developed to a greater and sometimes to a less extent.

§ 34. I have elsewhere called attention to the fact that the highly micaceous schists and the argillites of the Green Mountain and White Mountain series of rocks are, so far as known, wanting in the Laurentian, and with them the characteristic minerals of the latter series, staurolite, andalusite, and cyanite. There are, however, beds of a highly micaceous rock in the Laurentian which contain an unctuous magnesian mica with a pyroxenic admixture; these are very unlike the mica-schists composed of a non-magnesian mica and quartz, with orthoclase, which abound in the White Mountain rocks. These magnesian beds belong to the calcareous horizons in the Laurentian series, at which also occur the most numerous veins and the principal minerals of economic value. It is also along these horizons, marked by softer rocks, that the valleys and the arable lands of the Laurentian areas are chiefly found, and for this reason, also, the mineralogy of these parts is better known than that of the harder gneissic portions. The above observations on the lithological character of the stratified rocks are important on account of the relations between these and the included veins, in which the characteristic minerals of the gneissic and calcareous rocks are often found associated.

§ 35. The history of these veins, as seen in the Laurentian rocks of the Laurentides in Canada, the Adirondacks of northern New York, and the Highlands of southern New York and New Jersey, has been discussed at length by the author in an essay on The Mineralogy of the Laurentian Limestones, in the Report of the Geological Survey of Canada for 1863 – 66, pages 181 – 223.* In this essay, which will be frequently referred to in the present paper, the vein-stones found in the Laurentian rocks have been noticed under three heads: First, metalliferous veins carrying galenite, blende, pyrite, and chalcopyrite in a gangue of calcite, sometimes with celestine and fluorite; these, which are of palæozoic age or still younger, cut the Potsdam sandstone, the Calciferous sand-rock, and probably also the overlying Trenton limestones. Second, quartzo-feldspathic veins with muscovite, tourmaline, zircon, etc. These veins I have described as passing by insensible gradations into the third class, in which calcite and apatite, with pyroxene, phlogopite, and other calcareous and magnesian silicates predominate, though frequently accompanied by quartz and orthoclase. These veins are older than the Potsdam sandstone, which rests upon their eroded outcrops, and sometimes includes worn fragments of apatite derived from them.

§ 36. It is these last two classes which it is proposed to consider in the present paper under the name of granitic vein-stones. In justification of the extension of the term "granitic" to the whole of this series of veins, it must be repeated, that it is not possible to draw a line of distinction between those in which quartz and orthoclase are the characteristic minerals, and those in which calcite, apatite, pyroxene, and phlogopite prevail,

* This essay is reprinted, with some additions, in the Report of the Regents of the University of New York for 1867, Appendix E. The reader's attention is called to the note on the Hastings rocks, at the close of that reprint.

sometimes to the entire exclusion of quartz and feldspar, both of which minerals are, however, frequently intermixed with the preceding species in the same aggregate. In one example, in Burgess, Ontario, the sides of a large vein are occupied by a mixture of calcite and apatite, while the centre is filled by a vertical granite-like layer of reddish orthoclase, with a little quartz and green apatite. Of another vein in the township of Lake, in Ontario, one portion was found to consist of calcite with yellow phlogopite, while an adjacent part consisted of quartz, with brown tourmaline, bismuthine, native bismuth, and graphite.

§ 37. The resemblance between the minerals of these Laurentian vein-stones and the same species brought from Norway was noticed so long ago as 1827, by Dr. William Meade (American Journal Science (1), XII. 303). Daubrée, in his account of the metalliferous deposits of Scandinavia, published in 1843 (Annales des Mines (4), IV. 199, 282), has given us a careful description of the veins from which these minerals are derived. From this, together with the observations of Scheerer and Durocher, we are enabled to compare these vein-stones with those of the Laurentian rocks in North America, and show, as I have — in the essay above referred to — done in detail, and for each principal species, the great similarity which exists between the two. In the so-called Primitive Gneiss formation of Scandinavia these veins occur either in gneiss, or in a gneissoid rock consisting of various admixtures of pyroxene, hornblende, garnet, epidote, and mica, the whole associated with crystalline limestones. The veins which abound in the gneiss near the iron-mines of Arendal, in Norway, according to Daubrée, though occasionally containing calcite, apatite, hornblende, and scapolite, are sometimes destitute of all calcareous and magnesian minerals, and become granite-like aggregates of orthoclase and quartz. He hence describes these veins, as a whole, though frequently abounding in lamellar calcite, as essentially granitic in character. As already noticed in § 8, Daubrée agrees with Scheerer in regarding these vein-stones as produced by a process of secretion, in opposition to Durocher, who looked upon them as having been formed by igneous injection.

§ 38. The principal mineral species known in the corresponding vein-stones of the Laurentian rocks of North America are the following: *calcite, dolomite,* fluorite, *apatite, serpentine,* chrysolite, *chondrodite, wollastonite, hornblende, pyroxene, pyrallolite,* gieseckite, *scapolite,* petalite, *orthoclase,* oligoclase, *albite, muscovite, phlogopite,* seybertite, *tourmaline, garnet,* idocrase, *epidote,* allanite, zircon, spinel, chrysoberyl, corundum, *quartz, sphene,* rutile, menaccanite, *magnetite, hematite, pyrite,* and *graphite.* To which may be added some rarer species, such as tephroite, willemite, franklinite, zincite, warwickite, found in a few localities only, and others of less importance. Of the above list, those species whose names are in italics have been recognized as constituent minerals in the stratified rocks in which the veins occur.

The most important species in these vein-stones are calcite, quartz, orthoclase, phlogopite, pyroxene, apatite, and graphite, of which some one or more will generally be found to prevail in the veins in question. The greater part of the species named in the first list were observed by Daubrée in the veins near Arendal, and to these he adds axinite, gadolinite, and more rarely beryl and leucite;* while in the island of Utoë, associated with iron-ores, crystalline limestones, and hornblendic rocks passing into gneiss, are similar granitic vein-stones containing orthoclase and quartz, with tourmaline, cassiterite, and, in the middle of the veins, petalite, spodumene, and lepidolite. This association is the more worthy of notice, as the only other known locality of petalite (if we except the castor of Elba) is in the crystalline limestone of Bolton, Massachusetts, where it occurs, probably in a vein-stone, with scapolite, hornblende, pyroxene, chrysolite, spinel, apatite, and sphene.

* For a notice of the occurrence of leucite in these veins, and also in veins in Mexico, see the author's Contributions to Lithology (Amer. Journal Science, (2), XXXVII. 264). According to Garrigou, this rare species occurs both well crystallized and in compact porphyroid masses, in dioritic rocks (ophites), at Lusbé in the valley of Aspé, in the Pyrennees. (Bull. Soc. Geol. de Fr. (2), XXV. 727.)

§ 39. Evidences of the concretionary origin of these granitic vein-stones of the Laurentian rocks appear in their banded structure, their drusy cavities, the peculiar incrustations and modes of enclosure often observed in the crystals, and finally in the rounded forms of certain crystals, which show a process of partial solution succeeding that of deposition. A banded arrangement of the materials parallel to the sides of the vein is often well marked. Thus, while the walls may be coated with crystalline hornblende, or with phlogopite, the body of the vein will be filled with apatite, in the midst of which may be found a mass of loganite, or of crystalline orthoclase mixed with quartz, filling the centre of the vein, as already noticed in § 36. In other instances portions of the vein will be occupied by crystals of apatite, pyroxene, or phlogopite imbedded in calcareous spar, which in some other part of the breadth of the vein, or in its prolongation, will so far predominate as to give to the mass the aspect of a coarsely crystalline lamellar limestone. Prisms of apatite are often observed extending from either side toward the centre of the vein, which in some cases may be filled with calcite or another mineral, and in others is a vacant space lined with crystals. Drusy cavities of this kind, a foot in breadth and several feet in length and depth, are sometimes met with in these veins in Ontario.

§ 40. Further evidence of concretionary origin is seen in the manner in which the various minerals incrust each other. Thus small prisms of apatite are enclosed in large crystals of phlogopite, in pyroxene, in quartz, and even in massive apatite; crystals or rounded crystalline masses of calcite are imbedded in apatite and in quartz, and well-defined crystals of hornblende (pargasite) in pyroxene. In another example before me, small crystals of hornblende are implanted on a large crystal of pyroxene, and both, in their turn, are incrusted by small crystals of epidote. Crystalline graphite in like manner is enclosed alike in orthoclase, quartz, calcite, phlogopite, and pyroxene.

§ 41. Another noticeable evidence of the concretionary origin of these veins is the phenomenon already referred to in § 25, where the external skeleton or framework of a crystal is complete, while the space within either remains empty, or is filled with other minerals, often unsymmetrically arranged. This condition of things is rendered intelligible by the formation of similar hollow crystals during the cooling of certain saline solutions, as for example potash-nitrate. Small hollow prisms of red and green tourmaline, closely resembling the hollow nitre crystals, are common in the well-known granitic vein-stone of Paris, Maine. I have elsewhere referred to the formation of such moulds or skeleton-crystals as having taken place in vein-cavities, and as serving to explain many cases of enclosure of mineral species. (Address to the A. A. A. S., Indianapolis, 1871. Paper XIII. of the present volume.) In addition to the examples there cited, the Laurentian vein-stones afford some curious cases. Thus a prism of yellow idocrase half an inch in diameter from a vein in Grenville, Ontario, composed chiefly of orthoclase and pyroxene, is seen when broken across to consist of a thin shell of idocrase filled with a confused crystalline aggregate of orthoclase, which encloses a small crystal of zircon. In like manner large crystals of zircon from similar veins in St. Lawrence County, New York, are sometimes shells filled with calcite.

§ 42. The rounded forms of certain crystals in the Laurentian vein-stones were, I believe, first noticed by Emmons, who observed that crystals of quartz imbedded in carbonate of lime from Rossie, New York, have their angles so much rounded that the crystalline form is nearly or quite effaced, the surfaces being at the same time smooth and shining. This appearance is occasionally observed in other localities, and is not confined to quartz alone, crystals of calcite and of apatite sometimes exhibiting the same peculiarity. At the same time the orthoclase, scapolite, pyroxene, and zircon, which are associated with these rounded crystals, preserve all their sharpness of outline, as was observed by Emmons for the orthoclase in contact with the crystals of quartz just described. He suggested that the rounded outlines of these crystals were due to a partial fusion, although he did not overlook a fact which renders this explanation untenable, namely, that the species presenting rounded angles are much less fusible than those which, in contact with them, preserve their crystalline forms intact. (Geology of the First District of New York, pages 57, 58.) These facts are well shown in the apatite-veins of Elmsley and Burgess, Ontario, where the crystals of apatite rarely present sharp or well-defined forms, but (whether lining drusy cavities or imbedded in the calcite or other minerals of the vein-stone) are most frequently rounded or sub-cylindrical masses, while the pyroxene and sphene, which often accompany them, preserve their distinctness of form. This rounding of the angles of certain crystals appears to me nothing more than a result of the solvent action of the heated watery solutions from which the minerals of these veins were deposited; the crystals previously formed being partially redissolved by some change in the temperature or the chemical constitution of the solution. Heated solutions of alkaline silicate, as shown by Daubrée, are without action on feldspar, as might be expected from the fact observed by him of the production of crystals of feldspar, as well as of pyroxene, in the midst of such solutions. These liquids would, however, doubtless attack and dissolve apatite, which is in like manner decomposed by solutions of alkaline carbonate, and these latter at elevated temperatures dissolve crystallized quartz. That this solvent process has been repeated during the filling of the veins is seen by a specimen in my possession, which shows crystals of calcite previously rounded and enclosed in a large crystal of quartz, the angles of which are also nearly obliterated. From the alternations in the deposited mineral matters in many vein-stones, as well as from what we know of the changing composition of mineral springs, it is evident that the waters circulating in the fissures now occupied by veins must have been subject to periodical variations in composition.

§ 43. In the Geology of Canada (page 729) I have noticed an example of rounded quartz crystals in the veins at the Harvey Hill copper-mine in Leeds, Quebec. Large terminated prisms of limpid colorless quartz are there found imbedded in compact erubescite, their angles being much rounded, while their faces are concave, and have lost their polish, retaining only a somewhat greasy lustre. A thin shining green layer, apparently of a silicate of copper, covers the surfaces of the ore in contact with the crystals. From the mode of their arrangement in certain specimens, it is evident that these prisms of quartz, lining drusy cavities, were partially dissolved previous to the deposition of the metallic sulphide.

§ 44. Some of the more important types of Laurentian vein-stones may now be noticed. Those made up of quartz with orthoclase, muscovite, and black tourmaline, often with zircon, are not unfrequent in the Laurentian gneiss, though so far as yet observed less abundant than in the gneisses and mica-schists of the White Mountain series. It is true, as already pointed out, that, from the greater softness of the enclosing rocks, the veins of the latter series are often weathered into relief (§ 20), and are thus rendered more conspicuous than those in the harder Laurentian gneisses. Among other examples of this first type of granitic veins may be mentioned those in Yeo's Island among the Thousand Isles of the St. Lawrence, and the well-known vein in Greenfield, near Saratoga, remarkable for affording crystals of chrysoberyl. A frequent type among the Laurentian granitic veins is characterized by great cleavable masses of reddish or reddish-brown orthoclase, with quartz and but little mica. With these are sometimes associated equally large masses of white or pale-colored albite; these veins are sometimes of great size, one hundred feet or more in breadth. The perthite of Thompson, well known as a cleavable feldspar made up of alternate thin plates of reddish-brown orthoclase and white albite, forms with quartz a large granitic vein; while the peristerite of the same author is an opalescent or chatoyant white albite, with blue and green reflections, which occurs with quartz in another vein in the same region. Some of the veins of red orthoclase include large cleavable masses of dark green hornblende, occasionally with magnetite. A remarkable vein about eighty feet in width, in Buckingham, Quebec, is composed entirely of reddish-white orthoclase and cleavable magnetite, the latter in masses sometimes two or three inches in diameter, scattered through the feldspar.

§ 45. The veins hitherto noticed occur in gneiss, but on the river Rouge one consisting of large masses of quartz and albite is found in crystalline limestone. A remarkable vein described by Sir William Logan in Blythefield, Ontario, traverses alternate strata of limestone and gneiss, and has a breadth of not less than 150 feet. It consists in great part of a coarsely cleavable pale green pyroxene (sahlite), with a dark green hornblende, phlogopite, and calcite. Portions of the vein-stone, however, consist of an admixture of orthoclase, quartz, and black tourmaline, showing the transition from the calcareous to the feldspathic type of veins. In Ross, Ontario, a vein holds large isolated crystals of white orthoclase imbedded with black spinel, apatite, and fluorite in a base of lamellar pink carbonate of lime. One of the most remarkable of these composite veins is in Grenville, Quebec, and was formerly worked for graphite. It cuts a crystalline limestone, itself holding graphite and phlogopite, and has afforded not less than fourteen distinct mineral species, namely, calcite, apatite, serpentine, wollastonite, pyroxene, scapolite, orthoclase, oligoclase, garnet, idocrase, zircon, quartz, sphene, and graphite. An adjacent vein abounds in phlogopite, with pyroxene and zircon. A not less remarkable vein is that described by Blake in Vernon, New Jersey (this Journal (2), XIII. 116), in which calcite, fluorite, chondrodite, phlogopite, margarite, spinel, corundum, zircon, sphene, rutile, menaccanite, pyrite, and graphite occur. Some of these contain barytine, and in one case I have observed natrolite, both seemingly filling cavities, and of later origin than the other minerals. The remarkable zinciferous minerals, franklinite, zincite, dysluite, and willemite, found in the Laurentian limestones of New Jersey, appear from the descriptions of H. D. Rogers to belong to calcareous vein-stones. Granitic veins are found traversing the magnetic iron ore-beds of the Laurentian series. I have described one in Moriah, New York, which includes angular fragments of the magnetite which forms its walls, and consists of a cleavable greenish triclinic feldspar, with quartz crystals having rounded angles, octahedrons of magnite, allanite, and a soft greenish mineral resembling loganite.

§ 46. As regards the order of deposition of minerals in these veins, we find apatite enclosed alike in calcite, in quartz, in phlogopite, in spinel, in graphite, and in pyrite. On the other hand, apatite sometimes includes rounded crystals of calcite or of quartz; and graphite, which itself encloses apatite, is found included alike in quartz, in orthoclase, in pyroxene, and in calcite, in such a manner as to lead us to conclude that its crystallization was contemporaneous with that of all these minerals; while from the other facts mentioned it would appear that the order of deposition was subject to variation and to alternations. In a vein in Grenville large prisms of a white aluminous pyroxene (leucaugite) penetrate great crystals of phlogopite, while at the same time small crystals of a similar mica are completely imbedded in the crystallized pyroxene. Many facts relating to the association of various species in these vein-stones will be found in my essay, but the subject is one which still demands careful study. The banded structure of these veins is well shown in some of those which contain graphite. This mineral, though sometimes irregularly disseminated through the vein-stone, frequently occurs in sheets or layers alternating with orthoclase, quartz, or pyroxene, parallel to the walls of the vein and exhibiting a peculiar structure due to the formation of successive layers of crystalline lamellæ more or less nearly perpendicular to the plane of deposition.

§ 47. The veins hitherto noticed, whether feldspathic or calcareous, are generally vertical, or nearly so, and in most cases traverse the stratification. Of many of them which have been explored to some extent for apatite, mica, and graphite, it is noticed that they are subject to great changes in dimension as well as in mineral contents. They often appear to occupy short irregular fissures, and in some cases are to be described as more or less completely filled geode-like cavities rather than veins.

§ 48. In the reprint of my essay, already mentioned, several veins are noticed in the county of Hastings, Ontario, in rocks which were at that time referred by the Geological Survey of Canada to the Laurentian, but have since been found to belong to younger series. Such are the veins containing argentiferous fahlerz with mispickel, and that holding native gold with a quasi-anthracitic form of carbon, both from Madoc, and also the vein already noticed as occurring in the township of Lake (§ 36), which contains in one part bismuthine with tourmaline, quartz, and graphite, and in another part calcite with phlogopite. This latter vein occurs in an impure limestone, associated with quartzite and micaceous schists, and belonging to a series unconformably overlying the Laurentian, and resembling the rocks of the White Mountain series. It will be noticed that this vein is lithologically similar to those of the Laurentian, which are not improbably of the same age. Calcareous vein-stones like those already described are not unknown in the White Mountain rocks in Maine, where are found, on a small scale, aggregates of crystallized pyroxene, idocrase, and sphene, and others of calcite with hornblende, apatite, and graphite (§ 18), closely resembling the Laurentian vein-stones of New York and Canada.*

§ 49. The various minerals of these calcareous vein-stones are

[* In a note in the American Journal of Science for October, 1873, on The Copper Deposits of the Blue Ridge, I have described the occurrence in Virginia, North Carolina, and Tennessee of great concretionary veins in gneisses and mica-schists which I refer to the White Mountain series. These veins are sometimes transverse to the stratification, and at other times interbedded. An example of the latter is seen at the Ducktown copper-mine in Polk County, Tennessee, where there is a banded arrangement of the large masses parallel to the walls. The chief part of this vein is filled with pyrite, pyrrhotine, and chalcopyrite, rarely with galena, blende, mispickel, and molybdenite. These massive ores enclose large garnets, and are penetrated with prisms of zoisite, hornblende, and pyroxene, sometimes several inches in length. The hornblende crystals are bent and sometimes partially broken across, the transverse fissures being filled with sulphurets, which are also found between the cleavage planes of large pyroxene crystals. Other portions of the vein are of vitreous quartz, holding metallic sulphides and rarely garnets, while large masses of white cleavable pyroxene, and others of finely fibrous greenish or white hornblende, occur, besides masses of white cleavable calcite enclosing long prisms of green hornblende. This vein, with the exception of the abundance of metallic sulphurets, resembles closely in its contents the calcareous veins of the Laurentian rocks above described.]

generally described as occurring in crystalline limestones, though C. U. Shepard, H. D. Rogers, and W. P. Blake have each recognized the fact that these mineral species, with their calcareous gangue, belong to true veins. Emmons, however, failed to distinguish between these vein-stones and the stratified limestones of the series, which, as already stated, often contain disseminated many of the same species, though in a less perfectly crystallized condition than in the vein-stones. Since the latter are clearly seen to traverse the gneiss, like dikes, Emmons was led to look upon them as eruptive; and, generalizing from this, he declared that all the crystalline limestones of northern New York were non-stratified rocks of eruptive origin. (Geology of the First District of New York, 1842, pages 37 – 59.) This view of Emmons was, to a certain extent, adopted by Mather, who, while maintaining the stratified character of the crystalline limestones of southern New York, admitted the existence of eruptive limestones. Von Leonhard had already, in 1833, asserted that limestones have sometimes come from the interior of the earth in a liquid state, like other igneous rocks, and a similar view was at that time maintained by many other geologists. Among others we find Rozet asserting the eruptive origin of the crystalline limestones which are associated with gneiss in the mountains of the Vosges. (Bull. Soc. Geol. de France, III. 215 – 235.) In support of this view could be urged the dike-like form of the calcareous vein-stones, which other observers, like Emmons, confounded with the bedded limestones. The nature and origin of this misconception were, I believe, first pointed out by me in a communication to the American Association for the Advancement of Science in August, 1866 (Canadian Naturalist (2), III. 123), and subsequently more at length in the essay so often referred to. (Report Geol. Survey of Canada, 1863 – 66, p. 182.) It was there shown that many of these calcareous vein-stones are nearly free from foreign minerals, and so closely resemble in lithological characters the stratified limestones, that the different geognostical relations of the two alone enable us, in some examples, to distinguish between them. In this connection I called attention to the great dikes of granular limestones found traversing gneiss near Auerbach in the Bergstrasse, which Bischof has described as true vein-stones. These endogenous concretionary limestones are in fact to stratified limestones what endogenous granitic veins are to gneiss rocks.

XII.

THE ORIGIN OF METALLIFEROUS DEPOSITS.

This paper, unlike the others in this collection (with the exception of IV.), was a lecture to a general audience, given before the American Institute of New York, in May, 1872, and reported for their Proceedings. It is reprinted here because it states, though in a familiar manner, certain views which the author believes to be important. The following extract from a review of American Geology in the American Journal of Science for May, 1861 (a part of which is published as Essay V. of this volume), is prefixed as a concise statement of some of the points in the lecture.

"THE metals . . . . seem to have been originally brought to the surface in watery solutions, from which we conceive them to have been separated by the reducing agency of organic matters in the form of sulphurets or in the native state, and mingled with the contemporaneous sediments, where they occur in beds, in disseminated grains forming *fahlbands,* or are the cementing material of conglomerates. During the subsequent metamorphism of the strata these metallic matters, being taken into solution by alkaline carbonates or sulphurets, have been redeposited in fissures in the metalliferous strata, forming veins, or, ascending to higher beds, have given rise to metalliferous veins in strata not themselves metalliferous. Such we conceive to be, in a few words, the theory of metallic deposits; they belong to a period when the primal sediments were yet impregnated with metallic compounds which were soluble in the permeating waters. The metals of the sedimentary rocks are now, however, for the greater part in the form of insoluble sulphurets, so that we have only traces of them in a few mineral springs, which serve to show the agencies once at work in the sediments and waters of the earth's crust. The present occurrence of these metals in waters which are alkaline from the presence of carbonate of soda, is of great significance when taken in connection with the metalliferous character of certain dolomites, which, as we have shown, probably owe their origin to the action of similar alkaline springs upon basins of sea-water." (*Ante*, page 88.)

"The intervention of intense heat, sublimation, and similar hypotheses to explain the origin of metallic ores, we conceive to be uncalled for. The solvent powers of solutions of alkaline carbonates, chlorides, and sulphurets at elevated temperatures, taken in connection with the notions above enunciated, and with De Senarmont's and Daubrée's beautiful experiments on the crystallization of certain mineral species in the moist way, will suffice to form the basis of a satisfactory theory of metallic deposits." (*Ante*, page 25.)

There are about sixty bodies which chemists call elements; the simplest forms of matter which they have been able to extract from the rocky crust of our earth, its waters, and its atmosphere. These substances are distributed in very unequal quantities, and in very different manners. As regards the frequency of these elements in nature, neglecting for the present those which constitute air and water, and confining ourselves to the solid matters of the earth's crust, there are a few which are exceedingly abundant, making up nine tenths, if not ninety-five hundredths, of the rocks so far as known to us. The elements of which silica, alumina, lime, magnesia, potash, and soda are oxides are very common, and occur almost everywhere. There are others which are much rarer, being found in comparatively small quantities. Many of these rarer elements are, however, of great importance in the economy of nature. Such are the common metals and other substances used in the arts, which occur in nature in quantities relatively very minute, but which have been collected by various agencies, and thus made available for the wants of man. It is chiefly of the well-known metals, iron, copper, silver, and gold, that I propose to speak; but there are two other elements, not classed among the metals, which I shall notice for the reason that their history is extremely important, and will, moreover, enable us to comprehend more clearly some points in that of the metals themselves. I speak of phosphorus and iodine.

You all know the essential part which the former of these, combined as phosphate of lime, plays in the animal economy, in the formation of bones; and how plants require for their proper growth and development a certain amount of phosphorus. Ordinary soils contain only a few thousandths of this element, yet there are agencies at work in nature which gather this diffused phosphorus together in beds of mineral phosphates and in veins of crystalline apatite, which are now sought to enrich impoverished soils. Iodine, an element of great value in medicine and in the art of photography, is widely distributed, but still rarer than phosphorus; yet it abounds in certain mineral waters, and is, moreover, accumulated in marine plants. These extract it from the waters of the sea, where iodine exists in such minute quantities as almost to elude our chemical tests. (See the Appendix, page 237.)

There are probably no perfect separations in nature. We cannot, without great precautions, get any chemical element in a state of absolute purity, and we have reason to believe that even the rarest elements are everywhere diffused in infinitesimal quantities. The spectroscope, which we have lately learned to apply to the investigation alike of the chemistry of our own earth and of other worlds once supposed to be beyond the chemist's ken, not only demonstrates the very wide diffusion of various chemical elements here on the earth, but shows us that very many of them exist in the sun. If we accept, as most of us are now inclined to do, the nebular hypothesis, and admit that our earth was once, like the sun of to-day, an intensely heated vaporous mass; that it is, in fact, a cooled and condensed portion of that once great nebula of which the sun is also a part, — we might expect to find all the elements now discovered in the sun distributed throughout this consolidated globe. We may speculate about the condensation of some of these before others, and their consequent accumulation in the inner parts of the earth; but the fact that we have all the elements of the solar envelope (together with many more) in the exterior portions of our planet, shows that there was, at least, but a very partial concentration and separation of these elements during the period of cooling and condensation. The superficial crust of the earth, from which all the rocks and minerals which we know have been generated, must have contained, diffused through it, from the earliest time, all the elements which we now meet with in our study of the earth, whether still diffused, or accumulated, as we often find the rarer elements, in particular veins or beds.

The question now before us is, How have these elements thus been brought together, and why is it that they are not all still widely and universally diffused? Why are the compounds of iron in beds by themselves, copper, silver, and gold gathered together in veins, and iodine concentrated in a few ores and certain mineral waters? That we may the better discern the direction in which we are to look for the solution of this problem, let us premise that all of these elements, in some of their combinations, are more or less soluble in water. There are, in fact, no such things in nature as absolutely insoluble bodies, but all, under certain conditions, are capable of being taken up by water, and again deposited from it.* The alchemists sought in vain for a universal solvent; but we now know that water, aided in some cases by heat, pressure, and the presence of certain widely distributed substances, such as carbonic acid and alkaline carbonates and sulphides, will dissolve the most insoluble bodies; so that it may, after all, be looked upon as the long-sought-for alkahest or universal menstruum.

[* It is well known that many chemical compounds when first generated by double decomposition in watery solutions remain dissolved for a greater or less length of time before separating in an insoluble condition. The solubility of recently precipitated carbonate of lime in water holding certain neutral salts, as already described (*ante*, page 140), is a fact in the same order. In this connection may also be recalled the great solubility in water of silicic, titanic, stannic, ferric, aluminic, and chromic oxides when in what Graham has called their colloidal state. There is reason to believe that silicates of insoluble bases may assume a similar state, and it will probably one day be shown that for the greater number of those oxygenized compounds which we call insoluble there exists a modification soluble in water.]

Let us now compare the waters of rivers, seas, and subterranean springs, thus impregnated with various chemical elements, with the blood which circulates through our own bodies. The analysis of the blood shows it to contain albuminoids which go to form muscle, fat for the adipose tissues, phosphate of lime for the bones, fluorides for the enamel of the teeth, sulphur, which enters largely into the composition of the hair and nails, soda which accumulates in the bile, and potash, which abounds in the flesh-fluid. All of these are dissolved in the blood, and the great problem for the chemical physiologist is to determine how the living organism gathers them from this complex fluid, depositing them here and there, and giving to each part its proper material. This selection is generally ascribed to a certain vital force, peculiar to the living body. I shall not here discuss the vexed question of the nature of the force which determines the assimilation from the blood of these various matters for the needs of the animal organism, further than to say that modern investigations tend to show that it is only a subtler kind of chemistry, and that the study of the nature and relation of colloids and crystalloids, and of the phenomena of chemical diffusion, promises to subordinate all these obscure physiological processes to chemical and physical laws.

Let us now see how far the comparison which we have made between the earth and an animal organism will help us to understand the problem of the distribution of minerals in nature; how far water, the universal solvent, acting in accordance with known chemical and physical laws, will cause the separation of the mixed elements of the earth's crust, and their accumulation in veins and beds in the rocks. The subject is one of great importance to the geologist, who has to consider the genesis of the various rocks and ore-deposits, and the relations, which we are only beginning to understand, between certain metals and particular rocks, and between certain classes of ores and peculiar mineralogical and geological conditions. It is at the same time a vast one, and I can now only give you a few illustrations of the chemistry of the earth's crust, and of the laws of the terrestrial circulation, which I have compared to that of the blood distributing throughout the animal frame the elements necessary for its growth. The analogy is not altogether new, since a great French geologist, Elie de Beaumont, has already spoken of a terrestrial circulation in regard to certain elements in the earth's crust; though he has not, so far as I am aware, carried it out to the extent which I now propose to do in my attempt to explain some of the laws which have presided over the distribution of metals in the earth.

The chemist in his laboratory takes advantage of changes of temperature, and of the action of various solvents and precipitants, to separate, in the humid way, one element from another; but to these agencies, in the economy of nature, are added others which we have not yet succeeded in imitating, and which are exerted only in growing animals and plants. I repeat it; I do not wish to say that these latter processes are different in kind from those which we command in our laboratories, but rather that these organisms control a far finer and more delicate chemical and physical apparatus than we have yet invented. Plants have the power of selecting from the media in which they live the elements necessary for their support. The growing oak and the grass alike assimilate from the air and water the carbon, hydrogen, nitrogen, and oxygen which build up their tissues, and at the same time take from the soil a portion of phosphorus, which, though minute, is essential to the vegetable growth. The acorn of the oak and the grass alike become the food of animals, and the gathered phosphates pass into their bones, which are nearly pure phosphate of lime. In like manner the phosphates from organic waste and decay find their way to the sea, and through the agency of marine vegetation become at last the bony skeletons of fishes. These are, in turn, the prey of carnivorous birds, whose exuviæ form on tropical islands beds of phosphatic guano. A history not dissimilar will explain the origin of beds of coprolites and of some other deposits of mineral phosphates. [By whatever means the phosphates have been first concentrated, it appears from the recent studies of Sollas that the so-called coprolites of the green-sand in England result from a petrifaction of sponges by dissolved phosphates, and similar observations have been made by Edwards with regard to the guano of the Chincha Islands.]

But again, these plants or these animals may perish in the sea and be buried in its ooze. The phosphates which they have gathered are not lost, but become fixed in an insoluble form in the clayey matter; and when, in the revolutions of ages, these sea-muds, hardened to rock, become dry land, and crumble again to soil, the phosphates are there found ready for the wants of vegetation.

Most of what I have said of phosphates applies equally to the salts of potash, which are not less necessary to the growing plant. From the operation of these laws it results that neither of these elements is found in large quantities in the ocean. This great receptacle of the drainage from the land contains still smaller quantities of iodine; in fact, the traces of this element present in sea-water can scarcely be detected by our most delicate tests.* Yet marine plants have the power of separating this iodine, and accumulating it in their tissues, so that the ashes of these plants are not only rich in phosphates and in potash-salts, but contain so much iodine that our supplies of this precious element are almost wholly derived from this source, and that the gathering and burning of sea-weed for the extraction of iodine is in some regions an important industry. When this marine vegetation decays, the iodine which it contains appears, like the potash and phosphates, to pass into combination with metals, earths, or earthy phosphates, which retain it in an insoluble state, and in certain cases yield it to percolating saline solutions, which thus give rise to springs rich in iodine. (*Ante*, page 143.)

In all of these processes the action of organic life is direct and assimilative, but there are others in which its agency, although indirect, is not less important. I can hardly conceive of an accumulation of iron, copper, lead, silver, or gold, in the production of which animal or vegetable life has not either directly or indirectly been necessary, and I shall be-

* See the Appendix to this paper.

gin to explain my meaning by the case of iron. This, you are aware, is one of the most widely diffused elements in nature; all soils, all plants, contain it; and it is a necessary element in our blood. Clays and loams contain, however, at best, two or three hundredths of the metal, but so mixed with other matters that we could never make it available for the wants of this iron age of ours. How does it happen that we also find it gathered together in great beds of ore, which furnish an abundant supply of the metal? The chemist knows that the iron, as diffused in the rocks, exists chiefly in combination with oxygen, with which it forms two principal compounds: the first, or protoxide, which is readily soluble in waters impregnated with carbonic acid or other feeble acids; and the second, or peroxide, which is insoluble in the same liquids. I do not here speak of the magnetic oxide, which may be looked upon as a compound of the other two, neutral and indifferent to most natural chemical agencies. The combinations of the first oxide are either colorless or bluish or greenish in tint, while the peroxide is reddish-brown, and is the substance known as iron-rust. Ordinary brick-clays are bluish in color, and contain combined iron in the state of protoxide, but when burned in a kiln they become reddish, because this oxide absorbs from the air a further proportion of oxygen, and is converted into peroxide. But there are clays which are white when burned, and are much prized for this reason. Many of these were once ferruginous clays, which have lost their iron by a process everywhere going on around us. If we dig a ditch in a moist soil which is covered with turf or with decaying vegetation, we may observe that the stagnant water which collects at the bottom soon becomes coated with a shining, iridescent scum, which looks somewhat like oil, but is really a compound of peroxide of iron. The water as it oozes from the soil is colorless, but has an inky taste, from dissolved protoxide of iron. When exposed to the air, however, this absorbs oxygen, and the peroxide is formed, which is no longer soluble, but separates as a film on the surface of the water, and finally sinks to the bottom as a reddish ochre, or, under somewhat different conditions, becomes aggregated as a massive iron-ore. A process identical in kind with this has been at work at the earth's surface ever since there were decaying organic matters, dissolving the iron from the porous rocks, clays, and sands, and gathering it together in beds of iron-ore or iron-ochre. It is not necessary that these rocks and soils should contain the iron in the state of protoxide, since these organic products (which are themselves dissolved in the water) are able to remove a portion of the oxygen from the insoluble peroxide, and convert it into the soluble protoxide of iron, being themselves in part oxidized and converted into carbonic acid in the process.

We find in rock-formations of very different ages beds of sediments which have been deprived of iron by organic agencies, and near them will generally be found the accumulated iron. Go into any coal region, and you will see evidences that this process was at work when the coal-beds were forming. The soil in which the coal-plants grew has been deprived of its iron, and when burned turns white, as do most of the slaty beds from the coal-rocks. It is this ancient soil which constitutes the so-called fire-clays, prized for making bricks which, from the absence of both iron and alkalies, are very infusible. Interstratified with these we often find, in the form of ironstone, the separated metal; and thus from the same series of rocks may be obtained the fuel, the ore, and the fire-clay.

From what I have said it will be understood that great deposits of iron-ore generally occur in the shape of beds; although waters holding the compounds of iron in solution have, in some cases, deposited them in fissures or openings in the rocks, thus forming true veins of ore, of which we shall speak further on. I wish now to insist upon the property which dead and decaying organic matters possess of reducing to protoxide, and rendering soluble, the insoluble peroxide of iron diffused through the rocks; and reciprocally the power which this peroxide has of oxidizing and consuming these same organic matters, which are thereby finally converted into carbonic acid and water. This last action, let me say in passing, is illustrated by the destructive action of rusting iron bolts on moist wood, and the effect of iron stains in impairing the strength of linen fibre.

We see in the coal formation that the vegetable matter necessary for the production of the iron-ore beds was not wanting; but the question has been asked me, Where are the evidences of the organic material which was required to produce the vast beds of iron-ore found in the ancient crystalline rocks? I answer that the organic matter was, in most cases, entirely consumed in producing these great results; and that it was the large proportion of iron diffused in the soils and waters of these early times, which not only rendered possible the accumulation of such great beds of ore, but oxidized and destroyed the organic matters which in later ages appear in coals, lignites, pyroschists, and bitumens. Some of the carbon of these early times is, however, still preserved in the form of graphite, and it would be possible to calculate how much carbonaceous material was consumed in the formation of the great iron-ore beds of the older rocks, and to determine of how much coal or lignite they are the equivalents.

In the course of ages, however, as a large proportion of the once diffused iron-oxide has become segregated in the form of beds of ore, and thus removed from the terrestrial circulation, the conditions have grown more favorable for the preservation of the carbonaceous products of vegetable life. The crystalline magnetic and specular oxides, which constitute a large proportion of the ores of this metal, are almost or altogether indifferent to the action of organic matter. When, however, these ores are reduced in our furnaces, and the resulting metal is exposed to the oxidizing action of a moist atmosphere, it is again converted into iron-rust, which is soluble in water holding organic matters, and may thus be made to enter once more into the terrestrial circulation.

There is another form in which iron is frequently concentrated in nature, that of sulphide, and most frequently as the bisulphide, known as iron-pyrites. This substance is found both in the oldest and the newest rocks, and, like the oxide of iron, is even to-day forming in certain waters and in beds of mud and silt, where it sometimes takes a beautifully crystalline shape. What are the conditions in which the sulphide of iron is formed and deposited, instead of the oxide or carbonate of iron? Its production depends, like these, on decaying organic matters. The sulphates of lime and magnesia, which abound in sea-water, and in many other natural waters, when exposed to the action of decaying plants or animals, out of contact of air, are, like peroxide of iron, deoxidized, and are thereby converted into soluble sulphides; from which, if carbonic acid be present, sulphuretted hydrogen gas is set free. Such soluble sulphides, or sulphuretted hydrogen, are the reagents constantly employed in our laboratories to convert the soluble compounds of many of the common metals, such as iron, zinc, lead, copper, and silver, into sulphides, which are insoluble in water and in many acids, and are thus conveniently separated from a great many other bodies. Now, when in a water holding iron-oxide, sulphates are also present, the action of organic matter, deoxidizing the latter, furnishes the reagent necessary to convert the iron into a sulphide; which in some conditions, not well understood, contains two equivalents of sulphur for one of iron, and constitutes iron-pyrites. I may here say that I have found that the unstable protosulphide, which would naturally be first formed, may, under the influence of a persalt of iron, lose one half of its combined iron; and that from this reaction a stable bisulphide results. This subject of the origin of iron-pyrites is still under investigation.

The reducing action of organic matters upon soluble sulphates is well seen in the sulphuretted hydrogen which is evolved from the stagnant sea-water in the hold of a ship, and which coats silver exposed to it with a black film of sulphide of silver, and for the same reason discolors white-lead paint. The presence of sulphur in the exhalations from some other decaying matters is well known, and in all these cases a soluble compound of iron will act as a disinfectant, partly by fixing the sulphur as an insoluble sulphide. Silver coins brought from the ancient wreck of a treasure-ship in the Spanish Main were found to be deeply incrusted with sulphide of silver, formed in the ocean's depths by the process just explained, which is one that must go on wherever organic matters and sea-water are present, and atmospheric oxygen excluded.

The chemical history of iron is peculiar; since it requires reducing matters to bring it into solution, and since it may be precipitated alike by oxidation, and by further reduction provided sulphates are present. The metals, copper, lead, and silver, on the contrary, form compounds more or less soluble in water, from which they are not precipitated by oxygen, but only by reducing agents, which may separate them in some cases in a metallic state, but more frequently as sulphides. The solubility of the salts and oxides of these metals in water is such that they are found in many mineral springs, in the waters that flow from certain mines, and in the ocean itself, the waters of which have been found to contain copper, silver, and lead. Why, then, do not these metals accumulate in the sea, as the salts of soda have done during long ages? The direct agency of organic life comes again into play, precisely as in the case of phosphorus, iodine, and potash. Marine plants, which absorb these from the sea-water, take up at the same time the metals just named, traces of all of which are found in the ashes of sea-weeds. Copper, moreover, is met with in notable quantities in the blood of many marine molluscous animals, to which it may be as necessary as iron is to our own bodies. Indeed, the blood of man, and of the higher animals, appears never to be without traces of copper as well as of iron.

In the open ocean the waters are constantly aerated, so that soluble sulphides are never formed, and the only way in which these dissolved metals can be removed and converted into sulphides is by fixing them in organisms, either vegetable or animal. These, by their decay in the mud of the bottom, or the lagoons of the shore, generate the sulphides which fix their contained metals in an insoluble form, and thus remove them from the terrestrial circulation.

It is not, however, in all cases necessary to invoke the direct action of organisms to separate from water the dissolved metals. It often happens that the waters containing these, instead of finding their way to the ocean, flow into lakes or enclosed basins, as in the case of the drainage-waters of an English copper-mine, which have impregnated the turf of a neighboring bog to such an extent that its ashes have been found a profitable source of copper. Under certain conditions, not yet well understood, this metal is precipitated by organic matters in the metallic state, but if sulphates are present, a sulphide is formed. Thus, in certain mesozoic slates in Bohemia, sulphide of copper is found incrusting the remains of fishes, and in the sandstones of New Jersey we find it penetrating the stems of ancient trees. I have in my possession a portion of a small trunk taken from the mud of a spring in the province of Ontario, in which the yet undecayed wood of the centre is seen to be incrusted by hard and brilliant iron-pyrites. In like manner the trees found in the New Jersey sandstone became incrusted with copper-sulphide, which, as decay went on, in great part replaced the woody tissue. Similar deposits of sulphides of copper and of iron often took place in basins where the organic matter was present in such a condition or in such quantity as to be entirely decomposed, and to leave no trace of its form, unlike the examples just mentioned. In this way have been formed fahlbands, and beds of pyrites and other ores.

The fact that such deposits are associated with silver and with gold leads to the conclusion that these metals have obeyed the same laws as iron and copper. It is known that both persalts of iron and soluble sulphides have the power of rendering gold soluble, and its subsequent deposition in the metallic state is then easily understood.*

I have endeavored by a few illustrations to show you by what processes some of the more common metals are dissolved and again separated from their solution in insoluble forms. It now remains to say somewhat of the geological relations of

* See Appendix to this paper.

ore-deposits, which are naturally divided into two classes; the first including those which occur in beds, and have been formed contemporaneously with the enclosing earthy sediments. Such are the beds of iron-ores, which often hold embedded shells and other organic remains, and the copper-bearing strata already mentioned, in which the metal must have been deposited during the decay of the animal or plant which it incrusts or replaces. But there are other ore-deposits evidently of more recent formation than the rocky strata which enclose them, which have resulted from a process of infiltration, filling up fissures with the ore, or diffusing it irregularly through the rock. It is not always easy to distinguish between the two classes of deposits. Thus a fissure may in some cases be formed and filled between two sundered beds, from which may result a vein that may be mistaken for an interposed stratum. Again, a bed may be so porous that infiltrating waters may diffuse through it a metallic ore, or a metal, in such a manner as to leave it doubtful whether the process was contemporaneous with the deposition of the bed, or posterior to it. But I wish to speak of deposits which are evidently posterior, and occupy fissures in previously formed strata, constituting true veins. Whether produced by the great movements of the earth's crust, or by the local contraction of the rocks (and both of these causes have in different cases been in operation), such fissures sometimes extend to great lengths and depths; their arrangement and dimensions depending very much on the texture of the rocks which have been subjected to fracture. When a bone in our bodies is broken, nature goes to work to repair the fractured part, and gradually brings to it bony matter, which fills up the little interval, and at length makes the severed parts one again. So when there are fractures in the earth's crust, the circulating waters deposit in the openings mineral matters, which unite the broken portions, and thus make whole again the shattered rocks. Vein-stones are thus formed, and are the work of nature's conservative surgery.

Water, as we have seen, is a universal solvent, and the matters which it may bring and deposit in the fissures of the earth are very various. There is scarcely a spar or an ore to be met with in the stratified rocks that is not also found in some of these vein-stones, which are often very heterogeneous in composition. In certain veins we find the elements of limestone or of granite, and these often include the gems, such as tourmaline, garnet, topaz, hyacinth, emerald, and sapphire; while others abound in native metals or in metallic oxides or sulphides. The nature of the materials thus deposited depends very much on conditions of temperature and of pressure, which affect the solvent power of the liquid, and still more upon the nature of the adjacent rocks and of the waters permeating them. The chemistry of mineral veins is very complicated. Many of these fissures penetrate to a depth of thousands of feet of the earth's crust, and along the channels thus opened the ascending heated subterranean waters may receive in their course various contributions from the overlying strata. From these additions, and from the diminished solubility resulting from a decrease of pressure (*ante*, page 204), deposits of different minerals are formed upon the walls, and the slow changes in composition are often represented by successive layers of unlike substances. The power of these waters to dissolve and bring from the lower strata their contained metals and spars is probably due in great part to the alkaline carbonates and sulphides which these waters often hold in solution; but the chemical history of the deposition of the ores of iron, lead, copper, silver, tin, and gold, which are found in these veins, demands a lengthened study, and would furnish not less beautiful examples of nature's chemistry than those I have already laid before you.

The process of filling veins has been going on from the earliest ages; we know of some which were formed before the Cambrian rocks were deposited, while others are still forming, as the observations of Phillips have shown us in Nevada, where hot springs rise to the surface and deposit silica, with metallic ores, which incrusts the walls of the fissures. These thermal waters show that the agencies which in past times gave rise to the rich mineral deposits of our western regions, are still at work there.

Let us now consider the beneficent results of the process of vein-making. The precious metals, such as silver, are so sparsely distributed, that even the beds rich in the products of decaying sea-weed, which we have supposed to be deposited from the ocean, would contain too little silver to be profitably extracted. But in the course of ages these sediments, deeply buried, are lixiviated by permeating solutions, which dissolve the silver diffused through a vast mass of rock, and subsequently deposit it in some fissure, it may be in strata far above, as a rich silver-ore. This is nature's process of concentration.

We learn from the history which we have just sketched the important conclusion, that amid all the changes of the face of the globe the economy of nature has remained the same. We are apt, in explaining the appearances of the earth's crust, to refer the formation of ore-beds and veins to some distant and remote period, when conditions very unlike the present prevailed, when great convulsions took place, and mysterious forces were at work. Yet the same chemical and physical laws are now, as then, in operation: in one part dissolving the iron from the sediments and forming ore-beds, in another separating the rarer metals from the ocean's waters; while in still other regions the consolidated and buried sediments are permeated by heated waters, to which they give up their metallic matters, to be subsequently deposited in veins. These forces are always in operation, rearranging the chaotic admixture of elements which results from the constant change and decay around us. The laws which the First Great Cause imposed upon this material universe on the first day are still irresistibly at work fashioning its present order. One great design and purpose is seen to bind in necessary harmony the operations of the mineral with those of the vegetable and animal worlds, and to make all of these contribute to that terrestrial circulation which maintains the life of our mother earth.

While the phenomena of the material world have been looked upon as chemical and physical, it has been customary to speak of those of the organic world as vital. The tendency of modern investigation is, however, to regard the processes of animal and vegetable growth as themselves purely chemical and physical. That this is to a great extent true must be admitted, though I am not prepared to concede that we have in chemical and physical processes the whole secret of organic life. Still we are, in many respects, approximating the phenomena of the organic world to those of the mineral kingdom; and we at the same time learn that these so far interact and depend upon each other that we begin to see a certain truth underlying the notion of those old philosophers who extended to the mineral world the notion of a vital force, which led them to speak of the earth as a great living organism, and to look upon the various changes in its air, its waters, and its rocky depths, as processes belonging to the life of our planet.

APPENDIX.

ON IODINE AND GOLD IN SEA-WATER.

AFTER the above lecture was delivered, appeared the results of the researches of Sonstadt on the iodine in sea-water, which were published in the Chemical News for April 26, May 17, and May 24, 1872. According to him, this element exists in sea-water, under ordinary conditions, as iodate of calcium, to the amount of about one part of the iodate in 250,000 parts of the water. This compound, by decaying organic matter (and by most other reducing agents), is changed to iodide, from which, apparently by the action of carbonic acid, iodine is set free, and may be separated by agitating the water with bisulphide of carbon. The iodine thus liberated from sea-water by the action of dead organic matters, however, slowly decomposes water in presence of carbonate of calcium, and is reconverted into iodate, the oxygen of the air probably intervening to complete the oxidation, since, according to Sonstadt, iodides are readily converted into iodates under these conditions. He finds that the insolubility of the iodides of silver and of copper is so great that by the use of salts of these metals iodine may be separated from sea-water, without concentration, provided the iodate of calcium has first been reduced to iodide. By this property of iodine and its compounds to oxidize and be oxidized in turn, Sonstadt supposes them to perform the important function of consuming the products of organic decay, and so maintaining the salubrity of the ocean's waters. Their action would thus be very similar to that of the oxides of iron, as explained in the lecture.

Still more recently the same chemist has announced that the sea-water of the British coasts contains in solution, besides silver, an appreciable amount of gold, estimated by him at about one grain to a ton of water. This is separated by the addition of chloride of barium to the water, apparently as an aurate of baryta adhering to the precipitated sulphate, which yields by assay an alloy of about six parts of gold to four of silver. Other ways have been devised by him for separating these metals from their solution in sea-water. The agent which keeps the gold of the sea in a soluble and oxidized condition is, according to Sonstadt, the iodine liberated by the reaction already described. The views maintained by Lieber,

Wurtz, Genth, and Selwyn as to the solution and re-deposition of gold in modern alluvial deposits, seem to be well-grounded, and we are led to the conclusion that the circulation of this metal in nature is as easily effected as that of iron or of copper. The transfer of certain other elements, such as titanium, chrome, and tin, or at least their accumulation in concentrated forms, appears, on the contrary, to require conditions which are no longer operative, at least at the surface of the earth.

It should here be noticed, that Professor Henry Wurtz of New York, in a paper read before the American Association for the Advancement of Science in 1866, and published in the Journal of Mining in 1868, expressed the opinion that the ocean-waters contain gold, and urged experiments for its detection. According to his calculations, the total amount of gold hitherto extracted from the earth, and estimated at two thousand million dollars, would give only one dollar for two hundred and eighty million tons of sea-water ; while from the experiments of Sonstadt it would appear that the same quantity of gold is actually contained in twenty-five tons of water.

XIII.

THE GEOGNOSY OF THE APPALACHIANS AND THE ORIGIN OF CRYSTALLINE ROCKS.

The following address was delivered on retiring from the office of president of the American Association for the Advancement of Science at Indianapolis, August 16, 1871. It appears in the Proceedings of the Association and in the American Naturalist for October, and, with some abridgment of the first part, in Nature. A French translation of the entire address was also published in the Revue Scientifique. In reprinting it a few sentences have been substituted for the original references to the Cambrian rocks of Great Britian, and a fuller account of the Norian or Labrador series has been introduced, besides some minor additions in the first part. In the second part of the paper, also, important additions have been made. These new portions are distinguished by being enclosed in brackets.

In the American Journal of Science for February, 1872, appeared an adverse criticism of some parts of the address, by Professor J. D. Dana, to which the author in the same Journal for July, 1872, made a reply, which is here printed as an appendix to the paper; the short portion relating to geognosy being at the close. Professor Dana's rejoinder will be found in the same Journal for August, 1872.

In accordance with our custom it becomes my duty, in quitting the honorable position of president, which I have filled for the past year, to address you upon some theme which shall be germane to the objects of the Association. The presiding officer, as you are aware, is generally chosen to represent alternately one of the two great sections into which the members of the Association are supposed to be divided; namely, the students of the natural-history sciences on the one hand, and of the physico-mathematical and chemical sciences on the other. The arrangement by which, in our organization, geology is classed with the natural-history division, is based upon what may fairly be challenged as a somewhat narrow conception of its scope and aims. While theoretical geology or geogeny investigates the astronomical, physical, chemical, and biological laws which have presided over the development of our earth, and while practical geology or geognosy studies its natural history as exhibited in its physical structure, its mineralogy and its paleontology, it will be seen that this comprehensive science is a stranger to none of the studies which are included in the plan of our Association, but rather sits like a sovereign, commanding in turn the services of all.

As a student of geology, I scarcely know with which section of the Association I should to-day identify myself. Let me endeavor rather to mediate between the the two, and show you somewhat of the twofold aspect which geological science presents, when viewed respectively from the standpoints of natural history and of chemistry. I can hardly do this better than in the discussion of a subject which for the last generation has afforded some of the most fascinating and perplexing problems for our geological students; namely, the history of the great Appalachian mountain chain. Nowhere else in the world has a mountain system of such geographical extent and such geological complexity been studied by such a number of zealous and learned investigators, and no other, it may be confidently asserted, has furnished such vast and important results to geological science. The laws of mountain structure, as revealed in the Appalachians by the labors of the brothers Henry D. and William B. Rogers, of Lesley and of Hall, have given to the world the basis of a correct system of orographic geology,* and many of the obscure geological problems of Europe become plain when read in the light of our American experience. To discuss even in the most summary manner all of the questions which the theme suggests, would be a task too long for the present occasion; but I shall endeavor in the first place to bring before you certain facts in the history of the physical structure, the mineralogy, and the paleontology of the Appalachians; and, in the second place, to discuss some of the physical, chemical, and biological conditions which have presided over the formation of the ancient crystalline rocks that make up so large a portion of our great eastern mountain system.

* Amer. Jour. Sci. (2), XXX. 406; and *ante*, pages 49 - 58.

I. The Geognosy of the Appalachian System.

The age and geological relations of the crystalline stratified rocks of eastern North America have for a long time occupied the attention of geologists. A section across northern New York, from Ogdensburg on the St. Lawrence to Portland in Maine, shows the existence of three distinct regions of unlike crystalline schists. These are the Adirondacks to the west of Lake Champlain, the Green Mountains of Vermont, and the White Mountains of New Hampshire. The lithological and mineralogical differences between the rocks of these three regions are such as to have attracted the attention of some of the earlier observers. Eaton, one of the founders of American geology, at least as early as 1832 distinguished in his Geological Text-Book (2d edition) between the gneiss of the Adirondacks and that of the Green Mountains. Adopting the then received divisions of primary, transition, secondary, and tertiary rocks, he divided each of these series into three classes, which he named carboniferous, quartzose, and calcareous; meaning by the first, schistose, or argillaceous strata such as, according to him, might include carbonaceous matter. These three divisions, in fact, corresponded to clay, sand, and lime-rocks, and were supposed by him to be repeated in the same order in each series. This was apparently the first recognition of that law of cycles in sedimentation upon which I afterwards insisted in 1863.* Without, so far as I am aware, defining the relations of the Adirondacks, he referred to the lowest or carboniferous division of the primary series, the crystalline schists of the Green Mountains, while the quartzites and marbles at their western base were made the quartzose and calcareous divisions of this primary series. The argillites and sandstones lying still farther westward, but to the east of the Hudson River, were regarded as the first and second divisions of the transition se-

* Amer. Jour. Sci. (2), XXXV. 166. See, for an excellent presentation of this subject, with references to its literature, a paper by Dr. Newberry in the Proceedings of the American Association for the Advancement of Science for 1873, page 185.

ries, and were followed by its calcareous division, which seems to have included the limestones of the Trenton group; all of these rocks being supposed to dip to the westward, and away from the central axis of the Green Mountains. Eaton does not appear to have studied the White Mountains, nor to have considered their geological relations. They were, however, clearly distinguished from the former by Charles T. Jackson in 1844, when, in his report on the geology of New Hampshire, he described the White Mountains as an axis of primary granite, gneiss, and mica-schist, overlaid successively, both to the east and west, by what were designated by him Cambrian and Silurian rocks; these names having, since the time of Eaton's publication, been introduced by English geologists. While these overlying rocks in Maine were unaltered, he conceived that the corresponding strata in Vermont, on the western side of the granitic axis, had been changed by the action of intrusive serpentines and intrusive quartzites, which had altered the Cambrian into the Green Mountain gneiss, and converted a portion of the fossiliferous Silurian limestones of the Champlain valley into white marbles.* Jackson did not institute any comparison between the rocks of the White Mountains and those of the Adirondacks; but the Messrs. Rogers in the same year, 1844, published an essay on the geological age of the White Mountains, in which, while endeavoring to show their Silurian age, they speak of them as having been hitherto regarded as consisting exclusively of various modifications of granitic and gneissoid rocks, and as belonging "to the so-called primary periods of geologic time." † They, however, considered that these rocks had rather the aspect of altered palæozoic strata, and suggested that they might be, in part, at least, of the age of the Clinton division of the New York system; a view which was supported by the presence of what were at the time regarded by the Messrs. Rogers as organic remains. Subsequently, in 1847,‡ they announced that they no longer considered these to

* Geology of New Hampshire, 160-162.

† Amer. Jour. Sci. (2), I. 411.

‡ Ibid. (2), V. 116.

be of organic origin, without, however, retracting their opinion as to the palæozoic age of the strata. Reserving to another place in my address the discussion of the geological age of the White Mountain rocks, I proceed to notice briefly the distinctive characters of the three groups of crystalline strata just mentioned, which will be shown in the sequel to have an importance in geology beyond the limits of the Appalachians.

I. *The Adirondack or Laurentide Series.* — The rocks of this series, to which the name of the Laurentian system has been given, may be described as chiefly firm granitic gneisses, often very coarse grained, and generally reddish or grayish in color. They are frequently hornblendic, but seldom or never contain much mica, and the mica-schists (often accompanied with staurolite, garnet, andalusite, and cyanite), so characteristic of the White Mountain series, are wanting among the Laurentian rocks. They are also destitute of argillites, which are found in the other two series. The quartzites, and the pyroxenic and hornblendic rocks, associated with great formations of crystalline limestone, with graphite, and immense beds of magnetic iron-ore, give a peculiar character to portions of the Laurentian system.

II. *The Green Mountain Series.* — The quartzo-feldspathic rocks of this series are to a considerable extent represented by a fine-grained petrosilex or eurite, though they often assume the form of a true gneiss, which is ordinarily more micaceous than the typical Laurentian gneiss. The coarse-grained, porphyritic, reddish varieties common to the latter are wanting in the Green Mountains, where the gneiss is generally of pale greenish and grayish hues. [The quartziferous porphyries, which have been noticed *ante*, page 187, are supposed, in the present state of our knowledge, to belong to this series.] Massive stratified diorites, and epidotic and chloritic rocks, often more or less schistose, with steatites, dark-colored serpentines, and ferriferous dolomites and magnesites also characterize this gneissic series, and are intimately associated with beds of iron-ore, generally a slaty hematite, but occasionally magnetite. Chrome, titanium, nickel, copper, antimony, and gold are frequently met with in this series. The gneisses often pass into schistose micaceous quartzites, and the argillites, which abound, frequently assume a soft unctuous character, which has acquired for them the name of talcose or nacreous slates, though analysis shows them not to be magnesian, but to consist essentially of a hydrous micaceous mineral allied to paragonite. They are sometimes black and graphitic.

III. *The White Mountain Series.* — This series is characterized by the predominance of well-defined mica-schists interstratified with micaceous gneisses. These latter are ordinarily light colored from the presence of white feldspar, and, though generally fine in texture, are sometimes coarse grained and porphyritic. They are less strong and coherent than the gneisses of the Laurentian, and pass, through the predominance of mica, into mica-schists, which are themselves more or less tender and friable, and present every variety, from a coarse gneiss-like aggregate down to a fine-grained schist, which passes into argillite. The micaceous schists of this series are generally much richer in mica than those of the preceding series, and often contain a large proportion of well-defined crystalline tables belonging to the species muscovite. The cleavage of these micaceous schists is generally, if not always, coincident with the bedding; but the plates of mica in the coarser-grained varieties are often arranged at various angles to the cleavage and bedding-plane, showing that they were developed after sedimentation, by crystallization in the mass, a circumstance which distinguishes them from rocks derived from the ruins of these, which are met with in more recent series. The White Mountain rocks also include beds of micaceous quartzite. The basic silicates in this series are represented chiefly by dark-colored gneisses and schists in which hornblende takes the place of mica. These pass occasionally into beds of dark hornblende rock, sometimes holding garnets. Beds of crystalline limestone occur in the schists of the White Mountain series, and are sometimes accompanied by pyroxene, garnet, idocrase, sphene, and graphite, as in the corresponding rocks of the Laurentian, which this series, in its more gneissic portions, closely resembles, though apparently distinct geognostically. The limestones are intimately associated with the highly micaceous schists containing staurolite, andalusite, cyanite, and garnet. These schists are sometimes highly plumbaginous, as seen in the graphitic mica-schist holding garnets in Nelson, New Hampshire, and that associated with cyanite in Cornwall, Connecticut. To this third series of crystalline schists belong the concretionary granitic veins abounding in beryl, tourmaline, and lepidolite, and occasionally containing tinstone and columbite. (See Granites and Granitic Vein-Stones, *ante*, pages 194 – 199.) Granitic veins in the Laurentian gneisses frequently contain tourmaline, but have not, so far as yet known, yielded the other mineral species just mentioned.

Keeping in mind the characteristics of these three series, it will be easy to trace them southward by the aid of the concise and accurate descriptions which Professor H. D. Rogers has given us of the rocks of Pennsylvania. In his report on the geology of this State, he has distinguished three districts of various crystalline schists, which are by him included together under the name of gneissic or hypozoic rocks. Of these districts, the most northern, or the South Mountain belt, to the northwest of the Mesozoic basin, is said to be the continuation of the Highlands of New York and New Jersey, which, crossing the Delaware near Easton, is continued southward, through Pennsylvania and Maryland, into Virginia, where it appears in the Blue Ridge. The gneiss of this district in Pennsylvania is described as differing considerably from that of the southernmost district, being massive and granitoid, often hornblendic, with much magnetic iron, but destitute of any considerable beds of micaceous, talcose, or chloritic slate, which mark the rocks of the southern district. These characters are sufficient to show that the gneiss of this northern district is lithologically, as well as geognostically, identical with that of the Highlands, and belongs, like it, to the Adirondack, or Laurentian system of crystalline rocks. The gneiss of the middle district of Pennsylvania, to the south of the Mesozoic, but north of the Chester valley, is described by Rogers as resembling that of the South Mountain, or northern district, and to consist chiefly of white feldspathic and dark hornblendic gneiss, with very little mica, and with crystalline limestones.

The gneiss of the third or southern district (that lying to the south of the Montgomery and Chester valleys) comes from beneath the Mesozoic of New Jersey about six miles northeast of Trenton, and, stretching southwestward, occupies the southern border of Pennsylvania, extending into Delaware and Maryland. It is subdivided by Rogers into three belts. The first or most southern of these, passing through Philadelphia, consists of alternations of dark hornblendic and highly micaceous gneiss, with abundance of mica-slate, sometimes coarse grained, and at other times so fine grained as to constitute a sort of whet-slate. To the northwestward the strata become still more micaceous, with garnets and beds of hornblende slate, till we reach the second subdivision, which consists of a great belt of highly talcose and micaceous schists, with steatite and serpentine, and is in its turn succeeded by a third narrow belt, resembling the less micaceous members of the first or southernmost subdivision. The micaceous schists of this region abound in staurolite, garnet, cyanite, and corundum, and are traversed by numerous irregular granitic veins containing beryl and tourmaline. All of these characters lead us to refer the gneiss of this southern district to the third, or White Mountain series, with the exception of the middle subdivision, which presents the aspect of the second, or Green Mountain series.

Above the hypozoic gneisses Rogers has placed his azoic or semi-metamorphic series, which is traceable from the vicinity of Trenton to the Schuylkill, along the northern boundary of the southern hypozoic gneiss district. This series is supposed by Rogers to be an altered form of the primal sandstones and slates, and is described as consisting of a feldspathic quartzite, or eurite, containing in some cases porphyritic beds with crystals of feldspar and hornblende, together with various crystalline schists; including, in fact, the whole of the great serpentine belt of Montgomery, Chester, and Lancaster Counties, with its steatites, hornblendic, dioritic, chloritic, and micaceous schists (often garnet-bearing), together with a band of argillite, affording roofing-slates. With this great series are associated chromic and titanic iron, and ores of nickel and copper. Veins of albite with corundum also intersect this series near Unionville. We are repeatedly assured by Rogers that these rocks so much resemble the underlying hypozoic gneiss, as to be readily confounded with them; and when compared with the latter, as displayed in the southern district, it is difficult to believe that we have in this so-called azoic or metamorphic series of the Montgomery and Chester valleys anything else than a repetition of these same crystalline schists which have been described along their southern boundary, representing the Green Mountain and the White Mountain series. We thus avoid the difficulty of supposing that we have in this region two sets of serpentine rocks, and two of mica-schists, lithologically similar, but of widely different ages, — a conclusion highly improbable. It should be said that Rogers, in accordance with the notions then generally received, looked upon serpentine as an eruptive rock, which had altered the adjacent strata, converting the mica-schists into steatitic and chloritic rocks.

This so-called azoic series, according to Rogers, underlies the auroral limestone of Pennsylvania, thus apparently occupying the horizon of the primal palæozoic division. We find, however, in his report on the geology of the State, no satisfactory evidence of the identity of the two series of crystalline rocks. On the contrary, a very different conclusion would seem to follow from certain facts there detailed. The azoic or so-called metamorphic primal strata are said to have a very uniform nearly vertical dip, or with high angles to the southward, while the micaceous and gneissic strata of the northern subdivision of the southern district of so-called hypozoic rocks, limiting these last to the south, present either minute local contortions or wide gentle undulations, with comparatively moderate dips, for the most part to the northward.* From this, I think, we may infer that the nearly vertical strata must be, in truth,

* Rogers, Geology of Pennsylvania, I. pp. 69 - 74, and 154 - 158.

older underlying rocks belonging, not to the palæozoic system, but to our second series of crystalline schists. We conclude, then, that while the gneisses to the northwest, and probably those along the southeast rim of the mesozoic basin of Pennsylvania, are Laurentian, the great valley southward to the Delaware is occupied by the rocks of the Green Mountain and White Mountain series. The same two types of rocks, extending to the northeast, are developed about New York City, in the mica-schists of Manhattan and the serpentines of Staten Island and Hoboken; while in the range of the Highlands, the Laurentian gneiss belt of the South Mountain crosses the Hudson River.

The three series of gneissic rocks which we have distinguished in our section to the northward have, in southeastern New York, as in Pennsylvania, been grouped together in the primary system, and may thence all be traced into western New England. In Dr. Percival's Geological Report and Map of Connecticut, published in 1840, it will be seen that he refers to the gneiss of the Highlands two gneissic areas in Litchfield County; the one occupying parts of Cornwall and Ellsworth, and the other extending from Torrington, northward through Winchester, Norfolk, and Colebrooke into Berkshire County, Massachusetts. Further investigations may confirm the accuracy of Percival's identification, and show the Laurentian age of these New England gneisses, a view which is apparently supported by the mineralogical characters of some of the rocks in this region. Emmons informs us that primary limestones with graphite (perhaps Laurentian) are met with in the Hoosic range in Massachusetts east of the Stockbridge (Taconic) limestones.

The rocks of the second series are traceable from southwestern Connecticut northward to the Green Mountains in Vermont, and the micaceous schists and gneisses of the third, or White Mountain series are found both to the east and the west of the mesozoic valley in Connecticut and Massachusetts. They also occupy a considerable area in eastern Vermont, where they are separated from the White Mountain range by an outcrop of rocks of the second series. To the southeast of the White Mountains, along our line of section, the same mica-schists and gneisses, often with very moderate dips, extend as far as Portland, Maine, where they are interrupted by the outcropping of greenish chloritic and chromiferous schists, in nearly vertical beds, which appear to belong to the second series.

I find that the strata of the second series appear from beneath the Carboniferous at Newport, Rhode Island, in a nearly vertical attitude, and are also seen in the vicinity of Boston and Brighton, Saugus and Lynnfield. Their relations in this region to the gneisses with crystalline limestones of Chelmsford, etc., which I have referred to the Laurentian series,* have yet to be determined.

We have already mentioned that the crystalline rocks of Pennsylvania pass into Maryland and Virginia, where, as H. D. Rogers informs us, they appear in the mountains of the Blue Ridge. It remains to be seen whether the three types which we have pointed out in Pennsylvania are to be recognized in this region. A great belt of crystalline schists extends from Virginia through North and South Carolina, and into eastern Tennessee, where, according to Safford, these rocks underlie the Potsdam. It is easy, from the reports of Lieber on the geology of South Carolina, to recognize in this State the two types of the Green Mountain and White Mountain series. The former, as described by him, consists of talcose, chloritic, and epidotic schists, with diorites, steatites, actinolite-rock, and serpentines. It may be noted that he still adheres to the notion of the eruptive origin of the last three rocks, which the observations of Emmons, Logan, and myself in the Green Mountains have shown to be untenable. These rocks in South Carolina generally dip at very high angles. The great gneissic area of Anderson and Abbeville districts is described by Lieber as consisting of fine-grained gray gneisses with micaceous and hornblendic schists, and is cut by numerous veins of pegmatite, holding garnet, tourmaline, and beryl. These rocks, which

* American Journal of Science (2), XLIX. 75.

have the characters of the White Mountain series, appear, from the incidental observations to be found in Lieber's reports, to belong to a higher group than the chloritic and serpentine series, and to dip at comparatively moderate angles.*

Professor Emmons, whose attention was early turned to the geology of western New England, did not distinguish between the three types which we have defined, but, like Rogers in Pennsylvania, included all the crystalline rocks of that region in the primary system. It is to him, however, that we owe the first correct notions of the geological nature and relations of the Green Mountains. These, he has remarked, are often made to include two ranges of hills belonging to different geological series. The eastern range, including the Hoosic Mountain in Massachusetts and Mount Mansfield in Vermont, he referred to the primary; which he described as including gneiss, mica-schist, talcose slate, and hornblende, with beds and veins of granite, limestone, serpentine, and trap. He declared, moreover, that there is no clear line of demarcation among the various schistose primary rocks, and cited, as an illustration, the passage into each other of serpentine, steatite, and talcose schist. His description of the crystalline rocks of this range will be recognized as comprehensive and truthful.

[* My own observations have since shown me that the rocks of the White Mountain series are largely displayed, and rarely at high angles, in the Blue Ridge in Carroll County, Virginia, thence southwestward at least as far as Ashe County, North Carolina, and again in Polk County, Tennessee. The lithological study in these regions is rendered difficult by the fact that they are covered, often to a depth of a hundred feet or more, by the undisturbed products of their own decomposition, the protoxide bases having been removed by solution from the feldspar and the hornblende, and the whole rock, with the exception of the quartzose layers, reduced to a clayey mass, still, however, showing the inclined planes of stratification. The immense veins of pyritous copper-ores, which these rocks enclose (*ante*, page 217), have in like manner been changed, to as great depths, into hydrous peroxide of iron. I have already alluded to the significance, both chemical and geological, of this decomposition, and to its great antiquity (*ante*, page 10). The observations of C. A. White, in the northwest, show that such a decomposition of the Eozoic gneisses was anterior to the cretaceous period, while in Missouri, it appears from the studies of R. Pumpelly, confirmed by my own observations, that the quartziferous porphyries with which the iron-ores of that region occur, were thus decomposed before the deposition of the Cambrian sandstones.]

To the west of the hills of primary schist, he placed his Taconic system, named from the Taconic hills, which run from north to south along the boundary line of New York and Massachusetts, and form a range parallel with the Green Mountains. The lower portions of the Taconic system, according to Emmons, are schistose rocks made up from the ruins of the primary schists which lie to the east of them. Thus the talcose schists of Berkshire are said to be regenerated rocks, belonging to the newer system, but showing the color and texture of the older talcose schists from which they were formed. How far this is true of these particular strata may be a question, for there is reason to believe that Emmons included among his Taconic rocks some beds belonging to the older crystalline series of the Green Mountains; yet it is not less true that the possibility of derived rocks of this kind is one which has been too much overlooked by geologists.* Emmons elsewhere remarks that, while the talcose slates of the primary are associated with steatite and with hornblende, these are never found in the Taconic rocks, and also, that epidote, actinolite, titanium (rutile), etc., which are characteristic minerals of the primary, are wanting in the Taconic system.

The statements of Emmons on this point were sufficiently explicit; he included in the primary system all of the crystalline schists of the Green Mountains, except certain talcose and micaceous beds, which he supposed to be made up of the ruins of the similar strata in the primary, and to constitute, with a great mass of other rocks, the Taconic system; which was, in its turn, unconformably overlaid by the Potsdam sandstone and Calciferous sand-rock of the New York system. His views have, however, been misunderstood by more than one of his critics; thus, Mr. Marcou, while defending the Taconic system, makes it to include the three groups just mentioned, namely, 1. The Green Mountain gneiss; 2. The Taconic strata as defined by Emmons; and, 3. The Potsdam sandstone; † thus

* Some observations on this point will be found in Essay XIV.

† Proceedings of Boston Society of Natural History, November 6, 1861, and American Journal of Science (2), XXXIII. 282.

uniting in one system the crystalline schists and the overlying uncrystalline fossiliferous sediments, in direct opposition to the plainly expressed teachings of Emmons, as laid down in his report on the geology of the Northern District of New York, and later, in 1846,* in his memoir on the Taconic System.

In the geological survey of the State of New York, the rocks of the Champlain division (including the strata from the base of the Potsdam sandstone to the summit of the Loraine or Hudson River shales) had, by his colleagues, been looked upon as the lowest of the palæozoic system. Professor Emmons, however, was led to regard the very dissimilar strata of the Taconic hills as constituting a distinct and more ancient series. A similar view had been held by Eaton, who placed, as we have already seen, above the crystalline schists of the Green Mountains, his primary quartzose and calcareous formations, followed to the westward by transition argillites and sandstones, which latter appear to have corresponded to the Potsdam sandstone of New York. Emmons, however, gave a greater form and consisteney to this view, and endeavored to sustain it by the evidence of fossils, as well as by structure. The Taconic system, as defined by him, may be briefly described as a series of uncrystalline fossiliferous sediments reposing unconformably on the crystalline schists of the Green Mountains, and partly made up of their ruins ; while it is, at the same time, overlaid unconformably by the Potsdam and Calciferous formations of the Champlain division, and constitutes the true base of the palæozoic column.

Although he claimed to have traced this Taconic system throughout the Appalachian chain from Maine to North Carolina, it is along the confines of Massachusetts and New York that its development was most minutely studied. He separated it into a lower and an upper division, and estimated its total thickness at not less than thirty thousand feet, consisting, in the order of deposition, of the following members : 1. Granu-

* Loc. cit. p. 130, and Agriculture of New York, I. 53. This formed a part of the report by Emmons on the Agriculture of New York, but was also published separately.

lar quartz; 2. Stockbridge limestone; 3. Magnesian slate; 4. Sparry limestone; 5. Roofing-slate, graptolitic; 6. Silicious conglomerate; 7. Taconic slate; 8. Black slate. The apparent order of superposition differs from this, and it was conceived by Professor Emmons that during the accumulation of these Taconic rocks, the Green Mountain gneiss, which formed the eastern border of the basin, was gradually elevated so as to bring successively the older members above the ocean from which the sediments were being deposited. From this it resulted that the upper members of the system, such as the black slates, were confined to a very narrow belt, and never extended far eastward; although he admits that denudation may have removed large portions of these upper beds. At a subsequent period, a series of parallel faults, with upthrows on the eastern side, is supposed to have broken the strata, given them an eastward dip, and caused the newer beds to pass successively beneath the older ones, thus producing an *apparently inverted succession,* and making their present seeming order of superposition completely deceptive. In speaking of this supposed arrangement of the members of his Taconic system, Emmons alluded to them as "inverted strata"; while by Mr. Marcou, the strata were said to be "overturned on each side of the crystalline and eruptive rocks which occupy the centre of the chain, producing thus a fan-shaped structure," etc.* I have elsewhere shown that this notion, though to some extent countenanced by his vague and inaccurate use of terms, was never entertained by Emmons, whose own view, as defined in his Taconic System (p. 17),† is that just explained.

* Comptes Rendus de l'Académie, LIII. 804.

† See my further discussion of the matter, American Journal of Science (2), XXXII. 427; XXXIII. 135, 281. It is by an oversight that I have, in the latter volume (page 136), represented Barrande as sharing the misconception of Marcou, although his language, without careful scrutiny, would lead us to such a conclusion. In fact, in the Bull. Soc. Geol. de France ((2), XVIII. 261), in an elaborate study of the Taconic question, Barrande heads a section thus: "*Renversement conçu pour tout un système,*" and then proceeds to show that the *renversement* or *overturn* is only apparent, by explaining, in the language of Emmons, the view already set forth above.

The view of Emmons, that there exists at the western base of the Green Mountains an older fossiliferous series, underlying the Potsdam, met with general opposition from American geologists. In May, 1844, H. D. Rogers, in his address as president, before the American Association of Geologists, then met at Washington, criticised this view at length, and referred to a section from Stockbridge, Massachusetts, to the Hudson River, made by W. B. Rogers and himself, and by them laid before the American Philosophical Society in January, 1841. They then maintained that the quartz-rock of the Hoosic range was Potsdam, the Berkshire marble identical with the blue limestone of the Hudson valley, and the associated micaceous and talcose schists altered strata of the age of the slates at the base of the Appalachian system; that is to say, primal in the nomenclature of the Pennsylvania survey.

In 1843 Mather had asserted the Champlain age of the same crystalline rocks, and claimed that the whole of the division was there represented, including the Potsdam, the Hudson River group, and the intermediate limestones.* The conclusion of Mather was cited with approbation by Rogers, who apparently adopted it, and declared that Hitchcock held a similar view. It will be seen that these geologists thus united in one group the schists of the Hoosic range (regarded by Emmons as primary) with those of the Taconic range, and referred both to the age of the Champlain division, the whole of which was supposed to be included in the group.

In the same address Professor Rogers raised a very important question. Having referred to the Potsdam sandstone, which on Lake Champlain forms the base of the palæozoic system, he inquires, "Is this formation, then, the lowest limit of our Appalachian masses generally, or is the system expanded downward in other districts by the introduction beneath it of other conformable sedimentary rocks?" He then proceeded to state that from the Susquehanna River, southwestward, a more complex series appears at the base of the lower limestone than to the north of the Schuylkill, and in some parts of the Blue

* Geology of the Southern District of New York, p. 438.

Ridge he includes in the primal division (beneath the Calciferous sand-rock) "at least four independent and often very thick deposits, constituting one general group, in which the Potsdam or white sandstone (with Scolithus) is the second in descending order." This sandstone is overlaid by many hundred feet of arenaceous and ferriferous fucoidal slate, and underlaid by coarse sandy shales and flagstones; below which, in Virginia and East Tennessee, is a series of heterogeneous conglomerates, which rest on a great mass of crystalline strata. The accuracy of these statements is confirmed by Safford, who, in his report on the geology of Tennessee (1869), places at the base of the column a great series of crystalline schists, apparently representatives of those of southeastern Pennsylvania. (*Ante*, page 245.) Upon these repose what Safford designates as the Potsdam group, including, in ascending order, the Ocoee slates and conglomerates, estimated at 10,000 feet, and the Chilhowee shales and sandstones, 2,000 feet or more, with fucoids, worm-burrows, and Scolithus. These are conformably overlaid by the Knoxville division, consisting of fucoidal sandstones, shales, and limestones, the latter two holding fossils of the age of the Calciferous sand-rock. It is noteworthy that these rocks are greatly disturbed by faults, and that in Chilhowee Mountain the lower conglomerates are brought on the east against the Carboniferous limestone, by a vertical displacement of at least 12,000 feet. The general dip of all these strata, including the basal crystalline schists, is to the southeast.

The primal palæozoic rocks of the Blue Ridge were then by Rogers, as now by Safford, looked upon as wholly of Potsdam age, including the Scolithus sandstone as a subordinate member, so that the strata beneath this were still regarded as belonging to the New York system. Hence, while Rogers inquires whether the Taconic system "may not along the western border of Vermont and Massachusetts include also some of the sandy and slaty strata here spoken of as lying beneath the Potsdam sandstone,"* he would still embrace these lower strata in the Champlain division.

* American Journal of Science (1), XLVII. 152, 153.

Thus we see that at an early period the rocks of the Taconic system were, by Rogers and Mather, referred to the Champlain division of the New York system, a conclusion which has been sustained by subsequent observations. Before discussing these, and their somewhat involved history, we may state two questions which present themselves in connection with this solution of the problem. First, whether the Taconic system, as defined by Emmons, includes the whole or a part of the Champlain division; and, second, whether it embraces any strata older or newer than the members of this portion of the New York system. With reference to the first question it is to be remarked, that in their attempts to compare the Taconic rocks with those of the Champlain division as seen farther to the west, observers were led by lithological similarities to identify the upper members of the latter with certain portions of the Taconic. In fact, the Trenton limestone, with the Utica slates and the Loraine or Hudson River shales, making together the upper half of the Champlain division (in which Emmons, moreover, included the overlying Oneida and Medina conglomerates and sandstones), have in New York an aggregate thickness of not less than three or four thousand feet, and offer many lithological resemblances to the great mass of sediments at the western base of the Green Mountains, to which the name of Taconic had been applied. It is curious to find that Emmons, in 1842, referred to the Medina the Red sand-rock of the east shore of Lake Champlain, since shown to be Potsdam; and, moreover, placed the Sillery sandstone of the neighborhood of Quebec at the summit of the Champlain division, as the representative of the Oneida conglomerate; while at the same time he noticed the great resemblance which this sandstone, with its adjacent limestones, bore to similar rocks on the confines of Massachusetts, already referred by him to the Taconic system.*

This view of Emmons as to the Quebec rocks was adopted by Sir William Logan, when, a few years afterwards, he began to study the geology of that region. The sandstone of Sillery

* Geology of the Northern District of New York, pp. 124, 125.

was described by him as corresponding to the Oneida or Shawangunk conglomerate, while the limestones and shales of the vicinity, which were supposed to underlie it, were regarded as the representatives of the Trenton, Utica, and Hudson River formations.* By following these rocks along the western base of the Appalachians into Vermont and Massachusetts, they were found to be a continuation of the Taconic system, which Sir William was thus led to refer to the upper half of the Champlain division, as had already been done by Professor Adams in 1847.† As regards the crystalline strata of the Appalachians in this region, he, however, rejected the view of Emmons, and maintained that put forward by the Messrs. Rogers in 1841; namely, that these, instead of being older rocks, were but these same upper formations of the Champlain division in an altered condition; a view which was maintained during several years in all of the publications of those connected with the geological survey of Canada.

This conclusion, so far as regards the age of the unaltered fossiliferous rocks from Quebec to Massachusetts, was supposed to be confirmed by the evidence of organic remains found in them in Vermont. Mr. Emmons had described, as characteristic of the upper part of the Taconic system, two crustaceans, to which he gave the names of *Atops trilineatus* and *Elliptocephalus asaphoides*; the other fossils noticed by him being graptolites, fucoids, and what were apparently the marks of annelids. In 1847 Professor James Hall, in the first volume of his Paleontology, declared the Atops of Emmons to be identical with *Triarthrus* (*Calymene*) *Beckii*, a characteristic fossil of the Utica slate; while the Elliptocephalus was referred by him to the genus *Olenus*, now known to belong to the primordial fauna of Sweden, where it is found in slates lying beneath the orthoceratite limestone, and near the base of the palæozoic series. Although, as it now appears, the geological horizon of the Olenus slates was well known to Hisinger,

* Geological Survey of Canada, 1847-48, pp. 27, 57; and American Journal of Science (2), IX. 12.

† American Journal of Science (2), V. 108.

this author in his classic work, Lethæa Suecica, published in 1837, represents, by some unexplained error, these slates as overlying the orthoceratite limestone, which is the equivalent of the Trenton limestone of the Champlain division. Hence, as Mr. Barrande has remarked, Hall was justified by the authority of Hisinger's published work in assigning to the Olenus slates of Vermont a position above that limestone, and in placing them, as he then did, on the horizon of the Hudson River or Loraine shales. The double evidence afforded by these two fossil forms in the rocks of Vermont served to confirm Sir William Logan in placing in the upper part of the Champlain division the rocks which he regarded as their stratigraphical equivalents near Quebec; and which, as we have seen, had some years before been by Emmons himself assigned to the same horizon. The remarkable compound graptolites which occur in the shales of Pointe Levis, opposite Quebec, were described by Professor James Hall in the report of the Geological Survey of Canada for 1857, and were then referred to the Hudson River group; nor was it until August, 1860, that Mr. Billings described from the limestones of this same series at Pointe Levis a number of trilobites, among which were several species of Agnostus, Dikelocephalus, Bathyurus, etc., constituting a fauna whose geological horizon he decided to be in the lower part of the Champlain division.

Just previous to this time, in the report of the Regents of the University of New York for 1859, Professor Hall had described and figured by the name of Olenus two species of trilobites from the slates of Georgia, Vermont, which Emmons had wrongly referred to the genus Paradoxides. They were at once recognized by Barrande, who called attention to their primordial character, and thus led to a knowledge of their true stratigraphical horizon, and to the detection of the singular error in Hisinger's book, already noticed, by which American geologists had been misled.* They have since been separated from Olenus, and by Professor Hall referred to a new and

* For the correspondence on this matter between Barrande, Logan, and Hall, see American Journal of Science (2), XXXI. 210-226.

closely related genus, which he has named Olenellus, and which is now regarded as belonging to the horizon of the Potsdam sandstone, to which we shall presently advert.

Further studies of the fossiliferous rocks near Quebec showed the existence of a mass of sediments estimated at about 1,200 feet, holding a numerous fauna, and corresponding to a great development of strata about the age of the Calciferous and Chazy formations, or, more exactly, to a formation occupying a position between these two, and constituting, as it were, beds of passage between them. In this new formation were included the graptolites already described by Hall, and the numerous crustacea and brachiopoda described by Billings, all of which belong to the Levis slates and limestones. To these and their associated rocks Sir William Logan then gave the name of the Quebec group, including, besides the fossiliferous Levis formation, a great mass of overlying slates, sandstones, and magnesian limestones, hitherto without fossils, which have been named the Lauzon rocks, and the Sillery sandstones and shales, which he supposed to form the summit of the group, and which had afforded only an Obolella and two species of Lingula; * the volume of the whole group being about 7,000 feet.

The paleontological evidence thus obtained by Billings and by Hall, both from near Quebec and in Vermont, led to the conclusion that the strata of these regions, so much resembling the upper members of the Champlain division, were really a great development, in a modified form, of some of its lower portions. Their apparent stratigraphical relations were explained by Logan by the supposition of "an overturned anticlinal fold, with a crack and a great dislocation running along the summit, by which the Quebec group is brought to overlie the Hudson River group. Sometimes it may overlie the overturned Utica formation, and in Vermont points of the overturned Trenton appear occasionally to emerge from beneath the overlap." He, at the same time, declared that "from the physical structure alone, no person would suspect the break that must exist in

* See Billings, Palæozoic Fossils of Canada, p. 69.

the neighborhood of Quebec, and, without the evidence of fossils, every one would be authorized to deny it." *

The rocks from western Vermont, which had furnished to Hall the species of Olenellus, have long been known as the Red sand-rock, and, as we have seen, were by Emmons, in 1842, referred to the age of the Medina sandstone, — a view which the late Professor Adams still maintained as late as 1847.† In the mean time Emmons had, in 1855, declared this rock to represent the Calciferous and Potsdam formations, the brown sandstones of Burlington and Charlotte, Vermont, being referred to the latter.‡ This conclusion was confirmed by Billings, who, in 1861, after visiting the region and examining the organic remains of the Red sand-rock, assigned to it a position near the horizon of the Potsdam.§ Certain trilobites found in this Red sand-rock by Adams, in 1847, were by Hall recognized as belonging to the European genus *Conocephalus* (= *Conocephalites* and *Conocoryphe*), whose geological horizon was then undetermined.|| The formation in question consists in great part of a red or mottled granular dolomite, associated with beds of fucoidal sandstone, conglomerates, and slates. These rocks were carefully examined by Logan in Swanton, Vermont, where, according to him, they have a thickness of 2,200 feet, and include toward their base a mass of dark-colored shales holding Olenellus with Conocephalites, Obolella, etc.; *Conocephalites Teucer*, Billings, being common to the shales and the red sandy beds.¶ Many of these fossils are also found at Troy and at Bald Mountain, New York, where they accompany the Atops of Emmons, now recognized by Billings as a species of Conocephalites.

* Logan's letter to Barrande, American Journal of Science (2), XXXI. 218. The true date of this letter was December 31, 1860, but, by a misprint, it is made 1831.

† Adams, American Journal of Science (2), V. 108.

‡ Emmons, American Geology, II. 128.

§ American Journal of Science (2), XXXII. 232.

|| Ibid. (2), XXXIII. 374.

¶ Geology of Canada, 1863, p. 281; American Journal of Science (2), XLVI. 224.

A similar condition of things extends northeastward along the Appalachian region. On the south side of the St. Lawrence below Quebec a great thickness of limestones, sandstones, and slates, formerly referred to the Quebec group, is now regarded by Billings as, in part at least, of the Potsdam formation; while on the coast of Labrador and in northern Newfoundland the same formation, characterized by the same fossils as in Vermont, is largely developed, attaining in some parts, according to Murray, a thickness of 3,000 feet or more. Along the northern coast of the island it is nearly horizontal, and appears to be conformably overlaid by about 4,000 feet of fossiliferous strata representing the Calciferous sand-rock and the succeeding Levis formation.

Mr. Billings has described a section from the Laurentian of Crown Point, New York, to Cornwall, Vermont, from which it appears that to the eastward of a dislocation which brings up the Potsdam to overlie the higher members of the Champlain division, the Potsdam is itself overlaid, at a small angle, by a great mass of limestones representing the Calciferous, and having at the summit some of the characteristic fossils of the Levis formation. Next in ascending order are not less than 2,000 feet of limestones with Trenton fossils (embracing probably the Chazy division), while to the east of this the Levis again appears, including the white Stockbridge limestones.* We have here an evidence that the augmentation in volume observed in the lower members of the Champlain division in the Appalachian region extends to the Trenton, which to the west of Lake Champlain is represented, the Chazy included, by not more than 500 feet of limestone. The Potsdam, in the latter region, consists of from 500 to 700 feet of sandstone holding Conocephalites and Lingulella, and overlaid by 300 feet of magnesian limestone, the so-called Calciferous sand-rock. In the valley of the Mississippi these two formations in Iowa, Missouri, and Texas are represented by from 800 to 1,300 feet of sandstones and magnesian limestones; while in the Black Hills

* T. S. Hunt on the Geology of Vermont, American Journal of Science (2), XLVI. 227.

of Nebraska, according to Hayden, the only representative of these lower formations is about one hundred feet of sandstone holding Potsdam fossils.*

In striking contrast to this, it has been shown that along the Appalachian range from Newfoundland to Tennessee these lower formations are represented by from 8,000 to 15,000 feet of fossiliferous sediments. It has been suggested by Logan that these widely differing conditions represent deep-sea accumulations on the one hand, and the deposits from a shallow sea which covered a submerged continental plateau on the other; the sediments in the two areas being characterized by a similar fauna, though differing greatly in lithological characters and in thickness. To this we may add, that the continental area, being probably submerged and elevated at intervals, became overlaid with beds which represent only in a partial and imperfect manner the great succession of strata which were being accumulated in the adjacent ocean.†

In a paper which I hope to present to the geological section during the present meeting of the Association, it will be shown, from a study of the rocks of the Ottawa basin, that the typical Champlain division not only presents important paleontological breaks, but evidences of stratigraphical discordance at more

* American Journal of Science (2), XXV. 439; XXXI. 234. [Later observations show great variations in the thickness of these lower rocks in the West. In the Wahsatch Mountains are found, according to Bradley, from 1,500 to 2,000 feet of sandstones and conglomerates, regarded as Potsdam, overlaid by 3,000 feet of magnesian limestones and shales, holding fossils of the Levis, and, towards the summit, of Niagara and probably of Lower Helderberg age; the whole followed by 2,000 feet of Devonian sandstones and 3,000 feet of Carboniferous limestones. In the Teton Mountains, however, according to the same observer, this great thickness of Potsdam and Levis rocks is represented by only 700 feet of quartzites and limestones, overlaid by about 600 feet of magnesian limestones, probably of Niagara age, followed by 2,000 feet of Carboniferous limestones. In the Wind River Mountains, in western Wyoming, Professor Comstock has described a remarkable series, including Potsdam and Levis, followed by strata of Oriskany age, Carboniferous limestones, Triassic, Jurassic, and Cretaceous rocks, all apparently conformable, and resting at an angle of about 20° on the crystalline Eozoic rocks. Remains of the fauna of the Trenton period (Upper Cambrian) have moreover very recently been made known to us from the West.]

† Ibid. (2), XLVI. 225.

than one horizon over the continental area, which, as the result of widely spread movements, might be supposed to be represented in the Appalachian region. In the latter Logan has already observed that the absence of all but the highest beds of the Levis along the eastern limit of the Potsdam, near Swanton, Vermont (while the whole thickness of them appears a little farther westward), makes it probable that there is a want of conformity between the two; and I have in this connection insisted upon the entire absence, in this locality, of the Calciferous, which is met with a little farther south in the section just mentioned, as another evidence of the same unconformity.* There are also, I think, reasons for suspecting another stratigraphical break at the summit of the Quebec group,† in which case many problems in the geological structure of this region will be much simplified.

It should be remembered that the conditions of deposition in some areas have been such that accumulations of strata, corresponding to long geologic periods, and elsewhere marked by stratigraphical breaks, are arranged in conformable superposition; and moreover that movements of elevation and depression have even caused great paleontological breaks, which over considerable areas are not marked by any apparent discordance. Thus the remarkable break in the fauna between the Calciferous and the Chazy is not accompanied by any noticeable discordance in the Ottawa basin; and in Nebraska, according to Hayden, the Potsdam, Carboniferous, Jurassic, and Cretaceous formations are all represented in about 1,200 feet of conformable strata. ‡ In Sweden the whole series from the base of the Cambrian to the summit of the Silurian appears as a conformable sequence, while in North Wales, although there is no apparent discordance from the base of the Cambrian to the summit of the Lingula flags, stratigraphical breaks, according to Ramsay, probably occur both at the base and the summit

* American Journal of Science (2), XLVI. 225.

† See, for the evidence of this, Essay XV., Part Third.

‡ American Journal of Science (2), XXV. 440.

of the Tremadoc slates,* which are considered equivalent to the Levis formation.

We have seen that, according to Logan, a dislocation a little to the north of Lake Champlain causes the Quebec group to overlie the higher members of the Champlain division. The same uplift, according to him, brings up, farther south, the Red sand-rock of Vermont, which to the west of the dislocation rests upon the upturned and inverted strata of various formations from the Calciferous sand-rock to the Utica and Hudson River shales. These latter, according to him, are seen to pass for considerable distances beneath nearly horizontal layers of the Red sand-rock, the Utica slate, in one case, holding its characteristic fossil, *Triarthrus Beckii.* This relation, which is well shown in a section at St. Albans, figured by Hitchcock,† was looked upon by Emmons and by Adams as evidence that the Red sand-rock was the representative of the Medina sandstone of the New York system. When, however, the former had recognized the Potsdam age of the sand-rock, with its Olenellus, which he supposed to be Paradoxides, this condition of things was conceived to be an evidence of the existence beneath the Potsdam of an older and unconformable fossiliferous series already mentioned.

The objections made by Emmons to Rogers's view of the Champlain age of the Taconic rocks were threefold: first, the great differences in lithological characters, succession, and thickness between these and the rocks of the Champlain division as previously known in New York; second, the supposed unconformable infraposition of a fossiliferous series to the Potsdam; and, third, the distinct fauna which the Taconic rocks were supposed to contain. The first of these is met by the fact, now established, that, in the Appalachian region, the Champlain division is represented by rocks having, with the same organic remains, very different lithological characters, and a thickness tenfold greater than in the typical Champlain region of northern New York. The second objection has already

* Quar. Geol. Journal, XIX. p. 36.

† Geology of Vermont, p. 374.

been answered by showing that the rocks which, as in the St. Albans section, pass beneath the Potsdam are really newer strata belonging to the upper part of the division, and contain a characteristic fossil of the Utica slate. As to the third point, it has also been met, so far as regards the Atops and Elliptocephalus, by showing these two genera to belong to the Potsdam formation. If we inquire further into the Taconic fauna, we find that the Stockbridge limestone (the Eolian limestone of Hitchcock), which was placed by Emmons near the base of the Lower Taconic (while the Olenellus slates are near the summit of the Upper Taconic), is also fossiliferous, and contains, according to the determinations of Professor Hall, species belonging to the genera Euomphalus, Zaphrentis, Stromatopora, Chaetetes, and Stictopora.* Such a fauna would lead to the conclusion that these limestones, instead of being older, were really newer than the Olenellus beds, and that the apparent order of succession was, contrary to the supposition of Emmons, the true one. This conclusion was still further confirmed by the evidence obtained in 1868 by Mr. Billings, who found in that region a great number of characteristic species of the Levis formation, many of them in beds immediately above or below the white marbles,† which latter, from the recent observations of the Rev. Augustus Wing, in the vicinity of Rutland, Vermont, would seem to be among the upper beds of the Potsdam formation. Thus while some of the Taconic fossils belong to the Potsdam and Utica formations, the greater number of them, derived from beds supposed to be low down in the system, are shown to be of the age of the Levis formation. There is, therefore, at present, no evidence of the existence, among the unaltered sedimentary rocks of the western base of the Appalachians in Canada or New England, of any strata more ancient than those of the Champlain division,‡ to which, from

* Geology of Vermont, 419; and American Journal of Science (2), XXXIII. 419.

† American Journal of Science (2), XLVI. 227.

‡ See, on this point and on the possibly greater antiquity of the rocks called Potsdam, Essay XV., Part Third.

their organic remains, the fossiliferous Taconic rocks are shown to belong.

Mr. Billings has, it is true, distinguished provisionally what he has designated an upper and a lower division of the Potsdam, and has referred to the latter the Red sand-rock with the Olenellus slates of Vermont, together with beds holding similar fossils at Troy, New York, and along the Strait of Bellisle in Labrador and Newfoundland; the upper division of the Potsdam being represented by the basal sandstones of the Ottawa basin and of the Mississippi valley.* In the present state of our knowledge of the local variations in sediments and in their fauna dependent on depth, temperature, and ocean currents, Billings, however, conceives that it would be premature to assert that these two types of the Potsdam do not represent synchronous deposits.

The base of the Champlain division, as known in the Potsdam formation of New York, of the Mississippi valley, and the Appalachian belt, does not, however, represent the base of the palæozoic series in Europe. The Alum slates in Sweden are divided into two parts, an upper or Olenus zone, and a lower or Conocoryphe zone, as distinguished by Angelin. The latter is characterized by the genus Paradoxides, which also occupies a lower division in the primordial palæozoic rocks of Bohemia (Barrande's stage C), the greater part of which are regarded as the equivalent of the Olenus zone of Sweden and the Potsdam of North America. The Lingula flags of Wales belong to the same horizon, and it is at their base, in strata once referred to the Lower Lingula flags, that the Paradoxides is met with. These strata, for which Hicks and Salter, in 1865, proposed the name of the Menevian group, are regarded as corresponding to the lower division of the Alum slates, and, like it, contain a fauna not yet recognized in the basal rocks of the New York system. [Beneath the Menevian lie the Llanberis and Harlech rocks (the Longmynd), which constitute the Lower Cambrian of Sedgwick; while above it are the great mass of the Lingula flags and the Tremadoc rocks, his Middle

* Report Geol. of Canada, 1863 – 66, p. 236.

Cambrian. To these succeed the Bala or Upper Cambrian, the equivalent of the Llandeilo and Caradoc rocks, to which Murchison gave the name of Lower Silurian. He at first claimed the Llandeilo as the base of his Silurian system, but subsequently endeavored to extend it downwards so as to include, under the name of Primordial Silurian, the Middle Cambrian of Sedgwick. To this Lyell objected, and while conceding to Murchison the Upper Cambrian as Lower Silurian, gave to the middle division of Sedgwick's series the name of Upper Cambrian. Hicks in a recent paper (1873) has adopted a similar compromise, including, however, in the Lower Silurian the Arenig group, and making the Tremadoc the upper member of the Upper Cambrian. For a discussion of the relations of Cambrian and Silurian the reader is referred to Essay XV. in this volume.] The same classification is now adopted by Linarsson, in Sweden, where, in Westrogothia, the Cambrian rocks (resting unconformably on the crystalline schists to be noticed further on) are overlaid conformably by the orthoceratite-limestones, which are by him regarded as forming the base of the Silurian, and as the equivalent of the Llandeilo rocks of Wales. The total thickness of these lower rocks in Sweden, including the representatives of the Lingula flags, the Menevian beds, and an underlying fucoidal (Eophyton) sandstone, is only three hundred feet, while the first two divisions in Wales have a thickness of five to six thousand, and the Harlech grits and Llanberis slates (including the Welsh roofing-slates beneath) amount to eight thousand feet additional. Recent researches show that these lower rocks in Wales contain an abundant fauna, extending downward some 2,800 feet from the Menevian to the very base of strata regarded as the representatives of the Harlech grits. The brachiopoda of the Harlech beds appear identical with those of the Menevian, but new species of *Conocephalites*, *Microdiscus*, and *Paradoxides* are met with, besides a new genus, *Plutonia*, allied to the last mentioned.* [The Upper

* Hicks, Geol. Mag., V. 306; and Rep. Brit. Assoc., 1868, p. 69; also Harkness and Hicks in Nature, Proc. Geol. Soc., May 10, 1871.

Cambrian, as defined by Sedgwick, is represented in North America by the upper portion of the Champlain division of New York, from the top of the Chazy, while the Middle and Lower Cambrian have their equivalents in the Quebec group, the Chazy, Calciferous, and Potsdam, and in the strata holding Paradoxides and other primordial forms in Massachusetts, New Brunswick, and Newfoundland. The precise relation of these to the Potsdam formation of New York is yet to be determined, as well as the question whether there exists in the Appalachians any palæozoic rocks belonging to a lower horizon than the Potsdam. For a further discussion of these questions the reader is referred to Essay XV. in the present volume.]

In May, 1861, I called attention to the fact that beds of quartzose conglomerate at the base of the Potsdam in Hemmingford, near the outlet of Lake Champlain, on its western side, contain fragments of green and black slates, "showing the existence of argillaceous slates before the deposition of the Potsdam sandstone." * The more ancient strata, which furnished these slaty fragments to the Potsdam conglomerate, have perhaps been destroyed, or are concealed, but they or their equivalents may yet be discovered in some part of the great Appalachian region. They should not, however, be called Taconic, but receive the prior designation of Cambrian, unless, indeed, it shall appear that the source of these slate fragments was the more argillaceous beds of the still older Huronian schists. Emmons regarded his Taconic system as the equivalent of the Lower (and Middle) Cambrian of Sedgwick; but when, in 1842, Murchison announced that the name of Cambrian had ceased to have any zoölogical significance, being identical with Lower Silurian,† Emmons, conceiving, as he tells us, that all Cambrian rocks were not Silurian, instead of maintaining Sedgwick's name which, with the progress of paleontological study, is assuming a great zoölogical importance, devised the name of Taconic, as synonymous with the Lower (and Middle) Cambrian of Sedgwick.‡

* American Journal of Science (2), XXXI. 404.

† Proc. Geol. Soc. London, III. 642.

‡ Emmons, Geol. N. District of New York, 162; and Agric. of New York, I. 49.

The crystalline strata to which the name of the Huronian series has been given by the Geological Survey of Canada, have sometimes been called Cambrian from their resemblance to certain crystalline rocks in Anglesea, which have been imagined to be altered Cambrian. The typical Cambrian rocks of Wales, down to their base, are, however, uncrystalline sediments, and, as pointed out by Dr. Bigsby in 1863,* are not to be confounded with the Huronian, which he regarded as equivalent to the second division of the so-called azoic rocks of Norway, the *Urschiefer* or primitive schists, which in that country rest unconformably on the primitive gneiss (*Urgneiss*), and are in their turn overlaid unconformably by the fossiliferous Cambrian strata. This second or intermediate series in Norway is characterized by eurites, micaceous, chloritic, and hornblendic schists, with diorites, steatite, and dark-colored serpentines, generally associated with chrome; and abounds in ores of copper, nickel, and iron. In its mineralogical and lithological characters, the Urschiefer corresponds with what we have designated the second series of crystalline schists. It is, in Norway, divided into a lower or quartzose division, marked by a predominance of quartzites, conglomerates and more massive rocks, and an upper and more schistose division. Macfarlane, who was familiar with the rocks of Norway, after examining both the Huronian of Lake Superior and the crystalline strata of the Green Mountains, had already, in 1862, declared his opinion that both of these were representatives of the Norwegian Urschiefer,† thus anticipating, from his comparative studies, the conclusions of Bigsby.

The crystalline rocks of Anglesea and the adjacent part of Caernarvon, which have been described and mapped by the British Geological Survey as altered Cambrian, are directly overlaid by strata of the Llandeilo or Upper Cambrian division, corresponding to the Trenton and Hudson River formations. If we consult Ramsay's report on the region, it will be found that he speaks of the lower rocks as "probably Cambrian,"

* Quar. Jour. Geol. Soc., XIX. 36.
† Canadian Naturalist, VII. 125.

and states as a reason for that opinion, that they are connected by certain beds of intermediate lithological characters with strata of undoubted Cambrian age.* These, however, as he admits, present great local variations, and, after carefully scanning the whole of the evidence adduced, I am inclined to see in it nothing more than the existence, in this region, of Cambrian strata made up from the ruins from the great mass of pre-Cambrian schists, which are the crystalline rocks of Anglesea. Such a phenomenon is repeated in numerous instances in our North American rocks, and is the true explanation of many supposed examples of passage from crystalline schists to uncrystalline sediments. The Anglesea rocks are a highly inclined and much contorted series of quartzose, micaceous, chloritic, and epidotic schists, with diorites and dark-colored chromiferous serpentines, all of which, after a careful examination of them in the collections of the Geological Survey of Great Britain, I consider identical with the rocks of the Green Mountain or Huronian series. A similar view of their age is shared by Phillips and by Sedgwick, in opposition to the opinion of the British survey. The former asserts that the crystalline schists of Anglesea are "below all the Cambrian rocks"; † while Sedgwick expresses the opinion that they are of "a distinct epoch from the other rocks of the district, and evidently older." ‡

Associated with the fossiliferous Devonian rocks of the Rhine is a series of crystalline schists, similar to those just noticed, seen in the Taunus, the Hundsrück, and the Ardennes. These, in opposition to Dumont, who regarded them as belonging to an older system, are declared by Römer to have resulted from a subsequent alteration of a portion of the Devonian sediments. §

Turning now to the Highlands of Scotland, we have a similar series of crystalline schists, presenting all the mineralogical

* Geol. of North Wales, pp. 145, 175.

† Manual of Geology (1855), 89.

‡ Geol. Journal for 1845, 449.

§ Naumann, Geognosie, 2d edition, II. 383.

characters of those of Norway and of Anglesea, which, according to Murchison and Giekie, are younger than the fossiliferous limestones of the western coast (about the horizon of the Levis formation of the Quebec group), which seem to pass beneath them. Professor Nicol, on the contrary, maintains that this apparent superposition is due to uplifts, and that these crystalline schists are really older than the lowest Cambrians, which appear to the west of them as uncrystalline sediments resting on the Laurentian. He does not, however, confound these crystalline schists of the Scottish Highlands with the Laurentian, from which they differ mineralogically, but regards them as a distinct series.* In the presence of the differences of opinion which have been shown in this controversy, we may be permitted to ask whether, in such a case, stratigraphical evidence alone is to be relied upon. Repeated examples have shown that the most skilful stratigraphists may be misled in studying the structure of a disturbed region where there are no organic remains to guide them, or where unexpected faults and overslides may deceive even the most sagacious. I am convinced that in the study of the crystalline schists, the persistence of certain mineral characters must be relied upon as a guide, and that the language used by Delesse, in 1847, will be found susceptible of a wide application to crystalline strata: "Rocks of the same age have most generally the same chemical and mineralogical composition, and, reciprocally, rocks having the same chemical composition and the same minerals, associated in the same manner, are of the same age." † In this connection the testimony of Professor James Hall is also to the point. Speaking of the crystalline schists of the White Mountain series, he says: —

"Every observing student of one or two years' experience in the collection of minerals in the New England States knows well that he may trace a mica-schist of peculiar but varying character from Connecticut, through central Massachusetts, and

* Quar. Jour. Geol. Soc.; Murchison, XV. 353; Giekie, XVII. 171; Nicol, XVII. 58, XVIII. 443.

† Bull. Soc. Geol. de Fr. (2), IV. 786.

thence into Vermont and New Hampshire, by the presence of staurolite and some other associated minerals, which mark with the same unerring certainty the geological relations of the rock as the presence of *Pentamerus oblongus, P. galeatus, Spirifer Niagarensis*, or *S. macropleura*, and their respectively associated fossils, do the relations of the several rocks in which these occur." *

I am convinced that these crystalline schists of Germany, Anglesea, and the Scotch Highlands will be found, like those of Norway, to belong to a period anterior to the deposition of the Cambrian sediments, and will correspond with the newer gneissic series of our Appalachian region. There exists, in the Highlands of Scotland, a great volume of fine-grained, thin-bedded mica-schists with andalusite, staurolite, and cyanite, which are met with in Argyleshire, Aberdeenshire, Banffshire, and the Shetland Isles. Rocks regarded by Harkness as identical with these of the Scottish Highlands also occur in Donegal and Mayo in Ireland. Through the kindness of the Rev. Professor Haughton of Trinity College, and Mr. Robert H. Scott, then of Dublin, I received some years since a large collection of the crystalline rocks of Donegal, which I am thus enabled to compare with those of North America, and to assert the existence, in the northwest of Ireland, of our second and third series of crystalline schists. The Green Mountain rocks are there exactly represented by the dark-colored chromiferous serpentines of Aghadoey, and the steatite, crystalline talc, and actinolite of Crohy Head; while the mica-schist of Loch Derg, with white quartz, blue cyanite, staurolite, and garnet, all united in the same fragment, cannot be distinguished from specimens found at Cavendish, Vermont, and Windham, Maine. The fine-grained andalusite-schists of Clooney Lough are exactly like those from Mount Washington; while the granitoid mica-slates from several other localities in Donegal are not less clearly of the type of the White Mountain series. Similar micaceous schists, with andalusite (chiastolite), occur on Skiddaw, in Cumberland, England, the relations of which have

* Paleontology of New York, Vol. III., Introduction, page 93.

been clearly defined by Sedgwick, who groups the rocks of Skiddaw into four divisions. The lowest of these, succeeding the granite, is a series of crystalline rocks, not described lithologically, with mineral veins, "having some resemblance to the rocks of Cornwall," and including, towards the summit, "chiastolite-schists and chiastolite-rocks." These are followed in ascending order by two great series of slates and grits, succeeded by a fourth division of schists, sometimes carbonaceous, holding in parts fucoids and graptolites, which are apparently overlaid discordantly by sundry trappean conglomerates and chloritic slates.* The graptolites of the Skiddaw slates are found to be identical with those of the Levis formation,† and it is worthy of notice that although Sedgwick places the mica-schists with andalusite (chiastolite) so far below the graptolitic beds, he elsewhere, in comparing the rocks of North Wales and Cumberland, states that the chloritic and micaceous rocks of Anglesea and Caernarvon are not represented in Cumberland, being distinct from the other rocks of North Wales, and much older.‡

In Victoria, Australia, the position of the chiastolite schists, according to Selwyn, is beneath the graptolitic slates. Boblaye, it is true, asserted in 1838 that the chiastolite-schists of Les Salles, near Pontivy in Brittany, include *Orthis* and *Calymene*;§ but when we remember that even experienced observers in the White Mountains for a time mistook for remains of crustacea and brachiopods, certain obscure forms, which they afterwards found not to be organic, and that Dana, in this connection, has called attention to the deceptive resemblance to fossils presented by some imperfectly developed chiastolite crystals in the same region,‖ we may well require a verification of Boblaye's observation, especially since we find that more recently D'Archiac and Dalimier agree with De Beaumont and Dufrenoy in placing

* Synopsis of British Palæozoic Rocks, p. lxxxiv, being an Introduction to McCoy's Brit. Pal. Fossils (1855).

† Harkness and Salter, Quar. Jour. Geol. Soc., XIX. 135.

‡ Geol. Journal (1845), IV. 583.

§ Bull. Soc. Geol. de Fr., X. 227.

‖ American Journal of Science (2), I. 415, V. 116.

the chiastolite-schists of Brittany at the very base of the transition sediments, marking the summit of the crystalline schists.*

With regard to the crystalline schists of Lakes Huron and Superior, to which the name of the Huronian system has been given, the observations of all who have studied the region concur in placing them unconformably beneath the sediments which are supposed to represent the base of the New York system; while, on the other hand, they rest unconformably on the Laurentian gneiss, fragments of which are included in the Huronian conglomerates. The gneissic series of the Green Mountains had, however, as we have seen, been, since 1841, regarded, by the brothers Rogers, Mather, Hall, Hitchcock, Adams, Logan, myself, and others, as Lower Silurian (Cambrian of Sedgwick). Eaton and Emmons had alone claimed for it a pre-Cambrian age, until, in 1862, Macfarlane ventured to unite it with the Huronian system, and to identify both with the crystalline schists of a similar age in Norway. Later observations in Michigan justify still further this comparison; for not only the more schistose beds of the Green Mountain series, but even the mica-schists of the third or White Mountain series, with staurolite and garnet, are represented in Michigan, as appears by the recent collections of Major Brooks of the Geological Survey of Michigan, kindly placed in my hands for examination. He informs me that these latter schists are the highest of the crystalline strata in the northern peninsula. (*Ante*, page 18.)

To the north of Lake Superior, as I have already shown elsewhere, the schists of this third series, which, as early as 1861, I compared to those of the Appalachians, are widely spread; while in Hastings County, forty miles north of Lake Ontario, rocks having the mineralogical and lithological characters both of the second and third series are found resting on the first or Laurentian series, the three apparently unconformable, and all in turn overlaid by horizontal Trenton limestone.†

We have shown, that in Pennsylvania, while some of these

* Bull. Soc. Geol. de Fr. (2), XVIII. 664.

† American Journal of Science (2), XXXI. 395, and L. 85.

schists of the second and third series were regarded as altered primal rocks by H. D. Rogers, others, lithologically similar, were referred by him to the older so-called azoic series, which we believe to be their true position. Professor W. B. Rogers has lately informed me that in Virginia a gneissic series, having the characters of the Green Mountain rocks, is clearly overlaid unconformably by the lowest primal palæozoic strata of the region. Coming northward, the uncrystalline argillites and sandstones holding Paradoxides, at Braintree, Massachusetts,* and St. John, New Brunswick, overlie unconformably crystalline schists of the second series; and in the latter region, in one locality, rocks which are by Bailey and Matthew regarded of Laurentian age. In Newfoundland, in like manner, a great series of crystalline schists, in which Mr. Murray recognizes the Huronian system as first studied and described by him in the West, is unconformably overlaid by a group of sandstones, limestones, and slates, holding Paradoxides. The peculiar gneisses and mica-schists of the White Mountain series appear to be developed to a great extent in Newfoundland, which led me to propose for them the name of the Terranovan system.†

From the part which the ruins of these rocks play in the production of succeeding sediments, it is not always easy to define the limits between the ancient mica-schists and the Cambrian strata in these northeastern regions. It is not impossible that the two may graduate into each other, as some have supposed, in Newfoundland and Nova Scotia; but until further light is thrown upon the subject, I am disposed to regard the relation between the two as one of derivation rather than of passage.

We have already alluded to the history of the rocks of the White Mountains, formerly looked upon as primary, and by Jackson described as an old granitic and gneissic axis uplifting the more recent Green Mountain rocks. Their manifest differences from the more ancient gneiss of the Adirondacks, and their apparent superposition to the Green Mountain series, then

* Hunt, Proc. Bost. Soc. Nat. Hist., October 19, 1870.

† American Journal of Science (2), L. 87.

regarded by the Messrs. Rogers as belonging to the Champlain division, led them, in 1846, to look upon the White Mountains as altered strata belonging to the Levant division of their classification, corresponding to the Oneida, Medina, and Clinton of the New York system. In 1848, Sir William Logan came to a somewhat similar conclusion. Accepting, as we have seen, the view of Emmons, that the strata about Quebec included a portion of the Levant division, and regarding the Green Mountain gneisses as the equivalents of these, he was induced to place the White Mountain rocks still higher in the geological series than the Messrs. Rogers had done, and expressed his belief that they might be the altered representatives of the New York system, from the base of the Lower Helderberg to the top of the Chemung; in other words, that they were not Middle Silurian, but Upper Silurian and Devonian. This view, adopted and enforced by me,* was further supported by Lesley in 1860, and has been generally accepted up to this time. In 1870, however, I ventured to question it, and in a published letter, addressed to Professor Dana, concluded, from a great number of facts, that there exists a system of crystalline schists distinct from, and newer than, the Laurentian and Huronian, to which I gave the provisional name of Terranovan [since called Montalban], constituting the third or White Mountain series, which appears not only throughout the Appalachians, but westward to the north of Lake Ontario, and around and beyond Lake Superior.† Although I have, in common with most other American geologists, maintained that the crystalline rocks of the Green Mountain and White Mountain series are altered palæozoic sediments, I find, on a careful examination of the evidence, no satisfactory proof of such an age and origin, but an array of facts which appear to me incompatible with the hitherto received view, and lead me to conclude that the whole of our crystalline schists of eastern North America are not only pre-Silurian but pre-Cambrian in age.

* Geological Survey of Canada, Report 1847-48, p. 58; also American Journal of Science (2), IX. 19.

† American Journal of Science (2), L. 83.

In what precedes I have endeavored to discuss briefly and impartially some of the points in the history of the older rocks, and of the views which during the past thirty years have been entertained as to their age and geological relations, both in America and in Europe. I have said some things which will provoke criticism, and at the same time, I trust, lead to further study of these rocks, a correct knowledge of which lies at the basis of geological science.

I cannot, however, conclude this part of my subject without referring to the views put forth in 1869 by Professor Hermann Credner, of Leipzig, in an essay on the Eozoic or pre-Silurian formations of North America.* With Macfarlane, he refers to the Huronian the gneissic series of the Green Mountains, but includes with it, as part of the Huronian system, the so-called Lower Taconic rocks of Vermont, "with remains of annelids and crinoids." Credner thus falls into the very error against which Emmons warned American geologists, namely, the confounding in one system the ancient crystalline schists with the newer fossiliferous sediments. Resting unconformably on these, he places, first, the Upper Taconic, corresponding, according to him, to a part of the Quebec group; and, second, the Potsdam sandstone. In this he has copied, for the most part, Marcou, who, however, groups the whole of these various divisions in the Taconic system; while Credner, rejecting the name, unites a portion of the Taconic of Emmons with the Huronian system, and refers the other portion, together with the Potsdam, to the Silurian. These same views are set forth in a more recent paper, by the same author, on the Alleghany system, which is accompanied with sections and a geologically colored map.† In this, not content with including in the Huronian both the fossiliferous strata of the Levis formation and the crystalline schists of the Green Mountains, he refers the gneisses and mica-schists of the White Mountains to the same system; while the broad area of similar rocks from their base to the sea at Port-

* Die Gliederung der Eozoischen Formationsgruppe, u. s. w., p. 53. Halle, 1869.

† Petermann's Geographische Mittheilungen. 2 Heft, 1871.

land is regarded as Laurentian. This, on Credner's map, is also made to include, with the exception of the White Mountains themselves, all the rocks of the third or White Mountain series, which cover so large a part of New England. Those who have followed the historical sketch already given can see how widely these notions of Credner differ from those of Emmons, and from all other American geologists, and how much they are at variance with the present state of our knowledge. It is much to be regretted that so good a geologist and lithologist should, from a too superficial study, have fallen into these errors, which can only retard the progress of comparative geognosy, for which he has done so much. In England, again, Credner confounds the Cambrian and Huronian, referring to the latter system the whole of the Longmynd rocks with their characteristic Cambrian fauna, — a view which is supported only by the conjectured Cambrian age of the crystalline schists of Anglesea, which are pre-Cambrian and probably Huronian, like the Urschiefer of Scandinavia, which Credner correctly refers to the latter system, as Macfarlane and Bigsby had done before him. He, moreover, recognizes in the similar crystalline schists of Scotland, the Urals, and various parts of Germany, including those of Bavaria and Bohemia, a newer system, overlying the primary or Laurentian gneiss, and corresponding to the Huronian or Green Mountain series of North America; while he suggests a correspondence with similar rocks in Japan, Bengal, and Brazil. In a collection of rocks brought from the latter country by Professor C. F. Hartt, I have found, as elsewhere stated,* what appear to be representatives of the three types of crystalline schists which have been distinguished in eastern North America.

[I have not in the preceding discussion alluded to the Norian series, otherwise called the Labradorian or Upper Laurentian, for the reason that although largely developed in the southern part of the Adirondack region, it does not occur on our line of section, and, moreover, was not certainly known in the Appalachians. Subsequent observations of the Geological Survey of

* The Nation, December 1, 1870; and Hartt's Geology of Brazil, p. 550.

New Hampshire having, however, shown the existence of rocks supposed to belong to this series in the region of the White Mountains, a brief history of it will not be out of place; while for further details the student is referred to a paper by the present writer in the American Journal of Science for February, 1870 ((2) XLIX. 180). The rocks of this series were recognized by Emmons in Essex County, New York, and described by him in 1842, in the geology of the Northern District of that State (page 27). They were by him correctly regarded as identical with the hypersthene rock of the Western Islands of Scotland, described by MacCulloch, and were looked upon as intrusive. Similar rocks in erratic masses abound in the valley of the St. Lawrence, but were first found in place by Logan, and described by me in the Report of the Geological Survey of Canada for 1852 (page 167). They were shown by Logan to belong to a great stratified series, which was at first included by him in the Laurentian. Subsequent investigation, however, showed that these rocks rest unconformably on the Laurentian gneiss, and he therefore called them Upper Laurentian. Inasmuch as they are largely displayed in Labrador, and moreover consist in great part of labradorite feldspar, the name of the Labradorian series was also given to them. In 1870 it was shown by me, in the paper above referred to, that these rocks were apparently identical with the norites of Esmark, found in Norway under conditions very like those of the Labradorian rocks of North America, and that this name of norite, given in allusion to that country, has the right of priority. I therefore propose to speak of them by that name, and moreover to designate as the Norian series the great formation of crystalline stratified rocks of which the norites make up so large a part. The typical norites consist chiefly of a triclinic feldspar, varying in composition from anorthite to andesine, but generally near labradorite in composition. The color of these rocks is ordinarily some shade of blue, — from bluish-black or violet to bluish-gray, smoke-gray, or lavender, more rarely passing into flesh-red, and occasionally greenish-blue, greenish or bluish white. The weathered surfaces are opaque white. These norites are sometimes nearly pure feldspar, but often include small portions of hypersthene, pyroxene, or hornblende, — the former two being sometimes associated in the same specimen, and in contact with each other. A black mica (biotite), red garnet, epidote, chrysolite, and menacannite (titanic iron) are frequently present in these rocks; quartz, however, is rarely seen, and then only in small quantities. Through an admixture of the first-named minerals these norites pass into hyperite, diabase, and diorite. The norites vary in texture, being sometimes coarsely granitoid, and at other times fine grained and nearly impalpable. The coarser varieties often present large cleavable masses, showing the striæ characteristic of the polysynthetic macles of the triclinic feldspars, and sometimes exhibit a fine play of colors, as in the well-known specimens from Labrador. A gneissic structure is well marked in many of the less coarse-grained varieties of norite, and the lines of bedding are shown by the arrangement of the various foreign minerals. Although norites predominate in the Norian series, they are found in the area of these rocks which is seen to the north of Montreal to be interstratified with beds of micaceous orthoclase-gneiss, quartzite, and crystalline limestone, not unlike those met with in the Laurentian and White Mountain series. It was from their distribution in this region that Sir William Logan was enabled to show that the rocks of the Norian series rest unconformably upon the gneisses and limestones of the Laurentian. Further evidence of the same kind was obtained by Mr. Richardson, in 1869, on the north side of the Gulf of St. Lawrence, where rocks of the Norian series were found to lie in discordant stratification, and at moderate angles on the nearly vertical Laurentian gneiss. The norites may be readily studied in Essex County, New York, where they reach the shore of Lake Champlain just above the town of Westport, and include the great deposits of titanic iron ores of this region. The titanic ores of Bay St. Paul, Lake St. John, and the Bay of Seven Islands, in Canada, also occur in Norian rocks. In all of these localities they appear to be directly superposed on the Laurentian; but in the vicinity of St. John, New Brunswick, a small area of norites is found to occupy a position in contact with rocks regarded as belonging to the Huronian and the White Mountain series. The rocks which are referred to the Norian series in the White Mountain region, according to Hitchcock, rest upon the gneisses and mica-schists of the White Mountains; while these overlie unconformably a more ancient series of granitoid gneiss, supposed to represent the Laurentian.

The hypersthene rock of Skye was by MacCulloch regarded as an eruptive rock; and Giekie, in his memoir on the geology of a part of Skye, published in 1858 (Quarterly Journal of the Geological Society, XIV. page 1), appears to include them with certain syenites and greenstones, which he vaguely speaks of as not intrusive, though eruptive after the manner of granites (*loc. cit.*, pp. 11 – 14). Specimens of these rocks from Loch Scavig, and others in MacCulloch's collection from that vicinity, which I have examined, are, however, identical with the North American norites, whose stratified character is undoubted. I called attention to these resemblances in the Dublin Quarterly Journal for July, 1863 (*ante*, page 33); and Professor Haughton, of Dublin, who in 1864 visited Loch Scavig, subsequently described and analyzed the norite from that locality; which is, according to him, evidently "a bedded metamorphic rock." (Dublin Quarterly Journal for 1865, page 94.)

The distribution of the crystalline rocks of the Norian, Huronian, and Montalban or White Mountain series would seem to show that these are remaining portions of great, distinct, and unconformable series, once widely spread out over a more ancient floor of granitic gneiss of Laurentian age; but that the four series thus indicated include the whole of the crystalline stratified rocks of New England is by no means certain. How many more such formations may have been laid down over this region, and subsequently swept away, leaving no traces, or only isolated fragments, we may never know; but it is probable that a careful study of the geology of New England and the adjacent British Provinces may establish the existence of many more than the four series above enumerated. When it is considered that we find within the limits of southern

New Brunswick alone small areas of palæozoic sediments which are shown by their organic remains to belong to not less than five periods, namely, Menevian, Lower Helderberg, Chemung, Lower Carboniferous, and Carboniferous, all perfectly well distinguished, and each reposing directly upon the ancient crystalline rocks, we are prepared for a history not less varied and complex for the rocks belonging to Eozoic time. (See the author's Address before the American Institute of Mining Engineers, in their Proceedings for February, 1873.)

Professor C. H. Hitchcock, from the results of the Geological Survey of New Hampshire, now in progress, announces, in 1873 and 1874, a large number of divisions in the crystalline rocks of this State. The Norian series there, according to him, rests unconformably upon ancient gneisses, which, as he suggests, belong perhaps to the Laurentian, the appearance of which in northeastern Massachusetts I pointed out in 1870. With the Norian he has however included a great series of granites and of compact felsites, some of which, from specimens, appear identical with the orthophyres of our eastern coasts, of Lake Superior, and Missouri. These, so far as my observations go, are in no way related to the Norian, but probably belong to the Huronian series. (*Ante*, page 187.) Besides these, he recognizes the White Mountain series of gneisses and andalusite-schists (Montalban). He describes, under the name of gneiss, the so-called granites of Concord and Fitzwilliam, which I had already, in 1870, declared to be gneisses associated with the mica-schists of the Montalban series. (*Ante*, page 188.) This series he supposes to be more ancient than the well-characterized Huronian rocks of the State; but admits in addition a second and more recent series of mica-schists with andalusite and staurolite, named the Coös group. Further researches in this disturbed region will be required to determine whether, besides this series of andalusite and staurolite-bearing mica-schists, which (associated with gneisses) occurs in other regions, as I have in the previous pages of this essay endeavored to show, above the Huronian, there is another and an older series of similar rocks, or whether the two are one and the same series, repeated by stratigraphical accidents.]

II. The Origin of Crystalline Rocks.

We now approach the second part of our subject, namely, the genesis of the crystalline schists whose history we have just discussed. The origin of the mineral silicates which make up a great portion of the crystalline rocks of the earth's surface is a question of much geological interest, which has been to a great degree overlooked. The gneisses, mica-schists, and argillites of various geological periods do not differ very greatly in chemical constitution from modern mechanical sediments, and are now, by the greater number of geologists, regarded as resulting from a molecular rearrangement of similar sediments formed in earlier times by the disintegration of previously existing rocks, not very unlike them in composition; the oldest known formations being still composed of crystalline stratified deposits presumed to be of sedimentary origin. Before these the imagination conceives yet earlier rocks, until we reach the surface of unstratified material which the globe may be supposed to have presented before water had begun its work. It is not, however, my present plan to consider this far-off beginning of sedimentary rocks, which I have elsewhere discussed. (*Ante*, page 63.)

Apart from the rocks just referred to, whose composition may be said to be essentially quartz and aluminous silicates, chiefly in the forms of feldspars and micas, there is another class of crystalline silicated rocks, which, though far less important in bulk than the last, is of great and varied interest to the lithologist, the mineralogist, the geologist, and the chemist. The rocks of this second class may be defined as consisting in great part of the silicates of the protoxide bases, lime, magnesia, and ferrous oxide, either alone, or in combination with silicates of alumina and alkalies. They include the following as their chief constituent mineral species: pyroxene, hornblende, chrysolite, serpentine, talc, chlorite, epidote, garnet, and triclinic feldspars, such as labradorite. The great types of this second class are not less well defined than the first, and consist of pyroxenic and hornblendic rocks, passing into diorites, diabases, ophiolites, and talcose, chloritic, and epidotic rocks. Intermediate varieties resulting from the association of the minerals of this class with those of the first, and also with the materials of non-silicated rocks, such as limestones and dolomites, show an occasional blending of the conditions under which these various types of rocks were formed.

The distinctions just drawn between the two great divisions of silicated rocks are not confined to stratified deposits, but are equally well marked in eruptive and unstratified masses, among which the first type is represented by trachytes and granites; and the second, by dolerites and diorites. This fundamental difference between acidic and basic rocks, as the two classes have been called, finds its expression in the theories of Phillips, Durocher, and Bunsen, who have deduced all silicated rocks from two supposed layers of molten matter within the earth's crust, consisting respectively of acidic and basic mixtures; the trachytic and pyroxenic magmas of Bunsen. From these, by a process of partial crystallization and eliquation, or by commingling in various proportions, those eruptive rocks which depart more or less from the normal types are supposed by the theorists of this school to be generated. (*Ante*, pages 3 and 23.) The doctrine that these eruptive rocks are not derived directly from a hitherto uncongealed nucleus, but are softened and crystallized sediments, in fact, that the whole of the rocks at present known to us have at one time been aqueous deposits, has, however, found its advocates. In support of this view, I have endeavored to show that the natural result of forces constantly in operation tends to resolve mechanical sediments into two classes: the one coarse, sandy, and permeable; the other fine, clayey, and impervious. The action of infiltrating atmospheric waters on the first and more silicious strata will remove from them lime, magnesia, iron-oxide, and soda, leaving behind silica, alumina, and potash, — the elements of granitic, gneissic, and trachytic rocks. The finer and more aluminous sediments (including the ruins of the soft and easily abraded silicates of the pyroxene group), resisting the penetration of the water, will, on the contrary, retain their alkalies, lime, magnesia, and iron, and thus will have the composition of the more basic rocks. [We find, in fact, in the sediments of various geological periods, not only beds of clay and marl corresponding to the second class, but strata made up in great part of mechanically disintegrated though chemically unchanged orthoclase, with quartz, the *débris* of granitic rocks, constituting what is called arkose. Beds of this kind, as will be seen in the following paper on the Geology of the Alps, have even been mistaken for granite and gneiss, and similar recomposed rocks occur in the mesozoic sandstones of New England and New Jersey. Such processes of disintegration and decay have probably been going on from very remote times, and the crystalline rearrangement of the resulting rocks may be supposed to give rise to true crystalline schists, or their aqueo-igneous fusion to eruptive rocks. (*Ante*, pages 14 and 23.)]

A little consideration will, however, show that this process is inadequate to explain the production of many of the varieties of crystalline silicated rocks. Such are serpentine, steatite, chrysolite, hornblende, diallage, chlorite, pinite, labradorite, and orthoclase, all of which mineral species form rock-masses by themselves, frequently almost without admixture. No geological student will now question that all of these rocks occur as members of stratified formations. Moreover, the manner in which serpentines are found interstratified with steatite, chlorite, argillite, diorite, hornblende, and feldspar rocks, and these, in their turn, with quartzites and orthoclase rocks, is such as to forbid the notion that all of these various materials have been deposited, with their present composition, as mechanical sediments from the ruins of pre-existing rocks of plutonic origin.

There are two hypotheses which have been proposed to explain the origin of these various silicated rocks, and especially of the less abundant, and, as it were, exceptional species just mentioned. The first of these supposes that the minerals of which they are composed have resulted from an alteration of previously existing minerals of plutonic rocks, often very unlike in composition to the present, by the taking away of certain elements and the addition of certain others. This is the theory of metamorphism by pseudomorphic changes, as they are called, and is the one taught by the now reigning school of chemical geologists, of which the learned and laborious Bischof, whose recent death science deplores, may be regarded as the great exponent. The second hypothesis supposes that the elements of these various rocks were originally deposited as, for the most part, chemically formed sediments, or precipitates; and that the subsequent changes have been simply molecular, or, at most, confined in certain cases to reactions between the mingled elements of the sediments, with the elimination of water and carbonic acid. It is proposed to consider briefly these two opposite theories, which seek to explain the origin of the rocks in question respectively by pseudomorphic changes in pre-existing crystalline plutonic rocks, and by the crystallization of aqueous sediments, for the most part chemically formed precipitates.

Mineral pseudomorphism, that is to say, the assumption by one mineral substance of the crystalline form of another, may arise in several ways. First of these is the filling up of a mould left by the solution or decomposition of an imbedded crystal, a process which sometimes takes place in mineral veins, where the processes of solution and deposition can be freely carried on. Allied to this is the mineralization of organic remains, where carbonate of lime or silica, for example, fills the pores of wood. When subsequent decay removes the woody tissue, the vacant spaces may, in their turn, be filled by the same or another species.* In the second place we may consider pseudomorphs from alteration, which are the result of a gradual change in the composition of a mineral species. This process is exemplified in the conversion of feldspar into kaolin by the loss of its alkali and a portion of silica, and the fixation of water, or in the change of chalybite into limonite by the loss of carbonic acid and the absorption of water and oxygen.

* Hunt on the Silicification of Fossils, Canadian Naturalist, New Series, I. 46.

The doctrine of pseudomorphism by alteration, as taught by Gustaf Rose, Haidinger, Blum, Volger, Rammelsberg, Dana, Bischof, and many others, leads them, however, to admit still greater and more remarkable changes than these, and to maintain the possibility of converting almost any silicate into any other. Thus, by referring to the pages of Bischof's Chemical Geology, it will be found that serpentine is said to exist as a pseudomorph after augite, hornblende, chrysolite, chondrodite, garnet, mica, and probably also after labradorite and even orthoclase. Serpentine rock or ophiolite is supposed to have resulted, in different cases, from the alteration of hornblende-rock, diorite, granulite, and even granite. Not only silicates of protoxides and aluminous silicates are conceived to be capable of this transformation, but probably also quartz itself; at least Blum asserts that meerschaum, a closely related silicate of magnesia, which sometimes accompanies serpentine, results from the alteration of flint; while according to Rose, serpentine may even be produced from dolomite, which we are told is itself produced by the alteration of limestone. But this is not all, — feldspar may replace carbonate of lime, and carbonate of lime, feldspar; so that, according to Volger, some gneissoid limestones are probably formed from gneiss by the substitution of calcite for orthoclase. In this way, we are led from gneiss or granite to limestone, from limestone to dolomite, and from dolomite to serpentine, or, more directly, from granite, granulite, or diorite to serpentine at once, without passing through the intermediate stages of limestone and dolomite, till we are ready to exclaim in the words of Goethe, —

"Mich ängstigt das Verfängliche
Im widrigen Geschwätz,
Wo Nichts verharret, Alles flieht,
Wo schon verschwunden was man sieht,"*

which we may thus translate: "I am vexed with the sophistry in their contrary jargon, where nothing endures, but all is fugitive, and where what we see has already passed away."

Chinesisch-Deutsche Jahres und Tages-Zeiten, XI.

By far the greater number of cases on which this general theory of pseudomorphism by a slow process of alteration in minerals has been based are, as I shall endeavor to show, examples of the phenomenon of mineral envelopment, so well studied by Delesse in his essay on Pseudomorphs,* and may be considered under two heads: first, that of symmetrical envelopment, in which one mineral species is so enclosed within the other that the two appear to form a single crystalline individual. Examples of this are seen when prisms of cyanite are surrounded by staurolite, or staurolite crystals completely enveloped in those of cyanite, the vertical axes of the two prisms corresponding. Similar cases are seen in the enclosure of a prism of red in an envelope of green tourmaline, of allanite in epidote, and of various minerals of the pyroxene group in one another. The occurrence of muscovite in lepidolite, and of margarodite in lepidomelane, or the inverse, are well-known examples, and, according to Scheerer, the crystallization of serpentine around a nucleus of olivine is a similar case. This phenomenon of symmetrical envelopment, as remarked by Delesse, shows itself with species which are generally isomorphous or homœomorphous, and of related chemical composition. Allied to this is the repeated alternation of crystalline laminæ of related species, as in perthite, the crystalline cleavable masses of which consist of thin, alternating layers of orthoclase and albite.

Very unlike to the above are those cases of envelopment in which no relations of crystalline symmetry nor of similar chemical constitution can be traced. Examples of this kind are seen in garnet crystals, the walls of which are shells, sometimes no thicker than paper, enclosing, in different examples, crystalline carbonate of lime, epidote, chlorite, or quartz. In like manner, crystalline shells of leucite enclose feldspar, hollow prisms of tourmaline are filled with crystals of mica or with hydrous peroxide of iron, and crystals of beryl with a granular mixture of orthoclase and quartz, holding small crystals of garnet and tourmaline, a composition identical with the

* Annales des Mines (5), XVI. 317 – 392.

enclosing granitic vein-stone.* Similar shells of galenite and of zircon, having the external forms of these species, are also found filled with calcite. In many of these cases the process seems to have been first the formation of a hollow mould or skeleton-crystal (a phenomenon sometimes observed in salts crystallizing from solutions), the cavity being subsequently filled with other matters. (*Ante*, page 212.) Such a process is conceivable in free crystals formed in veins, as, for example, galenite, zircon, tourmaline, beryl, and some examples of garnet, but is not so intelligible in the case of those garnets imbedded in mica-schist, studied by Delesse, which enclosed within their crystalline shells irregular masses of white quartz, with some little admixture of garnet. Delesse conceives these and similar cases to be produced by a process analogous to that seen in the crystallizations of calcite in the Fontainebleau sandstone; where the quartz grains, mechanically enclosed in well-defined rhombohedral crystals, equal, according to him, sixty-five per cent of the mass. Very similar to these are the crystals with the form of orthoclase, which sometimes consist in large part of a granular mixture of quartz, mica, and orthoclase, with a little cassiterite, and in other cases contain two thirds their weight of the latter mineral, with an admixture of orthoclase and quartz. Crystals with the form of scapolite, but made up, in a great part, of mica, seem to be like cases of envelopment, in which a small proportion of one substance in the act of crystallization compels into its own crystalline form a large portion of some foreign material, which may even so mask the crystallizing element that this becomes overlooked, as of secondary importance. The substance which, under the name of houghite, has been described as an altered spinel, is found by analysis to be an admixture of völlknerite with a variable proportion of spinel, which, in some specimens, does not exceed eight per cent, but to which, nevertheless, these crystalloids (to use the term suggested by Naumann) appear to owe their more or less complete octohedral form.†

* Report Geol. Survey of Canada, 1866, p. 189.

† Ibid., pp. 189, 213 ; American Journal of Science (3), I. 188.

The above characteristic examples of symmetrical and asymmetrical envelopment are cited from a great number of others which might have been mentioned. Very many of these are by the pseudomorphists regarded as results of partial alteration. Thus, in the case of associated crystals of andalusite and cyanite, Bischof does not hesitate to maintain the derivation from andalusite of the latter species by an elimination of quartz; more than this, as the andalusite in question occurs in a granite-like rock, he suggests that itself is a product of the alteration of orthoclase. In like manner the mica, which in some cases coats tourmaline, and in others fills hollow prisms of this mineral, is supposed to result from a subsequent alteration of crystallized tourmaline. So in the case of shells of leucite filled with feldspar, or of garnet enclosing epidote, or chlorite, or quartz, a similar transformation of the interior is supposed to have been mysteriously effected, while the external portion of the crystal remains intact. Again, the aggregates of cassiterile, quartz, and orthoclase, having the form of the latter, are, by Bischof and his school, looked upon as results of a partial alteration of previously formed orthoclase crystals. It needed only to extend this view to the crystals of calcite enclosing sand-grains, and regard these as the result of a partial alteration of the carbonate of lime. There is absolutely no proof that these hard crystalline substances can undergo the changes supposed, or can be absorbed and modified like the tissues of a living organism. It may, moreover, be confidently affirmed that the obvious facts of envelopment are adequate to explain all the cases of association upon which this hypothesis of pseudomorphism by alteration has been based. Why the change should extend to some parts of a crystal and not to others, why in some cases the exterior of the crystal is altered, while in others the centre alone is removed and replaced by a different material, are questions which the advocates of this fanciful hypothesis have not explained. As taught by Blum and Bischof, however, these views of the alteration of mineral species have not only been generally accepted, but have formed the basis of the generally received theory of rock-metamorphism.

Protests against the views of this school have, however, not been wanting. Scheerer, in 1846, in his researches in Polymeric Isomorphism,* attempted to show that iolite and aspasiolite, a hydrous species which had been looked upon as resulting from its alteration, were isomorphous species crystallizing together, and, in like manner, that the association of chrysolite and serpentine in the same crystal, at Snarum in Norway, was a case of envelopment of two isomorphous species. In both of these instances he maintained the existence of isomorphous relations between silicates in which 3HO replace MgO. He hence rejected the view of Gustaf Rose, that these serpentine crystals were results of the alteration of chrysolite, and supported his own by reasons drawn from the conditions in which the crystals occur. In 1853 I took up this question and endeavored to show that these cases of isomorphism described by Scheerer entered into a more general law of isomorphism, pointed out by me among homologous compounds differing in their formulas by nM_2O_2 ($M =$ hydrogen or a metal). I insisted, moreover, on its bearing upon the received views of the alteration of minerals, and remarked: "The generally admitted notions of pseudomorphism seem to have originated in a too exclusive plutonism, and require such varied hypotheses to explain the different cases, that we are led to seek for some more simple explanation, and to find it, in many instances, in the association and crystallizing together of homologous and isomorphous species."† Subsequently, in 1860, I combated the view of Bischof, adopted by Dana, that "regional metamorphism is pseudomorphism on a grand scale," in the following terms: —

"The ingenious speculations of Bischof and others, on the possible alteration of mineral species by the action of various saline and alkaline solutions, may pass for what they are worth, although we are satisfied that by far the greater part of the so-called cases of pseudomorphism in silicates are purely imaginary, and, when real, are but local and accidental phenomena. Bischof's notion of

* Pogg. Annal., LXVIII. 319.

† American Journal of Science (2), XVI. 218.

the pseudomorphism of silicates like feldspars and pyroxenes presupposes the existence of crystalline rocks, whose generation this neptunist never attempts to explain, but takes his starting-point from a plutonic basis."

I then asserted that the problem to be solved in regional metamorphism is the conversion of sedimentary strata, "derived by chemical and mechanical agencies from the ocean-waters and pre-existing crystalline rocks into aggregations of crystalline silicates. These metamorphic rocks, once formed, are liable to alteration only by local and superficial agencies, and are not, like the tissues of a living organism, subject to incessant transformations, the pseudomorphism of Bischof." *

I had not, at that time, seen the essay by Delesse on Pseudomorphs, already referred to, published in 1859, in which he maintained views similar to those set forth by me in 1853 and 1860, declaring that much of what had been regarded as pseudomorphism had no other basis than the observed associations of minerals, and that often "the so-called metamorphism finds its natural explanation in envelopment." These views he ably and ingeniously defended by a careful discussion of the whole range of facts belonging to the history of the subject.

My own expression of opinion on this question, in 1853, had been adversely criticised, and I had been charged with a want of comprehension of the question. It was, therefore, with no small pleasure, that I not only saw my views so ably supported by Delesse, but read the language of Carl Friedrich Naumann, who in 1861 wrote to Delesse as follows, referring to his essay just noticed: —

"You have rendered a veritable service to science in restricting pseudomorphs to their true limits, and separating what had been erroneously united to them. As you have remarked, envelopments have, for the most part, nothing in common with pseudomorphs, and it is inconceivable that they have been united by so many mineralogists and geologists. It appears to me, moreover, that they commit an analogous error, when they regard gneisses, amphibo-

* American Journal of Science (2), XXX. 135.

lites, etc., as being, all of them, the results of metamorphic epigenesis, and not original rocks. It is precisely because pseudomorphism has been so often confounded with metamorphism, that this error has found acceptance. I only admit a pseudomorph where there is some crystal the form of which has been preserved. There are very many metamorphic substances which are in no sense of the word, pseudomorphs. Had the name of *crystalloid* been chosen, instead of pseudomorph, this confusion would certainly have never found its way into the science. I think, with you, that the envelopment of two minerals is most generally explained by a *contemporaneous* and *original* crystallization. Secondary envelopments, however, exist, and such may be called pseudomorphs or crystalloids, if they reproduce exactly the form of the crystal enveloped, whether this last still remains, or has entirely disappeared." *

It is unnecessary to remark that the view of Delesse and Naumann—namely, that the so-called cases of pseudomorphism, on which the theory of metamorphism by alteration has been built, are, for the most part, examples of association and envelopment, and the result of a contemporaneous and original crystallization — is identical with the view suggested by Scheerer, and generalized by myself long before, when, in 1853, I sought to explain the phenomena in question by "the association and crystallizing together of homologous and isomorphous species."

Later, in 1862, I wrote as follows :—

"Pseudomorphism, which is the change of one mineral species into another by the introduction or the elimination of some element or elements, presupposes metamorphism (i. e. metamorphic or crystalline rocks), since only definite mineral species can be the subjects of this process. To confound metamorphism with pseudomorphism, as Bischof and others after him have done, is therefore an error. It may be further remarked, that, although certain pseudomorphic changes may take place in some mineral species, in veins and near the surface, the alteration of great masses of silicated rocks by such a process is as yet an unproved hypothesis." †

* Bull. Soc. Geol. de France (2), XVIII. 678.

† Descriptive Catalogue, Crystalline Rocks of Canada, p. 80, London Exhibition, 1862 ; also Canadian Naturalist, VII. 262 ; Dublin Quar. Journal, July, 1863; and American Journal of Science (2), XXXVI. 218.

Thus this unproved theory of pseudomorphism, as taught by Bischof, does not, even if admitted to its fullest extent, advance us a single step towards a solution of the problem of the origin of the various silicates which, singly or intermingled, make up beds in the crystalline schists. Granting, for the sake of argument, that serpentine results from the alteration of chrysolite or labradorite, and steatite or chlorite from hornblende, the origin of these anhydrous silicates, which are the subjects of the supposed change, is still unaccounted for. The explanation of this short-sightedness is not far to seek; as already remarked, Bischof, although a professed neptunist, starts from a plutonic basis.

[The notion of the plutonic origin of crystalline stratified rocks has in fact found many advocates, as may be seen by reference to pages of Naumann's Lehrbuch der Geognosie. This learned author himself speaks of them as "those enigmatical deepest-lying rocks which resemble sedimentary strata in possessing more or less perfect stratification, while resembling eruptive rocks in mineral composition and crystalline structure" (*loc. cit.*, Vol. II. p. 8, *et seq.*). He declares them to be neither sedimentary nor eruptive in the ordinary sense of those terms; and evidently leans to the notion, of which he speaks with favor, that they are in some way the first-solidified portions of the once molten globe. He elsewhere says that the solidification being from the surface downwards, the lowest of these rocks must be the newest, except so far as eruptive masses may break up through the crust. Tchitatchef, from his recent researches in Asia Minor, holds to Naumann's view as to the plutonic origin of the gneissic rocks of that region. The most recent and most explicit statement of this view of the plutonic origin of these rocks is that put forth by Macfarlane, in a learned essay on The Eruptive and Primary Rocks, in the Canadian Naturalist for 1864. He conceives that the structure in these rocks may have been generated by currents in the molten mass of the globe; and, further, that the once-formed crust may have had a different rate of rotation from the liquid below; from which also would result a stratiform arrangement in the elements of the solidifying layers, such as is seen in many slags, and in certain eruptive rocks. (*Ante*, page 186.) Add to this notion that of the separation of the fluid or, rather, viscid mass into two or more layers of different composition and density (*ante*, page 3), and we might have generated from them, by their solidification under the above conditions, the various types of stratiform feldspathic, hornblendic, and chrysolitic rocks, which would afterwards be penetrated by injections from the yet liquid portions below. If now we imagine the various plutonic rocks thus formed, both stratified and unstratified, to be the subjects of epigenic or pseudomorphous changes, by which some beds or masses were converted in serpentine or into steatitic or chloritic rocks, while others were changed into limestone, quartzite, or iron-oxide, we shall have as clear a conception as it is possible to form of the vaguely defined views of Naumann, Bischof, and their school, as to the origin of the crystalline rocks as we now find them.

Naumann, while denying the sedimentary origin of the great mass of crystalline schists, admitted, however, the conversion of younger uncrystalline sedimentary strata, in certain cases, into crystalline gneisses and mica-schists, resembling those of the primary formations, and like them subject to epigenic changes. That such crystalline rocks have ever been formed from the alteration of palæozoic or more recent sediments, except locally (pages 18, 298, and 310), is, however, more than doubtful, as will appear from the examination of the supposed examples of this conversion in the preceding pages of this paper, and also in the following one on the Geology of the Alps. In connection with these two papers are given the views of Gümbel (page 305) and of Favre on this important question.

These crystalline rocks, whatever their origin or mode of formation, appear to be in all cases older than the palæozoic sediments. They belong to at least three or four geognostically discordant series, and are moreover occasionally associated with fragmentary rocks, which render it impossible to admit for them any other than an aqueous sedimentary origin, in accordance with the view already defined on page 286.]

Whence, then, come these silicates of magnesia, lime, and iron, which are the sources of the serpentine, chrysolite, pyroxene, hornblende, steatite, and chlorite, which abound in these rocks? This is the question which I proposed in 1860, when, after discussing the results of my examinations of the tertiary rocks near Paris, containing layers of a hydrous silicate of magnesia, related to talc in composition, among unaltered limestones and clays, I remarked that it is evident "such silicates may be formed in basins at the earth's surface, by reactions between magnesian solutions and dissolved silica"; and, after some discussion, said "further inquiries in this direction may show to what extent certain rocks composed of calcareous and magnesian silicates may be directly formed in the moist way."* Subsequently, in a paper on The Origin of some Magnesian and Aluminous Rocks, printed in the Canadian Naturalist for June, 1860,† I repeated these considerations, referring to the well-known fact that silicates of lime, magnesia, and iron-oxide are deposited during the evaporation of natural waters, including those of alkaline springs and of the Ottawa River. Having described the mode of occurrence of the magnesian silicate, sepiolite, in the Paris basin, and the related quincite, containing some iron-oxide, and disseminated in limestone, I suggested that while steatite has been derived from a compound like sepiolite, the source of serpentine was to be sought in another silicate richer in magnesia; and, moreover, that chlorite (unless the result of a subsequent reaction between clay and carbonate of magnesia) was directly formed by a process analogous to that which, according to Scheerer, has, in recent times, caused the deposition from waters of neolite, — a hydrous alumino-magnesian silicate, approaching to chlorite in composition,‡ "the type of a reaction which formerly generated beds of chlorite, in the same way as those of sepiolite or talc." Delesse, subsequently, in 1861, in his essay on Metamorphism, insisted upon the sepiolite or so-called magnesian marls, as probably the

* American Journal of Science (2), XXIX. 284; also (2), XL. 49.

† Ibid. (2), XXXII. 286.

‡ Pogg. Annal., LXXI. 288.

source of steatite, and suggested the derivation of serpentine, chlorite, and other related minerals of the crystalline schists, from deposits approaching these marls in composition.* He recalled, also, the occurrence of chromic oxide, a frequent accompaniment of these magnesian minerals, in the hydrated iron ores of the same geological horizon with the magnesian marls in France. Delesse did not, however, attempt to account for the origin of these deposits of magnesian marls, in explanation of which I afterwards verified Bischof's observations on the sparing solubility of silicate of magnesia, and showed that silicate of soda, or even artificial hydrated silicate of lime, when added to waters containing magnesian chloride or sulphate, gives rise, by double decomposition, to a very insoluble magnesian silicate. (*Ante*, page 122.)

To explain the generation of silicates like the feldspars, scapolite, garnet, and saussurite, I suggested that double aluminous silicates, allied to the zeolites, might have been formed, and subsequently rendered anhydrous. The production of zeolitic minerals observed by Daubrée, at Plombières and Luxeuil, by the action of a silicated alkaline water on the masonry of ancient Roman baths, was appealed to by way of illustration. (*Ante*, pages 25 and 205.) It has been shown by Daubrée that the elements of the zeolites were derived in part from the waters, and in part from the mortar, and even the clay of the bricks, which had been attacked, and had entered into combination with the soluble matters of the water to form chabazite. I, however, at the same time pointed out another source of silicated minerals, upon which I had insisted since 1857, namely, the reaction between silicious or argillaceous matters and earthy carbonates in the presence of alkaline solutions. Numerous experiments showed that when solutions of an alkaline carbonate were heated with a mixture of silica and carbonate of magnesia, the alkaline silicate formed acted upon the latter, yielding a silicate of magnesia, and regenerating the alkaline carbonate; which, without entering into permanent combination, was the medium through which the union of the silica and the

* Etudes sur le Metamorphisme, quarto, pp. 91. Paris, 1861.

magnesia was effected. In this way I endeavored to explain the alteration, in the vicinity of a great intrusive mass of dolerite, of a gray palæozoic limestone, which contained, besides a little carbonate of magnesia and iron-oxide, a portion of very silicious matter, consisting apparently of comminuted orthoclase and quartz. In place of this, there had been developed in the limestone, near its contact with the dolerite, an amorphous greenish basic silicate, which had seemingly resulted from the union of the silica and alumina with the iron-oxide, the magnesia, and a portion of lime. By the crystallization of the products thus generated, it was conceived that minerals like hornblende, garnet, and epidote might be developed in earthy sediments, and many cases of local alteration explained. Inasmuch as the reaction described required the intervention of alkaline solutions, rocks from which these were excluded would escape change, although the other conditions might not be wanting. The natural associations of minerals, moreover, led me to suggest that alkaline solutions might favor the crystallization of aluminous silicates, and thus convert mechanical sediments into gneisses and mica-schists. The ingenious experiments of Daubrée on the part which solutions of alkaline silicates, at elevated temperatures, may play in the formation of crystallized minerals, such as feldspar and pyroxene, were posterior to my early publications on the subject, and fully justified the importance which, early in 1857, I attributed to the intervention of alkaline silicates, in the formation of crystalline silicated minerals.* (*Ante*, pages 6 and 25.)

[While we may not question the regeneration of feldspars and zeolites (which are but hydrated feldsdars) by the combination of silicates of alumina, like clay, with soluble alkaline or calcareous silicates, it is evident that this process is not the chief nor the primary one; since the existence of clay supposes the previous existence and decay of feldspars. The deposition of immense quantities, alike of orthoclase, albite, and oligoclase in veins which are evidently of aqueous origin, shows that conditions have existed in which the elements of these

* Proc. Royal Soc., May 7, 1857.

mineral species were abundant in solution. The relation between these endogenous deposits and the great beds of orthoclase and triclinic feldspar rocks is similar to that between veins of calcite and of quartz and beds of marble and travertine, of quartzite and hornstone. But while the conditions in which these latter mineral species are deposited from solution are perpetuated to our own time, those of the deposition of feldspars and many other species, whether in veins or in beds, appear to belong only to remote geological ages, and at best are represented in more recent time only by the production of a few zeolitic minerals. See in this connection the paper on Granites and Granitic Vein-Stones, XI. of the present volume, *passim*, but especially §§ 30, 31, and 49.]

While, however, there is good reason to believe that solutions of alkaline silicates or carbonates have been efficient agents in the crystallization and molecular rearrangement of ancient sediments, and have also played an important part in that local alteration of sedimentary strata which is often observed in the vicinity of intrusive rocks, it is clear to me that the agency of these solutions is less universal than once supposed by Daubrée and myself, and will not account for the formation of various silicated rocks belonging to the crystalline schists, such as serpentine, hornblende, steatite, and chlorite. When I commenced the study of these crystalline strata I was led, in accordance with the almost universally received opinion of geologists, to regard them as resulting from a subsequent alteration of palæozoic sediments, which, according to different authorities, were of Cambrian, Silurian, or Devonian age. Thus in the Appalachian region, as we have already seen, they have, on supposed stratigraphical evidence, been successively placed at the base, at the summit, and in the middle of the Champlain division of the New York system. A careful chemical examination among the unaltered palæozoic sediments, which in Canada were looked upon as the stratigraphical equivalents of the bands of magnesian silicates in these crystalline schists, showed me, however, no magnesian rocks, except certain silicious and ferruginous dolomites. From a consideration of reactions which I had observed to take place in such admixtures in presence of heated alkaline solutions, and from the composition of the basic silicates which I had found to be formed in silicious limestones near their contact with eruptive rocks, I was led to suppose that similar actions, on a grand scale, might transform these silicious dolomites of the unaltered strata into crystalline magnesian silicates.

Further researches, however, convinced me that this view was inapplicable to the crystalline schists of the Appalachians, since, apart from the geognostical considerations set forth in the previous part of this paper, I found that these same crystalline strata hold beds of quartzose dolomite and magnesian carbonate, associated in such intimate relations with beds of serpentine, diallage, and steatite, as to forbid the notion that these silicates could have been generated by any transformations or chemical rearrangement of mixtures like the accompanying beds of quartzose magnesian carbonates. Hence it was that already, in 1860, as shown above, I announced my conclusion that serpentine, chlorite, and steatite had been derived from silicates like sepiolite, directly formed in waters at the earth's surface, and that the crystalline schists had resulted from the consolidation of previously formed sediments, partly chemical and partly mechanical in their origin. The latter being chiefly silico-aluminous, took, in part, the forms of gneiss and mica-schists, while from the more argillaceous strata, poorer in alkali, much of the aluminous silicate crystallized as andalusite, staurolite, cyanite, and garnet. These views were reiterated in 1863,* and further in 1864, in the following language, as regards the chemically formed sediments: "Steatite, serpentine, pyroxene, hornblende, and in many cases garnet, epidote, and other silicated minerals, are formed by a crystallization and molecular rearrangement of silicates generated by chemical processes in waters at the earth's surface." † Their alteration and crystallization was compared to that of the

* Geology of Canada, pp. 577 - 581.

† American Journal of Science (2), XXXVII. 266; and XXXVII. 183.

mechanically formed feldspathic, silicious, and argillaceous sediments just mentioned.

The direct formation of the crystalline schists from an aqueous magma is a notion which belongs to an early period in geological theory. Delabeche in 1834 * conceived that they were thrown down as chemical deposits from the waters of the heated ocean, after its reaction on the crust of the cooling globe, and before the appearance of organic life. This view was revived by Daubrée in 1860. Having sought to explain the alteration of palæozoic strata of mechanical origin by the action of heated waters, he proceeds to discuss the origin of the still more ancient crystalline schists. The first precipitated waters, according to him, acting on the anhydrous silicates of the earth's crust, at a very elevated temperature, and at a great pressure which he estimated at two hundred and fifty atmospheres, formed a magma from which, as it cooled, were successively deposited the various strata of the crystalline schists.† This hypothesis, violating, as it does, all the notions which sound theory teaches with regard to the chemistry of a cooling globe, has, moreover, to encounter grave geognostical difficulties. The pre-Cambrian crystalline rocks belong to two or more distinct systems of different ages, succeeding each other in discordant stratification. The whole history of these rocks, moreover, shows that their various alternating strata were deposited, not as precipitates from a seething solution, but under conditions of sedimentation not unlike those of more recent times. In the oldest known of them, the Laurentian system, great limestone formations are interstratified with gneisses, quartzites, and even with conglomerates. All analogy, moreover, leads us to conclude that, even at this early period, life existed at the surface of the planet. Great accumulations of iron-oxide, beds of metallic sulphides and of graphite, exist in these oldest strata, and we know of no other agency than that of organic matter capable of generating these products. [" The presence of graphite, of native iron, and of sulphurets in most

* Researches in Theoretical Geology, pp. 297 – 300.

† Etudes et expériences synthétiques sur le metamorphisme, pp. 119 – 121.

aërolites, not to mention the hydrocarbonaceous matters which they sometimes contain, tells us in unmistakable language that these bodies come from a region where vegetable life has performed a part not unlike that which still plays on our globe, and even leads us to hope for the discovery in them of organic forms which may give us some notion of life in other worlds than our own." *]

Bischof had already arrived at the conclusion, which in the present state of our knowledge seems inevitable, that "all the carbon yet known to occur in a free state can only be regarded as a product of the decomposition of carbonic acid, and as derived from the vegetable kingdom." He further adds, "living plants decompose carbonic acid, dead organic matters decompose sulphates, so that, like carbon, sulphur appears to owe its existence in a free state to the organic kingdom." † As a decomposition (deoxidation) of sulphates is necessary to the production of metallic sulphides, the presence of the latter, not less than that of free sulphur and free carbon, depends on organic bodies; the part which these play in reducing and rendering soluble the peroxide of iron, and in the production of iron-ores, is, moreover, well known. It was, therefore, that, after a careful study of these ancient rocks, I declared in May, 1858, that a great mass of evidence "points to the existence of organic life, even during the Laurentian or so-called azoic period." ‡

This prediction was soon verified in the discovery of the *Eozoön Canadense* of Dawson, the organic character of which is now admitted by most zoölogists and geologists of authority. But with this discovery appeared another fact, which afforded a signal verification of my theory as to the origin and mode of deposition of serpentine and pyroxene. The microscopic and chemical researches of Dawson and myself showed that the calcareous skeleton of this foraminiferal organism was filled

* The Chemistry of the Earth, § 19, in the Report of Smithsonian Institution for 1869.

† Bischof, Lehrbuch, 1st ed., II. 95; English ed., I. 252, 344.

‡ American Journal of Science (2), XXV. 436.

with the one or the other of these silicates in such a manner as to make it evident that they had replaced the sarcode of the animal, precisely as glauconite and similar silicates have, from Cambrian times to the present, filled and injected more recent foraminiferal skeletons. I recalled, in connection with this discovery, the observations of Ehrenberg, Mantell, and Bailey, and the more recent ones of Pourtales, to the effect that glauconite or some similar substance occasionally fills the spines of Echini, the cavities of corals and millepores, the canals in the shells of Balanus, and even forms casts of the holes made by burrowing sponges (Clionia) and worms. The significance of these facts was further illustrated by showing that the so-called glauconites differ considerably in composition, some of them containing more or less alumina or magnesia, and one from the tertiary limestones near Paris being, according to Berthier, a true serpentine.*

These facts in the history of Eozoön were first made known by me in May, 1864, in the American Journal of Science, and subsequently more in detail, February, 1865, in a communication to the Geological Society of London.† They were speedily verified by Dr. Gümbel, who was then engaged in the study of the ancient crystalline schists of Bavaria, and soon recognized the existence, in the limestones of the old Hercynian gneiss, of the characteristic *Eozoön Canadense*, injected with silicates in a manner precisely similar to that observed by Dawson and myself.‡ Later, in 1869, Robert Hoffmann described the results of a minute chemical examination of the Eozoön from Raspenau, in Bohemia, confirming the previous observations in Canada and Bavaria. He showed that the calcareous shell of the Eozoön, examined by him, had been injected by a peculiar silicate, which may be described as related in composition both to glauconite and to

* American Journal of Science (2), XL. 360 ; Report Geol. Survey of Canada, 1866, p. 231 ; and Quar. Geol. Jour., XXI. 71.

† American Journal of Science (2), XXXVII. 431; Quar. Geol. Jour., XXI. 67.

‡ Proc. Royal Bavar. Acad. for 1866 ; and Can. Naturalist, new series, III. 81.

chlorite. The masses of Eozoön he found to be enclosed and wrapped around by thin alternating layers of a green magnesian silicate allied to picrosmine, and a brown non-magnesian mineral, which proved to be a hydrous silicate of alumina, ferrous oxide and alkalies, related to fahlunite, or more nearly to jollyte in composition.*

Still more recently, Dr. Dawson has detected a crystalline silicated mineral insoluble in dilute acids, injecting the pores of crinoidal stems and plates in a palæozoic limestone from New Brunswick, which is made up of organic remains. This silicate, which, in decalcified specimens, exhibits in a beautiful manner the intimate structure of these ancient crinoids, I have found by analysis to be a hydrous silicate of alumina and ferrous oxide, with magnesia and alkalies, closely related to fahlunite and to jollyte. The microscopic examinations of Dr. Dawson show that this silicate had injected the pores of the crinoidal remains and some of the interstices of the associated shell fragments before the introduction of the calcite which cements the mass. I have since found a silicate almost identical with this occurring under similar conditions in a Silurian limestone said to be from Llangedoç in Wales.†

Gümbel, meanwhile, in the essay on the Laurentian rocks of Bavaria, in 1866, already referred to, fully recognized the truth of the views which I had put forward, both with regard to the mineralogy of Eozoön and to the origin of the crystalline schists. His results are still further detailed in his Geognost. Beschreibung des östbayerisches Grenzegebirges, 1868, p. 833. Credner, moreover, as he tells us,‡ had already, from his mineralogical and lithological studies, been led to admit my views as to the original formation of serpentine, pyroxene, and similar silicates (which he cites from my paper of 1865, above referred to §), when he found that Gümbel had arrived at

* Jour. fur. Prakt. Chem., May, 1869; and American Journal of Science (3), I. 378.

† American Journal of Science (3), I. 379, and II. 57.

‡ Hermann Credner; die Gleiderung der Eozoischen Formationsgruppe Nord Amerikas. Halle, 1869.

§ That in the Quar. Geol. Jour., XXI. 67.

similar conclusions. The views of the latter, as cited by Credner from the work just referred to, are in substance as follows : the crystalline schists, with their interstratified layers, have all the characters of altered sedimentary deposits, and from their mode of occurrence cannot be of igneous origin, nor the result of epigenic action. The originally formed sediments are conceived to have been amorphous, and under moderate heat and pressure to have arranged themselves, and crystallized, generating various mineral species in their midst by a change, which, to distinguish it from metamorphism by an epigenic process, Gümbel happily designates *diagenesis.**

It is unnecessary to remark that these views, the conclusions from the recent studies of Gümbel in Germany and Credner in North America, are identical with those put forth by me in 1860.†

[* The following is extracted from an essay by the author in the Report of the Smithsonian Institution for 1869, on The Chemistry of the Earth, § 33 : "The gradual transformation of amorphous precipitates under water into crystalline aggregates, so often observed in the laboratory, appears to depend upon partial solution and re-deposition of the material, which must not be entirely insoluble in the surrounding liquid. If the solvent power of this be reduced, the dissolved portions are deposited on certain particles rather than others. By a subsequent exaltation of the solvent power of the liquid, solution of a further portion takes place, and this, in its turn, is deposited around the nuclei already formed, which are thus augmented at the expense of the smaller particles, until these at length disappear, being gathered to the crystalline centres. Such a process, which has been studied by H. Deville, suffices, under the influence of the changing temperature of the seasons, to convert many fine precipitates into crystalline aggregates, by the aid of liquids of slight solvent powers. A similar agency may be supposed to have effected the crystallization of buried sediments, and changes in the solvent power of the permeating water might be due either to variations of temperature or of pressure. Simultaneously with this process one of chemical union of heterogeneous elements may go on, and in this way, for example, we may suppose the carbonates of lime and magnesia become united to form dolomite or magnesian limestone."]

[† Since the first publication of the above address I have received in a private letter from Gümbel the following re-statement of his views as to the origin of crystalline rocks : "I have seen no occasion to change my opinions, which are, I believe, identical with your own. I do not maintain a metamorphic origin for the primitive rocks; for, although these are certainly much altered, there are no firm and consolidated rocks which are not so. They were formed like, for example, the limestones of more recent periods; these

At the early periods in which the materials of the ancient crystalline schists were accumulated, it cannot be doubted that the chemical processes which generated silicates were much more active than in more recent times. The heat of the earth's crust was probably then far greater than at present, while a high temperature prevailed at comparatively small depths, and thermal waters abounded. A denser atmosphere, charged with carbonic-acid gas, must also have contributed to maintain, at the earth's surface, a greater degree of heat, though one not incompatible with the existence of organic life. (*Ante*, page 46.) These conditions must have favored many chemical processes, which, in later times, have nearly ceased to operate. Hence we find that subsequently to the eozoic times, silicated rocks of clearly marked chemical origin are comparatively rare. In the mechanical sediments of later periods certain crystalline minerals may be developed by a process of molecular rearrangement, — diagenesis. These are, in the feldspathic and aluminous sediments, orthoclase, muscovite, garnet, staurolite, cyanite, and chiastolite, and in the more basic sediments, hornblendic minerals. It is possible that these latter and similar silicates may sometimes be generated by reactions between silica on the one hand and carbonates and oxides on the other, as already pointed out in some cases of local alteration. Such a case may apply to more or less hornblendic gneisses, for example, but no sediments, not of direct chemical origin, are pure enough to have given rise to the great beds of serpentine, pyroxene, steatite, labradorite, etc., which abound in the ancient crystalline schists. Thus, while the materials

were once pastes, magmas or muds, and so were the primitive rocks at the time of their origin, but during these first ages of the earth the consolidating and crystallizing forces (differing in degree only from those of the present time, and aided by a higher temperature) allowed the magma to assume the form of mineral species, more or less distinct. If we choose to call this change metamorphism, then the rocks thus formed are metamorphic; but so also are the limestones of later periods. The primitive rocks originated by way of sedimentation, the one after the other, constituting distinct formations, and there are no eruptive gneisses." See, in this connection, the Introduction to Essay III. of the present volume, and the statements of Favre in the Appendix to Essay XIV.]

for producing, by diagenesis, the aluminous silicates just mentioned are to be met with in the mud and clay-rocks of all ages, the chemically formed silicates, capable of crystallizing into pyroxene, talc, serpentine, etc., have only been formed under special conditions. [While the generation of various crystalline silicated minerals in rocks since the Eozoic age is theoretically not impossible, the accumulation of evidence goes to show that although such changes have taken place locally in the proximity of eruptive rocks, and by the invasion of thermal waters, there has been no wide-spread alteration or regional metamorphism, as it has been called, of these more recent sedimentary deposits.]

The same reasoning which led me to maintain the theory of an original formation of the mineral silicates of the crystalline schists, induced me to question the received notion of the epigenic origin of gypsums and magnesian limestones or dolomites. The interstratification of dolomites and pure limestones, and the enclosure of pebbles of the latter in a paste of crystalline dolomite, are of themselves sufficient to show that in these cases, at least, dolomites have not been formed by the alteration of pure limestones. The first results of a very long series of experiments and inquiries into the history of gypsum were published by me in 1859, and further researches, reiterating and confirming my previous conclusions, appeared in 1866. (*Ante*, page 80.) In these two papers it will, I think, be found that the following facts in the history of dolomite are established: namely, first, its origin in nature by direct sedimentation, and not by the alteration of non-magnesian limestones; second, its artificial production by the direct union of carbonate of lime and hydrous carbonate of magnesia, at a gentle heat, in the presence of water. As to the sources of the hydrous magnesian carbonate, I have endeavored to show that it is formed from the magnesian chloride or sulphate of the sea or other saline waters in two ways: first, by the action of the bicarbonate of soda found in many natural waters; this, after converting all soluble lime-salts into insoluble carbonate, forms a comparatively soluble bicarbonate of magnesia, from which a hydrous carbonate slowly separates; second, by the action of bicarbonate of lime in solution, which, with sulphate of magnesia, gives rise to gypsum; this first crystallizes out, leaving behind a much more soluble bicarbonate of magnesia, which deposits the hydrous carbonate in its turn. In this way, for the first time, in 1859, the origin of gypsums and their intimate relation with magnesian limestones were explained.*

It was, moreover, shown that, to the perfect operation of this reaction, an excess of carbonic acid in the solution during the evaporation was necessary to prevent the decomposing action of the hydrous mono-carbonate of magnesia upon the already formed gypsum. Having found that a prolonged exposure to the air, by permitting the loss of carbonic acid, partially interfered with the process, I was led to repeat the experiment in a confined atmosphere, charged with carbonic acid, but rendered drying by the presence of a layer of desiccated chloride of calcium. As had been foreseen, the process under these conditions proceeded uninterruptedly, pure gypsum first crystallizing out from the liquid, and, subsequently, the hydrous magnesian carbonate.† This experiment is instructive, as showing the results which must have attended this process in past ages, when the quantity of carbonic acid in the atmosphere greatly exceeded its present amount. (*Ante*, pages 43, 47, and 91.)

As regards the hypotheses put forward to explain the supposed dolomitization of previously formed limestones by an epigenic process, I may remark that I repeated very many times, under varying conditions, the often-cited experiment of Von Morlot, who claimed to have generated dolomite by the action of sulphate of magnesia on carbonate of lime, in the presence of water at a somewhat elevated temperature under pressure. I showed that what he regarded as dolomite was not such, but an admixture of carbonate of lime with anhydrous and sparingly soluble carbonate of magnesia; the conditions in which the carbonate of magnesia is liberated in this reaction not being favorable to its union with the carbonate of lime to form the double salt

* See the recent conclusions of Ramsay, noticed *ante*, page 92.

† Canadian Naturalist, new series.

which constitutes dolomite. The experiment of Marignac, who thought to form dolomite by substituting a solution of chloride of magnesium for the sulphate, I found to yield similar results, the greater part of the magnesian carbonate produced passing at once into the insoluble condition, without combining with the excess of carbonate of lime present. The process for the production of the double carbonate described by Charles Deville, namely, the action of vapors of anhydrous magnesian chloride on heated carbonate of lime, in accordance with Von Buch's strange theory of dolomitization, I have not thought necessary to submit to the test of experiment, since the conditions required are scarcely conceivable in nature. Multiplied geognostical observations show that the notion of the epigenic production of dolomite from limestone is untenable, although its re-solution and deposition in veins, cavities, or pores in other rocks is a phenomenon of frequent occurrence.

The dolomites or magnesian limestones may be conveniently considered in two classes: first, those which are found with gypsums at various geological horizons; and, second, the more abundant and widely distributed rocks of the same kind, which are not associated with deposits of gypsum. The production of the first class is dependent upon the decomposition of sulphate of magnesia by solutions of bicarbonate of lime, while those of the second class owe their origin to the decomposition of magnesian chloride or sulphate by solutions of alkaline bicarbonates. In both cases, however, the bicarbonate of magnesia, which the carbonated waters generally contain, contributes a more or less important part to the generation of the magnesian sediments. The carbonated alkaline waters of deep-seated springs often contain, as is well known, besides the bicarbonates of soda, lime and magnesia, compounds of iron, manganese, and many of the rarer metals in solution; and thus the metalliferous character of many of the dolomites of the second class is explained. The simultaneous occurrence of alkaline silicates in such mineral waters would give rise, as already pointed out, to the production of insoluble silicates of magnesia, and thus the frequent association of such silicates with dolomites and magnesian carbonates in the crystalline schists is explained, as marking portions of one continuous process. The formation of these mineral waters depends upon the decomposition of feldspathic rocks by subterranean or sub-aerial processes, which were doubtless more active in former ages than in our own. The subsequent action upon magnesian waters of these bicarbonated solutions, whether alkaline or not, is dependent upon climatic conditions; since, in a region where the rain-fall is abundant, such waters would find their way down the river-courses to the open sea, where the excess of dissolved sulphate of lime would prevent the deposition of magnesian carbonate. It is in dry and desert regions, with closed lake or sea basins, that we must seek for the production of magnesian carbonates; and I have argued from these considerations that much of northeastern America, including the present basins of the Upper Mississippi, Ohio, and St. Lawrence, must, during long intervals in the palæozoic period, have had a climate of excessive dryness, and a surface marked by shallow enclosed basins, as is shown by the widely spread magnesian limestones, and by the existence of gypsum and rock-salt at more than one geological horizon within that area.* (*Ante*, page 76.) The occurrence of serpentine and diallage at Syracuse, New York, offers a curious example of the local development of crystalline magnesian silicates in Silurian dolomitic strata, under conditions which are imperfectly known, and, in the present state of the locality, cannot be studied.†

Since the uncombined and hydrated magnesian mono-carbonate is at once decomposed by sulphate or chloride of calcium, it follows that the whole of these lime-salts in a sea-basin must be converted into carbonates before the production of carbonated magnesian sediments can begin. The carbonate of lime formed by the action of carbonates of magnesia and soda remains at first dissolved, either as carbonate (*ante*, page 140) or as bicarbonate, and is only separated in a solid form, when in excess,

* Geology of Southwestern Ontario, American Journal of Science (2), XLVII. 355.

† Geology of the Third District of New York, 108-110; and Hunt on Ophiolites, American Journal of Science (2), XXVI. 236.

or when required for the needs of living plants or animals, which are dependent for their supply of calcareous matter on the carbonate of lime produced, in part by the process just described, and in part by the action of carbonic acid on insoluble lime-compounds of the earth's solid crust. So many limestones are made up of calcareous organic remains, that a notion exists among many writers on geology that all limestones are, in some way, of organic origin. At the bottom of this lies the idea of an analogy between the chemical relations of vegetable and animal life. As plants give rise to beds of coal, so animals are supposed to produce limestones. In fact, however, the synthetic process by which the growing plant, from the elements of water, carbonic acid, and ammonia, generates hydrocarbonaceous and azotized matters, has no analogy with the assimilative process by which the growing animal appropriates alike these organic matters and the carbonate and phosphate of lime. Without the plant, the synthesis of the hydrocarbons would not take place; while, independently of the existence of coral or mollusk, the carbonate of lime would still be generated by chemical reactions, and would accumulate in the waters until, these being saturated, its excess would be deposited as gypsum or rock-salt are deposited. Hence, in such waters, where, from any causes, life is excluded, accumulations of pure carbonate of lime may be formed. In 1861 I called attention to the white marbles of Vermont, which occur intercalated among impure and fossiliferous beds, as apparently examples of such a process.*

It is by a fallacy similar to that which prevails as to the organic origin of limestones, that Daubeny and Murchison were led to appeal to the absence of phosphates from certain old strata, as evidence of the absence of organic life at the time of their accumulation.† Phosphates, like silica and iron-oxide, were doubtless constituents of the primitive earth's crust, and the production of apatite crystals in granitic veins or in crystalline schists is a process as independent of life as the forma-

* American Journal of Science (2), XXXI. 402.

† Siluria, 4th ed., pp. 28 and 537.

tion of crystals of quartz or of hematite. Growing plants, it is true, take up from the soil or the waters dissolved phosphates, which pass into the skeletons of animals, — a process which has been active from very remote periods. I showed, in 1854, that the shells of Lingula and Orbicula, both those from the base of the palæozoic rocks and those of the present time, have (like Conularia and Serpulites) a chemical composition similar to the skeletons of vertebrate animals.* The relations of both carbonate and phosphate of lime to organized beings are similar to those of silica, which, like them, is held in watery solution, and by processes independent of life, is deposited both in amorphous and crystalline forms, but in certain cases is appropriated by diatoms and sponges, and made to assume organized shapes. In a word, the assimilation of silica, like that of phosphate and carbonate of lime, is a purely secondary and accidental process; and where life is absent, all of these substances are deposited in mineral and inorganic forms.

* American Journal of Science (2), XVII. 236.

APPENDIX.

REPLY TO MR. DANA'S CRITICISMS.

In the American Journal of Science for February, 1872, Professor Dana has criticised certain points in my address, On the Geognosy of the Appalachians and the Origin of Crystalline Rocks, given in August, 1871, at Indianapolis, before the American Association for the Advancement of Science. I am charged by him with rejecting, for many mineral silicates, the view that they are pseudomorphs; that is to say, crystals chemically altered without loss of external form. I have denied that crystals of serpentine having the shape of chrysolite, pyroxene, dolomite, etc., and crystals of pinite having the shapes of nepheline or scapolite, are results of a chemical change of these species, nothwithstanding this view is now held by most mineralogists, on the grounds of similarities of geometrical form and the existence of what are regarded as intermediate stages in the process of transmutation; and I have maintained another and a very different view, which, in my opinion, is more rational. Until we can watch the transmutation of one of these species into another, the argument from supposed intermediate forms is worth no more in the mineral than in the organic world; the reasoning of the transmutationists, in the one case and the other, resting upon somewhat similar considerations. In either case we may say, with Professor Warrington Smyth, that in these intermediate forms "lie the materials for a history"; while we venture, with him, to express a doubt whether, from a series of specimens supposed to show a transition from chrysolite to serpentine, or from hornblende to chlorite, "we are obliged to conclude that there has been, *historically speaking*, an actual transition from the one to the other." (See his anniversary address, as President of the Geological Society of London, in 1867.)

Professor Dana says that Scheerer is the only one who shares my peculiar views on this question. I have, however, asserted in my address that Delesse has maintained the views of Scheerer and myself, as opposed to the popular doctrine of epigenesis, and shall endeavor to make good my assertion. In his essay on Pseudomorphs, published in 1859 (Ann. des Mines (5), XVI. 317 – 392),

Delesse begins his argument by remarking that since, in some cases, a mineral is found to be surrounded by another clearly resulting from its alteration (as, for example, anhydrite by gypsum), certain mineralogists have supposed that wherever one mineral encloses another there has been epigenesis or pseudomorphous alteration. Such, he says, may sometimes be the case, but it is easy to see that it is not so habitually. A crystallized mineral species frequently includes a large and even a predominating portion of another, and the combination is then considered by many as an example of partial pseudomorphous alteration. In such instances, remarks Delesse, the question arises whether we have to do with the results of envelopment or of chemical alteration ; to resolve which it becomes necessary to study carefully the problem of envelopment. He then proceeds to show that the enveloped substance is, in some cases, crystalline (and arranged either symmetrically or asymmetrically with regard to the enveloping mass) ; while in other cases it is amorphous, and enclosed like the sand-grains which predominate in the calcite crystals of Fontainebleau. The difficulty in deciding whether we have to do with envelopment or with epigenesis increases when the enveloped mineral becomes so abundant as to obscure the enveloping species, or when it becomes mixed with it in so intimate a manner as to seem one with the latter (*se fondre insensiblement avec lui*). The proportions of the enveloped and the enveloping mineral, we are told, may so far vary that the one or the other is no longer recognizable. "As the forces which determine crystallization have a great energy, the enveloping mineral is sometimes found in so small a quantity as to be entirely masked by the enveloping species." "When minerals have crystallized simultaneously, they have been able to become associated with each other and to envelop each other in all proportions" (loc. cit., pages 338, 339, 341, 353).

Our author then proceeds to tell us that, having carefully studied in numerous specimens the supposed mica-pseudomorphs of iolite, andalusite, cyanite, pyroxene, hornblende, etc., he regards them as, in all cases, examples of envelopment, and expresses the opinion that we must omit from our lists a great number of the so-called pseudomorphous minerals, especially among the silicates. The final result of the process of envelopment is, according to Delesse, this, — to give rise to mixed mineral aggregates, owing their external forms to the crystallizing energy of one of the constituents, which may be present in so small a quantity as to be completely obscured by the other matter present. From this condition of things result crystalline forms which, though totally different in their origin from the products of chemical alteration or substitution, are emphatically pseudomorphs.

From this process of mechanical and more or less heterogeneous envelopment, Delesse next proceeds to consider the crystallizing together of isomorphous or homœomorphous species, in relation to the generally received notion of epigenic pseudomorphism. He declares that "isomorphism explains very well facts which are often attributed to pseudomorphism," and that many "minerals which are still considered pseudomorphs are in reality examples of isomorphism" (pages 364, 365). Referring to the well-known investigations of Mitscherlich upon the crystallizing together, in all proportions, of isomorphous species, and of the symmetrical crystallization of one salt around a nucleus of another isomorphous with it, Delesse suggests that the different forms and varieties of hornblendic and pyroxenic minerals afford many examples of the kind. He then adds, "If, as Scheerer has remarked, water plays in silicates the part of a base, anhydrous silicates may crystallize at the same time with hydrated silicates, and, moreover, be isomorphous with them." In this way, he suggests, we may explain by isomorphism or homœomorphism, the association with pyroxene of the hydrous species, schiller-spar, as well as that "of various anhydrous and hydrated minerals" (pages 357, 358).

In further illustration of the words just quoted from Delesse, we may cite from Scheerer, as examples of what he called polymeric isomorphism, the association (in the same crystals) of iolite and aspasiolite, and of chrysolite and serpentine. If these and similar species crystallize together because they are isomorphous, it is evident that they may each crystallize separately; and thus the crystals of serpentine with the form of chrysolite, and those of aspasiolite and other so-called hydrous iolites, may be regarded as examples, not of epigenesis, but of isomorphism.

We have thus endeavored to set forth, chiefly in his own words, the views enunciated in 1859 by Delesse, according to whom the phenomena of so-called pseudomorphism among mineral silicates are to be explained, for the most part, not by chemical alterations of pre-existing species, but by envelopment and by isomorphism. That the above are really his views, and are, moreover, regarded by himself as contrary to those of the school which I oppose, Delesse does not permit us to doubt; for, after having set them forth as his own (*après avoir exposé notre manière de voir*), he says, "We hasten to add that these facts may also be explained in a manner altogether different (*peuvent aussi s'interpreter d'une manière toute différente*); and some *savans* of Germany, notably G. Rose, Haidinger, Blum, G. Bischof, and Rammelsberg, have sought their explanation in pseudomorphism. Their example has been followed by most mineralogists, etc." (pages 358, 359).

That the "pseudomorphism" of the authors just named is chemical alteration or epigenesis, it is not necessary to remind the reader, who will now be able to judge whether it is Professor Dana or myself who has misrepresented or misunderstood Delesse. Let us, however, add that the long and somewhat diffuse memoir of the latter, from which we have quoted, is wanting in unity of plan and purpose, and that parts of it, if we may hazard a conjecture, seem to have been written while he still inclined to the views of the opposite school. From the table of pseudomorphs which he has given, and from many passages in the text, it might be inferred that he then held the notions of Rose, Haidinger, etc., which he elsewhere, in the same paper, speaks of as being entirely different from his own. The views of Delesse, about this time, underwent a great change, which has a historic importance in connection with those which I advocate. When, in 1857 and 1858, he published the first and second parts of his admirable series of studies on metamorphism, Delesse held, in common with nearly every geologist of the time, to the eruptive origin of serpentine and the related magnesian rocks. Serpentine was then classed by him with other "trappean rocks"; and he elsewhere asserted that "granitic and trappean rocks" undergo in certain cases a change near their contact with the enclosing rock, by which they lose silica, alumina, and alkalies, and acquire magnesia and water, being thus changed into a magnesian silicate, which may take the form of saponite, serpentine, talc, or chlorite (Ann. des Mines (5), XII. 509; XIII. 393, 415). It would be difficult to state more distinctly the view, which he then held, of the origin of these magnesian rocks and minerals by the chemical alteration of plutonic (granitic and trappean) rocks. This was in 1858, and in 1859 appeared the memoir on pseudomorphs, already noticed, in which, in place of the theory of epigenic pseudomorphism, or chemical alteration of various mineral silicates, taught by the German school, he brought forward, in explanation of the facts upon which this was based, another theory, which was only an extension of that already maintained by Scheerer and myself.

It was not until 1861 that Delesse published the last part of his studies on metamorphism, which appeared in the Memoirs of the Academy of Sciences of France (Vol. XVII.); and in it we find that, consistently with the new views adopted by him in 1859, the old doctrine of the epigenic origin of serpentine and the related magnesian rocks from the alteration of plutonic rocks is abandoned. In its stead, it is here suggested by Delesse that all these magnesian rocks result from the crystallization of the sepiolites or so-called magnesian clays, which are frequent in many sedimentary deposits. These, according to him, by a molecular rearrangement of their elements, may give rise to serpentine, talc, chlorite, and their various associated and related minerals. The rocks thus generated are still declared to pass insensibly into plutonic rocks; but, instead of maintaining, as in 1858, that they are derived from the latter, Delesse, in 1861, asserts, on the contrary, that "the plutonic rocks are formed from the metamorphic rocks, and represent the maximum of intensity, or extreme limit of metamorphism."

This recognition of the notion that the great masses of serpentine, with their constantly associated hornblendic, talcose and chloritic rocks, have been directly formed from the molecular rearrangement or *diagenesis* of aqueous magnesian sediments, and not from the chemical alteration or *epigenesis* of erupted plutonic masses, marks a complete revolution in our views of the history of the crystalline rocks. The new doctrine did not, however, originate with Delesse, but was previously put forward by myself in a paper, On some Points of Chemical Geology, read before the Geological Society of London in January, 1859, appearing in abstract in the Philosophical Magazine for February, and published at length in the Geological Journal for November, in the same year. I there maintained that serpentines were "undoubtedly indigenous rocks, resulting from the alteration of silico-magnesian sediments"; and moreover asserted that the final result of heat, aided by water, on such rocks, would be their softening, and, in certain cases, their extravasation as plutonic rocks; which were regarded "as, in all cases, altered and displaced sediments." When this paper was written, in 1858, I still supposed that the reactions between the elements in beds of silicious magnesian carbonates (which, I had shown, may give rise to certain magnesian silicates in immediate proximity to eruptive rocks) might serve to explain the origin of great areas of serpentine and related crystalline magnesian silicates; but my studies of the silicates deposited during the evaporation of natural waters, and of the magnesian sediments of the Paris basin, soon led me to seek the origin of these rocks in the alteration of previously formed uncrystalline magnesian silicates. This view was set forth by me in the American Journal of Science for March, 1860 ((2), XXIX. 284), and more fully in the Canadian Naturalist for June, 1860 (also in the American Journal (2), XXXII. 286), where it was pointed out that steatite, chlorite and serpentine were probably derived from sediments similar to the magnesian silicates found among the tertiary beds in the vicinity of Paris, the so-called magnesian clays.

We have seen that these various novel views, put forth by me in 1859 and 1860, though totally different from those taught by Delesse in 1858, were integrally adopted by him in 1861. These dates are circumstantially given in my address of last year, and yet Professor Dana, in his review of it, charges me with "following nearly Delesse" as to the origin of serpentine. He also asserts that I "make Delesse the author of the theory of envelopment," when I have there declared that the view of Delesse — "that the so-called cases of pseudomorphism, on which the theory of metamorphism by alteration has been built, are, for the most part, examples of association and envelopment, and the result of a contemporaneous and original crystallization — is identical with the view suggested by Scheerer in 1846, and generalized by myself, when, in 1853, I sought to explain the phenomena in question by the association and crystallizing together of homologous and isomorphous species." To Delesse, therefore, belongs the merit, not of having suggested the notion of envelopment in this connection, but of having pointed out the bearing of the envelopment of heteromorphous and amorphous species on the question before us.

Professor Dana moreover asserts that, while Scheerer is the only one who maintains similar views to myself, I, in common with all other chemists, reject the chemical speculations which lie at the base of his views. On the contrary, unlike most chemists, who have failed to see the great principle which underlies Scheerer's doctrine of polymeric isomorphism, I have maintained (American Journal of Science (2), XV. 230; XVI. 218) that it enters into a general law, in accordance with which bodies whose formulas differ by nM_2O_2 or nH_2O_2 may (like those differing by nH_2C_2) have relations of homology, and moreover be isomorphous. (See, further, Paper XVII. of the present volume.) The existence of these same relations was further maintained and exemplified in a paper on Atomic Volumes, read by me before the French Academy of

Sciences and published in the Comptes Rendus of July 9, 1855. This doctrine, which I have never repudiated, is reiterated in my address last year (*ante*, page 291), and declared to include the polymeric isomorphism of Scheerer.

Professor Dana next says that, in asserting that "the doctrine of pseudomorphism by alteration, as taught by G. Rose, Haidinger, Blum, Volger, Rammelsberg, Dana, Bischof, and many others, leads them . . . . to maintain the possibility of converting almost any silicate into any other," I have "grossly misrepresented the views of at least Rose, Haidinger, Blum, Rammelsberg, and Dana" ; and that I "complete the caricature" by this sentence, to be found in my address : "In this way we are led from gneiss or granite to limestone, from limestone to dolomite, and from dolomite to serpentine ; or more directly from granite, granulite, or diorite, to serpentine at once, without passing through the intermediate stages of limestone and dolomite" ; — "part of which transformations," says Professor Dana, "I, for one, had never conceived ; and Rose, Haidinger, Rammelsberg, and probably Blum, and the 'many others,' would repudiate them as strongly as myself." The "many others," as he rightly remarks, are "other writers on pseudomorphism," among whom it would be unjust not to name their progenitor, Breithaupt, Von Rath, and Müller, at the same time with Volger and Bischof. According to Professor Dana, I "add to the misrepresentation by means of the strange conclusion that, because such writers hold that crystals may undergo certain alterations in composition, therefore they believe that rocks of the same constitution may undergo the same changes." This "strange conclusion" I have always supposed to be Professor Dana's own. No one has perhaps asserted it so clearly or so broadly as himself, and I shall therefore quote his own words in my justification. As early as 1845, in an article entitled Observations on Pseudomorphism (American Journal of Science (1), XLVIII. 92), he wrote : "The same process which has altered a few crystals to quartz has distributed silica to fossils without number, scattered through rocks of all ages. The same causes that have originated the steatitic scapolites occasionally picked out of the rocks, have given magnesia to whole rock-formations, and altered, throughout, their physical and chemical characters. If it be true that the crystals of serpentine are pseudomorphous crystals, altered from chrysolite, it is also true, as Breithaupt has suggested, that the beds of serpentine containing them are likewise altered, though often covering square leagues in extent, and common in most primary formations. The beds of steatite, the still more extensive talcose formations, contain everywhere evidence of the same agents." Again, in 1854, in his Mineralogy, 4th edition (page 226), Professor Dana, after a complete list of pseudomorphs, compiled from the writers of the school in question, says: "These examples of pseudomorphism should be understood as cases not simply of alteration of crystals, but in many instances of changes in beds of rock. Thus all serpentine, whether in mountain-masses or the simple crystal, has been formed through *a process of pseudomorphism, or in more general language, of metamorphism;* the same is true of other magnesian rocks, as steatitic, talcose, or chloritic slates. *Thus the subject of metamorphism, as it bears on all crystalline rocks, and of pseudomorphism, are but branches of one system of phenomena.*" If there could be any doubt as to the meaning of the words which I have italicized in quoting them from Professor Dana, it is removed by his language in 1858. Then, as now, adversely criticising my views on this question, he refers to the statements above cited, made in 1845 and 1854, as expressions of his doctrine, mentioning especially the first one, in which he says, "*metamorphism is spoken of as pseudomorphism on a broad scale.*" (American Journal of Science (2), XXV. 445.) I confess that I do not understand Professor Dana, when in his last criticism of me, fourteen years after the one just quoted, he reproaches me with having charged him with holding the doctrine that "*regional metaphorphism is pseudomorphism on a grand scale*"; and declares that he makes no such remark, neither expresses the sentiment in his Mineralogy of 1854.

With these citations before us, and remembering the views of Scheerer, and the later ones of Delesse, together with the language of the latter in his essay on Pseudomorphs, let us notice the words of Naumann, addressed to Delesse in 1861, in allusion to the essay in question: "Permit me to express to you my satisfaction for the ideas enunciated in your memoir on Pseudomorphs, — ideas which my friend Scheerer will doubtless share with myself" (*idées que mon ami M. Scheerer partagera sans doute comme moi-même*). Then follows the language which I have quoted in my address, in which he combats the error of those who hold that gneisses, amphibolites, and other crystalline rocks are "the results of metamorphic epigenesis, and not original rocks," and adds, "It is precisely *because pseudomorphism has so often been confounded with metamorphism* that this error has found acceptance." (Bull. Soc. Geol. de France (2),

XVIII. 678.) The reader must now judge whose opinions it is that are here denounced as erroneous, and whether Naumann was on the side of Professor Dana, or, with Delesse, on the side of Scheerer and myself. I insist the more strongly on this matter, because Professor Dana not only declares that Delesse and Naumann have always avowed the doctrines of the transmutationist school, and do not in any way whatever countenance my views, but implies that I have dealt unfairly with these authorities.

Professor Dana says, " If there was any occasion for a notice of my opinions, a critic of 1871 should have referred to the formal expression of them in my Manual of Geology, first published in 1863. The reader will there find the *diagenesis* of Gümbel, which Mr. Hunt takes occasion to commend, ... with but a brief allusion to pseudomorphism." The doctrine of diagenesis, it is hardly necessary to say, I have never attributed to Gümbel, nor does he claim it. It is the old doctrine of Hutton, Playfair, and Bouë, is taught by Bischof (Chemical Geology, III. 318, 325, 342), and pervades my papers of 1859 and 1860, already referred to. But while it has been generally admitted that what, in my address, I have called the first class of crystalline rocks (consisting chiefly of quartz and aluminous silicates) might result from the molecular rearrangement of the elements of clay and sand-rocks, I maintained in those papers that what I have called the crystalline rocks of the second class (in which protoxide-silicates predominate) have been generated, by a similar process, from aqueous deposits of chemically formed silicates. This view, advanced by me in 1860, having been adopted by Delesse and by Gümbel to explain the origin of the various magnesian silicated rocks hitherto generally regarded as the product of epigenesis, the latter has proposed to designate the process as diagenesis; a term which I adopt, as one well fitted to denote the generation of all kinds of crystalline rocks through a molecular rearrangement of sedimentary deposits, of whatever origin. Professor Dana, in common with most other geologists, admits in his Manual of Geology the old conception of the production by diagenesis from mechanical sediments of the rocks of the first class, but in the case of serpentine and steatite declares them to have been formed by epigenic pseudomorphism or chemical alteration of pyroxenic and other crystalline rocks; *the origin of which is by him left entirely unexplained.* It is true that his allusions to pseudomorphism in that volume are confined to very brief notices on pages 704 and 710; a fact which is the more noticeable, when we recall that the author had formerly expressed the belief "that pseudomorphism will soon constitute one of the most important chapters in geological treatises." (American Journal of Science (1), XLVIII. 66.) That Professor Dana has receded from the extreme views on this subject which he maintained from 1845 to 1858, and which I have constantly opposed, seems probable; but until he formally rejects them, the student of geology will not unnaturally suppose that he still gives the sanction of his authority to the doctrine which he so long taught without any qualification, but now repudiates, that "*metamorphism is pseudomorphism on a broad scale.*"

[In the Neues Jahrbuch für Mineralogie for November, 1872 (page 865), appeared a note from the venerable Carl Friedrich Naumann (who has since died at an advanced age), in which he comments upon my interpretation of his letter to Delesse. He begins by saying that I have, in my address in 1871, cited some passages from that letter, of which he then proceeds to repeat the substance, and adds: "Although I am still strongly opposed to the excesses of the metamorphic doctrine, I cannot explain how Professor Sterry Hunt can, from the extracts of my letter to Delesse, conclude that I regard those cases of pseudomorphism upon which the theory of metamorphism is grounded as in great part only examples of association and development, and also as a result of a simultaneous and original crystallization, and that my view is identical with his own, which he first put forth in the year 1853."

Upon this I have to remark that, instead of citing in my address extracts from his published letter to Delesse, I gave therein a translation of the whole letter, with the exception of the first three lines, which are, however, given above, with some other extracts, in my reply to Dana's criticisms. From this language I conclude that Naumann knew my address only through these misleading criticisms and my reply thereto. In the next place, it is not clear what were the excesses of the metamorphic doctrine which he still condemned in 1872. He, as we have shown from his Lehrbuch (*ante*, page 294), regarded gneisses and similar rocks as, for the most part, in some unexplained way, of plutonic origin, though he admitted their production in certain cases by the alteration of sediments, agreeable to the Huttonian view of diagenesis; while in the letter above mentioned he characterizes as erroneous the very different notion that all "gneisses, amphibolites, etc.," are "the results of metamorphic epigenesis." From his language in 1872, however, it would appear that, in his opinion, such rocks, once formed, may become the subjects of epigenic pseudomorphism, and be metamorphosed, as supposed by Bischof, Dana, and others, into serpentines, steatites, etc. In this case his implied sympathy, in 1861, with the teachings of Scheerer, who, in denying the epigenic origin of the serpentine associated with chrysolite and many similar cases, had struck a blow, in the language of Naumann, at "those cases of pseudomorphism upon which the theory of metamorphism is grounded"; and finally, his congratulations to Delesse (who had just declared that often "the so-called metamorphism finds its natural explanation in envelopment," and asserted the view of Scheerer and myself that much of what has been regarded as pseudomorphism has no other basis than the observed associations of mineral species) could, in my opinion, admit of no other interpretation than the one which I in 1871 gave to it. There is a confusion, not to say a contradiction, in these expressed views of the venerable teacher, which it is not easy to explain.

Nothing has been further from my intention than to misrepresent the views either of Naumann or of Dana; and my error, if I have fallen into one, arises from the difficulty of knowing their real opinions upon the matters in discussion. Let Professor Dana define, as clearly as I have done, his present views as to the origin of magnesian rocks, both those made up of chrysolite and pyroxenic minerals, and those composed of serpentine, steatite and chlorite, which he has supposed to come from an epigenesis of the former; let him tell us whether he holds the doctrine of pseudomorphic metamorphism which he taught in 1845, 1854 and 1858, and which, as I have shown, was held by Delesse as late as 1857, or that doctrine so long maintained by me, which the latter adopted in 1861. Such a definition would be eminently satisfactory to those who look to him as a teacher in science, and would prevent any further misconception or unintentional misrepresentation of his views.]

Professor Dana, having clearly defined the proposition that the chemical alterations which are recognized in individual crystals are to be conceived as extending to rock-masses, and having, moreover, asserted that the principle of the identity of metamorphism and pseudomorphism "bears on all crystalline rocks," is logically committed to all the deductions as to the changes of rocks which the transmutationist school has drawn from the supposed alteration of minerals. By reference to the table of pseudomorphs in the fourth edition of Dana's Mineralogy, it will be seen that each one of the metamorphoses of rocks mentioned in the above extract from my address is based upon an asserted epigenic change or conversion of the constituent species. I shall, however, show, in addition, that in each case the application of the principle to rock-masses has been recognized by one or more of the authorities already named, and that the so-called caricature has been drawn by their own hands. It would be easy, did space permit, to extend greatly this list of supposed transmutations. The various associations of rocks and minerals in nature, when interpreted according to the canons of this school, seem, in fact, as remarked by Professor Warrington Smyth, in his address already quoted, "to offer a premium to the ingenious for inventing an almost infinite series of possible combinations and permutations." Before proceeding further it is to be noted that no distinction can, in many cases, be established between the results of alteration (or partial replacement) and substitution (or complete replacement); since successive alterations may give the same product as direct substitution. Thus, for example, quartz might be directly replaced by calcite, or else first altered to a silicate of lime, which, in its turn, might be changed to carbonate. The alteration of quartz to a silicate of magnesia, and that of both pyroxene and pectolite to calcite, is maintained by the writers of the present school.

Metamorphosis of granite or gneiss to limestone: — Calcite, we are told, is pseudomorphous of quartz, of feldspar, of pyroxene, and of garnet, besides other species; it moreover replaces both orthoclase and albite "by some process of solution and substitution." (Dana's Mineralogy, 5th edition, 361.) Since quartz, orthoclase, and albite can be replaced by calcite, the transmutation of granite or gneiss into limestone presents no difficulty. [In the opinion of Messrs. King and Rowney, the crystalline limestones of Tyree in the Hebrides, those of Aker in Sweden, and similar limestones in the Laurentian of North America, were at one time beds of gneiss, diorite, and other silicated rocks, which have been changed by an epigenic process. (Annals and Magazine of Natural History for 1869, Vol. XIII. page 390.) Volger has also asserted a similar origin for certain gneissoid limestones.]

Metamorphosis of limestone to dolomite: — This change is maintained by Von Buch, Haidinger, and many others. I am blamed for mentioning in connection with this school the name of Haidinger, who, Professor Dana says, "never wrote upon the subject of the alteration of rocks." It will, however, be noticed, that his name has been quoted by Delesse with those of Bischof, Blum, and others as a disciple of this school, and it has never before been questioned that Haidinger was the first, if not to suggest, to clearly set forth, the theory of the supposed conversion of limestone into dolomite by the action of magnesian solutions, aided by heat and pressure, — a theory which I have elsewhere refuted. (Bischof, Chem. Geol., III. 155, 158; Zirkel, Petrographie, I. 246; Liebig and Kopp, Jahresbericht, 1847–48, 1289; and American Journal of Science (2), XXVIII. 376).

Metamorphosis of dolomite to serpentine: — This change is maintained by G. Rose (Bischof, Chem. Geol., II. 423), and by Dana (American Journal of Science (3), III. 89).

Metamorphosis of granite, granulite, and eclogite directly into serpentine, chlorite, and talc: — These transmutations are maintained by Müller, and adopted by Bischof. (Chem. Geol., II. 424, 434.)

Metamorphosis of limestone to granite or gneiss: — This is taught by Blum and Volger. (Chem. Geol., II. 186; III. 431.)

Having thus given the authorities for the examples cited in my address, I may notice some further illustrations of the doctrine from the pages of Bischof's work already quoted. Metamorphosis of diorite, hornblende-rock, and labradorite to serpentine; G. Rose, Breithaupt, Von Rath (II. 417, 418): diorite and hornblende-slate to talc-slate and chlorite-slate; G. Rose (III. 312): mica-slate to talc-slate and steatite, and mica to serpentine, steatite, and talc; Blum, C. Gmelin (II. 405, 468): quartz-rock to steatite; Blum (II. 468).

[That the extravagant views of the transmutationists, as set forth in the preceding pages, though now denied by Professor Dana, are still maintained by others, is well shown by two recent publications. In one of these, just referred to, Messrs. King and Rowney have gone even further than their predecessors. Not content with teaching the conversion of feldspar, quartz, hornblende, pyroxene, and chondrodite into calcite, they imagine that serpentine, which, according to Dana and others, results in all cases from the alteration of silicated or carbonated species, may itself become the subject of epigenic change, and be converted into calcite. The ophicalce rocks, which are mixtures of serpentine and carbonate of lime, have, according to King and Rowney, been formed in this manner from serpentine; and they further imagine this process to have been so guided as to leave the unchanged portions of the serpentine with the forms of a foraminiferal organism, the *Eozoon Canadense* of Dawson. This singular supplement to the hypothesis of epigenic change recalls the notion of the older naturalists, who, rather than admit the organic origin of shells found in the rocks, imagined them to have been generated by a plastic force. It is evident that it makes little difference what mineral species is taken as a starting-point for these transformations, and Dr. Genth has assumed corundum. In a recent paper (Proceedings of the American Philosophical Society, September 19, 1873) he has discussed various facts observed in the association and envelopment of the minerals associated with it, and concludes that there have been formed from corundum, by epigenesis, spinel, tourmaline, fibrolite, cyanite, paragonite, damourite and other micas, chlorite, and probably various feldspars. According to him, great beds of micaceous and chloritic schists have resulted from the transformation of corundum, and even the beds of bauxite, a mixture of hydrous aluminic and ferric oxides, allied to limonite, which abounds in certain tertiary deposits, were once corundum or emery, from which this amorphous hydrate is supposed to have been derived by a retrograde metamorphosis ; a striking example of the strange conclusions to which this doctrine of epigenic pseudomorphism may lead. The corundum-bearing vein-stones present close resemblance in the grouping and association of minerals to the granitic and calcareous vein-stones described in Essay XI. of the present volume. See, further, the author's criticisms on this subject, Proceedings Boston Society of Natural History, March 4, 1874.]

Coming now to his criticism of the first part of my address, with regard to New England rocks, Professor Dana asserts that "there are gneisses, mica-schists, and chloritic and talcoid schists in the Taconic series." I have, however, shown in my address that Emmons, the author of the Taconic system, expressly excluded therefrom the crystalline rocks, which he included in an older primary system ; excepting, however, certain micaceous and talcose beds, which he declared to be recomposed rocks, made up from the ruins of the primary schists, and distinguished from these by the absence of the characteristic crystalline minerals which belong to the Green Mountain primary schists.

Again, Professor Dana states that I make the crystalline schists of the White Mountains a newer series than the Green Mountain rocks. Such a view of their geognostical relations has been maintained for the last generation by the Messrs. Rogers, Logan, and many others, all of whom assigned the crystalline schists of the White Mountains to a higher geological horizon than those of the Green Mountains. In support of this view of their relative antiquity, I have, however, brought together observations from South Carolina, Pennsylvania, Michigan, Ontario, and Maine, all of which point to the same conclusion ; and I might now add similar evidence from New Brunswick and from Nova Scotia. My "chronological arrangement" of New England crystalline rocks, as it is called by Professor Dana, so far as it is my own, is limited to my affirmation that they are all of pre-Cambrian age ; in proof of which it need only be mentioned that the crystalline schists of both the types in question are, in southern New Brunswick, directly overlaid by uncrystalline shales, sandstones and conglomerates, made up in part of the ruins of these, and holding a Cambrian (Menevian) fauna.

As regards the mica-schists with staurolite, cyanite, andalusite, and garnet, I have in my address pointed out the fact that they appear to belong to a great series of rocks, very constant in character, which have a continuous outcrop from the Hudson River to the St. John, a distance of five hundred miles, and in the latter region are clearly pre-Cambrian. I have, moreover, brought together the evidence of observers in other parts of North America, in Great Britain, in continental Europe, and in Australia, showing that similar crystalline schists, holding these same minerals, always occupy, in these regions, a similar geological horizon. Professor Dana hereupon inquires whether any one has yet proved that these mineral characters are restricted to rocks of a certain geological period. I answer, that in opposition to these facts, it has not yet been proved that they belong to any later geological period than the one already indicated ; and that it is only by bringing together observations, as I have done, that we can ever hope to determine the geological value of these mineral fossils. In no other way did William Smith prove, in Great Britain, the value of organic fossils, and thus lay the foundations of paleontological geology.

XIV.

THE GEOLOGY OF THE ALPS.

This review appeared in the American Journal of Science for January, 1872, and serves to throw much light upon many important and still debated points of geology. I have added as an appendix to the present reprint the recent conclusions of Favre, and the statements of Pillet, which serve to confirm certain positions assumed in the review, and elsewhere in this volume.*

SINCE the days of De Saussure, the Alps have been the object of constant study. No other portion of Europe offers so many problems of interest to the geologist and the physical geographer as this great mountain-chain, whether we consider its lakes, glaciers, and moraines, its curiously disturbed and inverted fossiliferous strata, which seem, at first sight, arranged for the confusion alike of paleontologists and stratigraphists, or the crystalline rocks which form its highest summits. To give a list of the various investigators who have contributed their share to the elucidation of this region would, of itself, be no slight task, and would besides be foreign to our present purpose; which is to call attention to the learned work of Professor Alphonse Favre of Geneva, in which he has given us the results of more than twenty-five years of labor in the study of Alpine geology, chiefly in Savoy and the adjacent parts of Piedmont and Switzerland, embracing Mont Blanc and its vicinity. It is now twelve years since the present writer had occasion to review, in the American Journal of Science ((2), XXIX. 118), some points in Alpine geology raised by our author in his memoir "Sur les terrains liassique et keuperien de

* Recherches Géologiques dans les parties de la Savoie, du Piémont, et de la Suisse voisines du Mont Blanc, avec un Atlas de 32 planches, par Alphonse Favre, Professeur de Géologie à l'Académie de Genève. 3 Vols. 8vo. Paris. 1867.

la Savoie," published in 1859. Since that time the views then maintained by Favre have, in spite of much opposition, gained ground, and are set forth at length in the present work, supported by an amount of evidence which seems convincing. We shall endeavor from its pages to present a condensed summary of our present knowledge of the structure of Mont Blanc and the adjacent regions.

The crystalline rocks of the Alps, as first shown by Studer, do not form a continuous chain, but appear as distinct masses, separated from each other by uncrystalline sedimentary deposits, generally fossiliferous. According to Desor, there are between Nice and the plains of Hungary not less than thirty-four such areas, standing up like islands from out of the sedimentary rocks, and presenting for the most part a fan-like structure (*en eventail*). Of these masses of crystalline rock, Mont Blanc is the most remarkable, and is described by Elie de Beaumont as "rising through a solution of continuity in the secondary strata, which may be compared to a great button-hole." The length of this area of crystalline rock, measured from the Col du Bonhomme on the southwest to Saxon in the Valais on the northeast, is fifty-nine kilometres, while its breadth, from Chamonix on the northwest to Entrèves near Courmayeur on the southeast, is fourteen kilometres. The length of the central mass of protogine is, however, only twenty-seven kilometres. Of the numerous peaks in this area the highest attains an elevation of 4,810 metres above the level of the sea, being 3,760 metres above the valley of Chamonix, and 3,520 metres above the valley of Entrèves. This great mass is described by Favre as supported at the four corners by as many buttresses rising from the surrounding valleys, and known as the Cols de Balme, de Voza, de la Seigne, and de Ferret. The distance between the two valleys just named is only 13,500 metres, and the boldness with which the mountain rises from them is strikingly apparent if we take the Col de l'Aiguille du Midi and the Col du Géant, which are about 3,460 metres above the sea, and distant from each other 5,000 metres, giving a slope of about 30°. A still greater inclination is obtained if we choose, instead of these, the summits of the Aiguilles which bear the same names, and, although now isolated, represent portions of the former mass of Mont Blanc.

The crystalline rocks of this region present two types: first, the protogines which form the centre; and, second, the crystalline schists which occupy the flanks and form the Aiguilles Rouges. These schists are also found at a great elevation on the mountain; at the Grands Mulets (4,666 metres) the rocks are talcose and quartzose schists with graphite, hornblende, epidote, talc, and asbestus, and similar rocks and minerals are found from thence to the summit. The protogines themselves, according to the evidence of nearly all who have studied them, are stratified rocks, gneissic in structure, and pass in places into more schistose varieties, though Favre regards the distinction between these and the crystalline schists proper as one clearly marked. The outlines presented by the weathering of the protogine are very unlike the rounded forms assumed by true granite rocks. According to Delesse, the rock to which Jurine gave the name of protogine is a talco-micaceous granite or gneiss, made up of quartz, generally more or less grayish or smoky in tint, with orthoclase, grayish or reddish in color, and a white or greenish oligoclase with characteristic striæ, often penetrated with greenish talc. The mica (biotite), which some previous observers had mistaken for chlorite, is dark green in color, becoming of a reddish bronze by exposure. It is binaxial, nearly anhydrous, and contains a large portion of ferric oxide. The composition of the protogine rock, as a whole, differs from that of ordinary granite, according to Delesse, only in the presence of one or two hundredths of iron-oxide and magnesia. The name of arkesine was given by Jurine to a variety of protogine containing chlorite with hornblende, and sometimes sphene. Among the other crystalline rocks of the Alps are various talcose and chloritic schists, with steatites, chromiferous serpentines, diallage rocks, diorites, and euphotides, associated with beds of petrosilex or eurite, frequently porphyritic. Highly micaceous schists, often quartzose, and holding garnet, staurolite, and cyanite, are also met with among the crystalline rocks of the Alps. A great belt of serpentine and chloritic schists, traced for a long distance, may be seen at the base of the Montanvert overlaid by the euritic porphyries, into which they appear to graduate; the whole series, here supposed to be inverted, dipping at about 60° from the valley of Chamonix toward Mont Blanc, and overlaid by the more massive gneiss or protogine. The chloritic and talcose schists of the Alps have close resemblances with those of the Urals, and, as Damour has shown, contain a great many mineral species in common with them. Favre has, moreover, remarked the strong likeness between the chloritic and talcose schists and the mica-schists with staurolite of the western Alps and those found in Great Britain.

Granite, though not abundant in the vicinity of Mont Blanc, occurs in several localities, the best known of which is Valorsine, where a porphyroid granite with black mica forms considerable masses, and sends large veins into the adjacent gneiss. These, with others found at the Col de Balme and in the Aiguilles Rouges, appear to be true eruptive granites. Numerous small veins met with among the crystalline schists in the gorge of Trient appear, however, to belong to what I have described as endogenous granites. (*Ante*, page 193.) Favre has himself maintained that they are the results of aqueous infiltration, and has noticed the fact of a joint running longitudinally through the middle of many of them as an evidence of this mode of formation.

The uncrystalline strata in the region around Mont Blanc include representatives of the carboniferous, triassic, jurassic, neocomian, cretaceous, and tertiary. The existence of an apparently carboniferous flora, and its intimate association with a liassic fauna, has long been a well-known fact in Alpine geology. In 1859, Favre pointed out the existence of a zone of triassic rocks in this region represented by red and green shales, with sandstones, gypsum, and a cavernous magnesian limestone (cargneule). These rocks had long before been referred to this period by Buckland and Bakewell, but their horizon was established by the discovery of Favre that their position is intermediate between the carboniferous and the strata containing *Avicula contorta* (the Kössen beds, or the Rhætic beds of Gümbel), which are recognized as forming a passage between the trias and the lias, at the base of the jurassic system. To these, to the northwest of Mont Blanc, succeed the higher members of the system, followed by the neocomian, the cretaceous, and the nummulitic strata of the eocene, with overlying sandstones and shales, the flysch of some Alpine geologists.

Few questions in geology have been more keenly debated, or given rise to more often-repeated examinations, than the association of a carboniferous flora with liassic belemnites in the districts of Maurienne and Tarentaise, to the southwest of Mont Blanc. As seen at Petit-Cœur, the schists, with impressions of ferns and beds of anthracite, were so long ago as 1828 described by Elie de Beaumont as apparently intercalated in the jurassic system. Scipion Gras, and Sismonda after him, have agreed in regarding the rocks as constituting one great system, which according to Gras is of carboniferous age, but with a jurassic fauna; while De Beaumont and Sismonda, on the contrary, regarded it as of jurassic age, but with a carboniferous flora, and imagined that by some means there had been in this region a local survival of the vegetation of the palæozoic period. These conclusions were accepted by many geologists, though rejected by not a few. A brief account of the controversy up to that date will be found in the American Journal of Science for January, 1860, page 120; and in the work of Favre now before us the whole matter is discussed at great length in Chapter XXX. The anthracitic system of the Alps, as recognized by Gras, was by him estimated to have a thickness of from 25,000 to 30,000 feet, and included, besides the dolomites and gypsums now referred by Favre to the trias, coal-plants and layers of anthracite, together with limestones holding belemnites of jurassic age. Included in this great system were, moreover, gneissic, micaceous, and talcose rocks, with graphite, serpentine, euphotide, etc., all of which were regarded by Gras as formed by the local alteration of portions of the anthracitic system. To this was added in 1860 the discovery by Pillet of nummulitic beds intercalated in the same series near St. Julien in Maurienne. This fact was, however, in accordance with the conclusion previously reached by Sismonda from an examination of Taninge, that "the plants of the carboniferous period were still flourishing while the seas were depositing the rocks of the nummulitic period."

The question involved in this controversy had more than a local interest, since it touched the very bases of paleontology, by pretending that in the Alps the laws of succession which elsewhere prevail were suspended, and that the same types of vegetation had continued unchanged from the palæozoic to the tertiary period. Already, in 1841, Favre had brought forward the suggestion of Voltz, that these apparent anomalies might be explained by inversions of the strata; but this notion was rejected by De Mortillet and Murchison, as inadmissible for the section at Petit-Cœur. The recognition by Favre, in 1861, of the true age and position of the cargneules and their associated rocks, however, threw a new light on the question, for it was shown that these triassic rocks were interposed at Petit-Cœur between the limestones holding belemnites and the schists with coal-plants. In 1861, the Geological Society of France held its extraordinary session at St. Jean in Maurienne, and there also the succession was made clearly evident, as follows: nummulitic, liassic, infra-liassic, triassic, and carboniferous; the last resting on crystalline schists.

Attempts had been made to sustain the supposed jurassic age of the so-called anthracitic formation, by maintaining that some at least of the coal-plants were jurassic forms; but Heer, who had long maintained the contrary, published in 1863 a further study of the fossil flora of Switzerland and Savoy, in which he showed that of sixty species fourteen are peculiar to these regions, while forty-six belong to the carboniferous flora of Europe, and twenty-seven are common with that of North America. One species only has been identified as of liassic age, namely, *Odontopteris cycadea* Brongn., and is found in a locality near jurassic belemnites, but associated with no other plant.

Both Lory and Pillet now admit with Favre that the supposed paleontological anomalies of this region have no existence, and that this anthracitic system includes carboniferous, jurassic, and nummulitic strata inverted and folded upon themselves; nor is it without reason that Lory in this connection remarks upon "the illusions without number to which a purely stratigraphical study of the Alps may give rise." To this we may add the judgment of Dumont, in discussing the disturbed and inverted anthracite system of the Ardennes, that for regions thus affected "we cannot establish the relative age of the rocks from their inclination or their superposition."

These conclusions were not, however, admitted by Sismonda, who, in 1866, presented to the Royal Academy of Sciences of Turin an elaborate memoir on the anthracite system of the Alps.* In this, while admitting at Petit-Cœur the existence of evidence of more or less contortion, rupture, and overriding (*enchevauchement*) of the strata, he still maintains that the anthracitic system of Maurienne and Tarentaise is one great continuous series of jurassic age, from the fundamental gneiss and protogine, upon which it immediately rests, to the upper member in which occur thick beds of anthracite, with an abundant carboniferous flora, which he assigns, however, to the middle oölite (Oxfordian); the great mass of strata below being referred to the lias. He then particularly indicated the line of the great Mont Cenis tunnel, which, commencing in the upper anthracitic member, should pass downward through the quartzites and gypsums, thence through talcose schists and limestones, as far as Bardonecchia. These schists and limestones, according to him, are in "a very advanced stage of metamorphism," and include eruptive serpentines, with euphotide, steatite, and other magnesian rocks.

Since the completion of the tunnel, Messrs. Sismonda and Elie de Beaumont have presented to the Academy of Sciences of Paris an extended report on the geological results obtained in this great work. It is accompanied by a description of 134 specimens of the rocks collected at intervals throughout the en-

* Memoirs of the Acad., Second Series, XXIV 333.

tire distance of the tunnel, which, it will be remembered, passes from near Modane in Savoy to Bardonecchia in Piedmont (about fifteen miles to the southwest of Mont Cenis), a distance of 12,220 metres. The direction of the tunnel is N. 14° W., and the dip of the strata throughout nearly uniform, N. 55° W., at an angle of about 50°. From this we deduce by calculation that the vertical thickness of the strata is equal to nearly 60 per cent of the distance traversed, or in round numbers about 7,000 metres. Of this not less than 5,831 metres, beginning at the southern extremity, are occupied by lustrous and more or less talcose schists with crystalline micaceous limestones, often cut by veins of quartz with chlorite and calcite. Above there are 515 metres in thickness of alternations of anhydrous sulphate of lime (karstenite) with talcose schist and crystalline limestone. The anhydrite enclosed lamellar talc in irregular nodules, with dolomite, crystallized quartz, sulphur, and masses of rock-salt. This was overlaid by 220 metres of quartzite, occasionally alternating with greenish talcose schists, and enclosing veins and masses of anhydrite. A considerable break occurs in the series of specimens above this, but for the distance of 1,707 metres from the northern entrance to the tunnel, corresponding to a vertical thickness of 1,024 metres, we have principally sandstones, conglomerates, and argillites, occasionally with anthracite. The serpentines and euphotides which appear among the crystalline schists at Bramant, near the line of the tunnel, were not met with, nor was the underlying gneiss encountered. The work terminated at Bardonecchia among the crystalline limestones.

According to Sismonda and Elie de Beaumont, there is throughout this entire section no evidence of inversion, dislocation, or repetition in the series of 7,000 metres of strata, a conclusion which they support by very cogent arguments. Lory, on the contrary, while he agrees with the observers just mentioned in looking upon the crystalline strata as altered mesozoic, conceives them to include both trias and lias, and to be placed beneath the true carboniferous by a great inversion of the whole succession. This series of crystalline rocks is very conspicuous along the southeast side of Mont Blanc, extending into the Valais, and is regarded by Lory as a peculiar modification of the trias and lias, so enormously thickened and so profoundly altered as to be very unlike these formations to the northwest of Mont Blanc. In this view he is followed by Favre (§§ 666, 753). The serpentines and related rocks of this series are by De Beaumont, Sismonda, and Lory considered to be eruptive. The latter speaks of these as eruptions contemporaneous with the deposition of the strata, probably accompanied by emanations which effected the alteration of the sediments. According to Favre, they are clearly interstratified with the lustrous argillo-talcose schists, micaceous limestones and quartzites of the great series, and are by him placed in the trias. He has particularly described those of Mont Joret and those of the Val de Bruglié, near the Petit St. Bernard, where they are immediately interstratified with greenish schists, and associated with steatite, hornblendic and gneissic strata. The serpentines of Taninge in the Chablais, to the northwest of Mont Blanc, he also classes with these in the trias. The conclusions of Lory and Favre as to the geological age of these crystalline schists and limestones appear to us untenable in the light of Sismonda's investigations. If we admit with the latter that the whole section of the tunnel represents an uninverted series, and with Favre that its uppermost and uncrystalline portion at Modane is truly of carboniferous age, it is clear that the great mass of crystalline schists which underlie the latter should correspond more or less completely to the pre-carboniferous crystalline strata to the northwest of Mont Blanc. Among these latter, in fact, as observed by Favre, there occur at Col Joli and Taninge crystalline limestones and talcose schists like those of Maurienne. According to this view, which harmonizes the conflicting opinions, and makes the crystalline schists and limestones of the southeast pre-carboniferous, the anhydrites, with limestones, talcose slates, and quartzites seen in the Mont Cenis tunnel, are not the equivalents of the gypsum and cargneule of the trias, but may correspond to the anhydrites which, with gypsum, dolomite, serpentine and chloritic slate, are met with in the primitive schists of Fahlun in Sweden.

The existence of great and perplexing inversions of strata in many parts of the Alps is well known. One of the most striking cases is that figured by Murchison in his remarkable paper on the geology of the Alps in 1848 (Quar. Jour. Geol. Soc., V. 246), as occurring at the pass of Martinsloch in the canton of Glarus, 8,000 feet above the sea. Here nummulitic beds, dipping S. S. E. at a high angle, are regurlaly overlaid by the succeeding sandstone (flysch), resting unconformably and in a nearly horizontal attitude upon the edges of which are 150 feet of hard jurassic limestone, overlaid in its turn by talcose and micaceous schists, which are by Escher regarded as similar to those which underlie these limestones in the valley below. This mass of flysch appears near by to dip beneath these limestones, which, in their turn, are regularly overlaid by neocomian and cretaceous strata. This remarkable superposition of secondary and older crystalline rocks to tertiary is explained by Murchison, in accordance with the suggestion of H. D. Rogers, as the probable result of fracture and displacement along an anticlinal. Many striking examples of inversion are described by Favre in the vicinity of Mont Blanc. The mountain of the Voirons, near Geneva, shows at its base tertiary overlaid by cretaceous rocks, upon which jurassic strata are superimposed. Similar phenomena are met with along the north side of the Alps from Geneva to Austria, and at various localities on the southern side, in Lombardy. This inversion, moreover, is by no means confined to secondary and tertiary strata. In the valley of Chamonix the secondary limestones dip at a high angle toward Mont Blanc, and plunge beneath its crystalline schists. Other examples of the superposition of crystalline schists to the fossiliferous sediments have been pointed out by Elie de Beaumont in the mountains of Oisans, and confirmed by Lory and Dausse, while similar cases have been recognized by Morlot and Von Hauer in the eastern Alps, and by Ramond, De Boucheporn, and others in the Pyrenees. All of these cases are by Favre regarded as examples of the same process of inversion already noticed in so many instances among the secondary and tertiary strata of the region. He proceeds to contrast these examples with that of the gneisses with chloritic and micaceous schists, which in western Scotland, according to Murchison, overlie fossiliferous Lower Silurian beds, and are by him regarded as younger. This, upon the authority of Murchison, Favre regards as a singular and anomalous fact. It should, however, be said that this view of Murchison is rejected by Nicoll, who explains the appearances as the result of dislocation and oversliding of older crystalline schists upon the newer fossiliferous beds, in which case the western Highlands will form no exception to the general law of similar appearances in the Alps and Pyrenees. (*Ante*, page 271.)

The fact that the jurassic rocks in the valley of Chamonix pass beneath the crystalline schists of Mont Blanc was first noticed by De Saussure, and was afterwards observed by Bergmann and by Bertrand, who argued from this that the limestones were older than the gneiss. Bertrand's paper, as noticed by Favre, occurs in the Journal des Mines, VII. 376 (1797 – 1798). Later, in 1824, we find Keferstein inquiring whether these overlying gneisses and protogines might not be altered flysch (that is, eocene), a view which he subsequently maintained. Similar views have found favor among later geologists; we find Murchison asserting the eocene age of certain Alpine gneisses, mica-schists, and granites; while Lyell has suggested that the protogines, gneisses, etc., of the Alps may have resulted from the alteration both of secondary and tertiary strata. (Anniversary Address to the Geological Society, 1850.) Studer has taught that the flysch of the Grisons has been changed into crystalline gneiss, while Rozet and Fournet, with Lory and Sismonda, have assigned to the jurassic period the great system of gneisses, with talcose and micaceous schists, which make up Monts Cenis and Pelvoux, and much of the mountains on the frontier of Piedmont and in the Valais.

Hutton, as early as 1788, had taught that what he called the primary schists were sediments, the ruins of earlier rocks altered by heat, but it does not appear that he attempted to fix the relative age of any such altered rocks. In fact, the notion of geological periods, based upon the study of fossils, was not as yet fully recognized. The suggestions of Bergmann and Bertrand, that the crystalline rocks of the Alps are newer than the fossiliferous limestones which pass beneath them, seems to have been the first attempt to give to Hutton's view a definite and special application, and the inception of that hypothesis with which we have since become familiar, which supposes the conversion of mountain masses of palæozoic, mesozoic, and even cenozoic sediments, in the Alps and elsewhere, into gneisses and other crystalline rocks.* Numerous sections in the vicinity of Mont Blanc show the sedimentary strata in their normal attitude, resting unconformably upon the crystalline schists, while in some localities the whole succession from the carboniferous to the eocene, both inclusive, is met with. In many parts, however, the carboniferous is wanting, and the trias forms the base of the column, while elsewhere the infra-liassic beds repose on the crystalline schists, and in the Bernese Alps no fossiliferous beds lower than the oölite are observed. These variations would appear to be connected with the movement of subsidence which permitted the deposition of marine limestones above the carboniferous strata; and Favre has further pointed out, in the vicinity of Dorenaz, a want of conformity between these and the succeeding formations.

To the carboniferous belongs the well-known conglomerate of Valorsine, which includes pebbles of gneiss, quartzite, talcose, and micaceous schist, and of quartz veins with tourmaline. The paste, which is reddish, talcose, and micaceous, seems identical with many of the pebbles, so that it is sometimes difficult to distinguish these from the matrix. A thin fibrous envelope often surrounds the pebbles (§ 521). Although the alternation of these beds with others holding plants shows them to be of carboniferous age, it is often, says Favre, difficult to fix the lower limit of this formation, on account of the great resemblance between certain of the carboniferous sandstones and

[* Already, before Hutton, Von Trebra, in 1785, had taught a somewhat similar doctrine. He supposed that a slow change under the influence of heat and water, which he compared to a fermentation, is constantly going on in the interior of the rocks, and may in time convert mountains of granite into gneiss, and of gray wacke into clay-slate. (Erfahrungen von Innern der Gebirge, page 48.)]

portions of the older crystalline schists, which, in cases where the former are destitute of pebbles, makes it impossible to distinguish between the two. Necker, in like manner, asserted that it was impossible to draw a line of demarcation, and was hence led to assert a passage from the one to the other. The same close resemblance was noticed by De Saussure, and is testified to by De Mortillet and by Sismonda, who says of the feldspathic sandstone (grès) near St. Jean in Maurienne, that "unless we take care we run the risk of being deceived, and of confounding it with gneiss"; while elsewhere similar rocks assume the aspect of granite from the predominance in them of feldspar. Hence it has happened that observers like Dolomieu and Bakewell placed the anthracites of the Alps in the mica-slate formation, and that Berger described as a "veined granite" the Aiguille des Posettes, which, according to Favre, consists of nearly vertical beds of carboniferous sediments. In illustration of this condition of things, Favre cites the observation of Boulanger, according to whom the triassic sandstones of the department of Allier are made up of quartz, feldspar, and mica, so united as to give rise to a sandstone which would be taken for a primitive rock but for the occasional presence of a rolled pebble of granite.* The paste of this Valorsine conglomerate, which seems identical with certain of the enclosed pebbles, appears, according to Favre, to have undergone a certain rearrangement, so that the beds of these "pretendus schists cristallins" of the carboniferous are with difficulty distinguished from the "vrais schists cristallins" upon which they rest unconformably. I insist the more upon these details, because in the earlier notice of Favre's investigations I erroneously represented him as including in the carboniferous a great mass of the older crystalline schists.

In this connection we may cite the observation of Sedgwick, who cites similar cases of recomposed rocks in Scotland, "which it is not always possible to distinguish from the parent rock," and remarks that "a mechanical rock may appear highly crys-

* See Favre, Terrains liassique et keuperien, etc. (1859), pp. 78, 79, to which, in this work, he refers the reader for further explanation on this point.

talline because it is composed of crystalline parts derived from some pre-existing crystalline rock." * Emmons also has called attention to the existence of secondary or recomposed beds of talcose, chloritic, and micaceous schists in the Taconic hills of western New England, which, according to him, have been confounded with the older parent rocks. (*Ante*, page 251.) It would hardly seem necessary to call attention to facts which are familiar to all field-geologists who have worked much among newer deposits in the vicinity of older crystalline rocks, were it not that their significance is so great in connection with Alpine geology.

This deceptive resemblance to the older crystalline rocks in the Alps, as might be supposed, is not confined to the carboniferous. Similar cases are noticed by Favre in the trias, while at the Cols du Bonhomme and des Fours are crystalline aggregates also noticed by Saussure as closely resembling the older crystalline rocks, which are shown by the fossils of interstratified beds to be of infra-liassic age. Studer, in opposition to Murchison, maintained that the apparently granitic layers in the flysch (eocene) near Interlaken are but the débris of an older crystalline rock, while the crystalline schists of the Bolghen mountain in the eastern Alps, supposed by Murchison to be in some way interposed in the flysch, are both by Studer and by Boué regarded as merely masses of the older crystalline rocks in a tertiary conglomerate.†

In discussing the age of the "true crystalline schists" of the Alps, to make use of his expression already quoted, Favre, as we have seen, places them beneath the carboniferous, and in opposition to the suggestion of Murchison and the opinion of Gueymard, that they may be of Cambrian and Silurian age, concludes that we have no proof of the existence of representatives of these systems in the western Alps (§ 808). In this connection we may note with Favre the presence at Dienten, in the Tyrol, of a Silurian fauna, intercalated in beds of gray and green chloritic schists (§ 697 *b*). The gneiss of Mettenbach,

* Geol. Transactions (1835), III. 479.

† Ibid., III. 334; Geol. Jour., V. 210.

near the Jungfrau, has afforded to Favre a pale green ophicalce resembling that of the Laurentian, in which he has detected *Eozoon Canadense* (§ 697 *a*). Having thus declared his conviction of the great antiquity of the crystalline schists, whose ruins enter into the composition of the conglomerate of Valorsine, he proceeds to remark that "the part played by the Alps of Savoy by that mysterious force called metamorphism, to which the formation of the crystalline schists is often attributed, has been greatly exaggerated." He adds, "I have always been surprised to find in the Alps so few traces of this pretended action," and suggests that the question has been complicated by the resemblances already noted between the crystalline schists and the recomposed rocks of the coal measures (§ 697 *c*). In the same spirit he declared in 1859 that there are "scarcely any evidences of alteration after the Valorsine conglomerate"; in the paste of which he admits a crystalline rearrangement, by no means improbable.* It appears inconsistent with these expressions of opinion to find our author admitting with Lory the triassic and jurassic age of the great mass of lustrous schists and micaceous limestones which are overlaid by the carboniferous at Modane, and at various localities, as we have seen, include serpentines, steatites, etc. Our author feels this to be a difficulty, and speaks of these serpentines, unlike those of the Montanvert, the Aiguilles Rouges, etc., as belonging to non-crystalline formations, a character which can hardly be ascribed to them. If, however, Sismonda be correct in placing them below rocks which are, according to Favre, true coal measures, these serpentines and steatites, with their accompanying schists and limestones, are, as we have already shown, in the same horizon with the crystalline schists to the north of Mont Blanc.

The origin of the fan-like structure attributed to the Alps by nearly all observers since the time of De Saussure, and correctly represented in the sections published by Studer in 1851, and by Favre in 1859, is explained by the latter in accordance with the view put forward by Lory in 1860.† He supposes

* Terrains liassique et keuperien, page 77.

† Lory, Description géologique du Dauphiné, p. 180.

that the underlying crystalline rocks, forced by great lateral pressure, formed an elevated anticlinal arch, which, breaking on the crown, from the excess of curvature, shows the lowest rocks in the centre of the rupture, flanked on either side by the overlying strata. These, in their upper part, are subjected to a comparatively feeble lateral pressure, while the deeper portions are forcibly compressed by the smaller folds on either side, from which results the fan-like or sheaf-like structure of the mass. The newer strata in the synclinals are by this process arranged in troughs, and are more or less overlaid by the older rocks. Such a synclinal exists in the valley of Chamonix, between the two ruptured and eroded anticlinals represented by Mont Blanc and the Brevent. In illustration of this structure Favre has given a grand section commencing to the northwest in the mountain known as Les Fiz, which, overlooking the Col d'Anterne, rises to a height of 3,180 metres, and displays all the Alpine formations from the sandstones of Taviglionaz, overlying the nummulitic beds, down to the carboniferous, which is seen resting on the crystalline schists. These appear in the height of Pormenaz, and in the Brevent, at the northwest base of which the carboniferous rocks are arranged in a sharp fold dipping beneath the crystalline strata. The latter, to the northeast, rise in the Aiguilles Rouges, which are steep hills of vertical beds including hornblendic, chloritic, and talcose rocks, with petrosilex, eclogite, and serpentine. The highest of the Aiguilles rises 2,944 metres above the sea, and consequently 1,892 metres above the valley of Chamonix. This summit, which was visited by Favre, was found to be capped by horizontal strata, consisting at the top of about thirty-seven metres of jurassic beds, with belemnites and ammonites, underlaid by infraliassic strata with cargneules, sandstones, and schists, the whole resting upon vertical strata of unctuous mica-schists, which enclosed a bed of saccharoidal limestone. From thence we pass over the valley of Chamonix, which holds enfolded in crystalline schists triassic and jurassic strata, and over the summit of Mont Blanc, to find the same folding repeated between the base of the latter and the protogines of Mont Chétif. The fan-like structure attributed to this last is questioned by Lory, according to whom the strata of this mountain dip uniformly to the southeast, and are overlaid by the great mass of crystalline talcose schists and micaceous limestones assigned by him to the trias; but apparently, as we have endeavored to show, a portion of the pre-carboniferous crystalline schists. These rocks are well displayed further on in the mountain of Cramont, and are regarded by Favre as identical with those of Mont Cenis.* Lory conceives that the attitude of the rocks of Mont Chétif to the jurassic strata in the trough at the southeast base of Mont Blanc is due to a great fault with an uplift, which has brought these older rocks to overlie the jurassic beds.

With the facts before us, we can with Favre trace the history of Mont Blanc from the time when over a partially submerged region of gneiss and crystalline schists the carboniferous strata with their beds of coal and their plant-remains were being deposited; many of the strata being made up from the partially disintegrated crystalline schists and now scarcely distinguishable from them. After some disturbance, the secondary formations were laid down unconformably alike over the carboniferous and the older strata, followed by the nummulitic beds and their overlying sandstones; the whole, from the base of the trias, having in this region an aggregate thickness of about 1,250 metres. Subsequently to this occurred the great movements which threw into folds all of these strata, enclosing, as in the Tarentaise, the nummulites, with jurassic and carboniferous fossils, among the folds of the crystalline schists. This was followed by great denudation, which removed from the broken anticlinals the secondary rocks, leaving, however, in the horizontal jurassic beds which still cap the Aiguilles Rouges, an evidence of the former spread of these formations, which once extended over what is now the summit of Mont Blanc. It is worthy of note that the highest portions of this latter do not exhibit the underlying gneiss, but are capped by crystalline schists, which may be supposed to rest upon it, as do the sec-

* See in this connection Hebert, Bull. Soc. Geol. de France (2), XXV. 356.

ondary strata upon the schists of the Aiguilles Rouges. These elevated points are evidences of the enormous erosion in this region, the results of which have contributed to build up in the lower regions of the Alps, and in the Jura, the great masses of miocene sediment known as the molasse, — a formation partly marine and partly lacustrine, which attains in some parts a thickness of more than 2,000 metres. This period was followed by other movements which have raised the beds of molasse to a vertical attitude, and in some cases inverted them, so that they appear dipping beneath the nummulitic formation. It is worthy of note that the molasse near Geneva includes in its upper part a lacustrine limestone, followed by marls with gypsum, and by lignites.

That the nature of the fan-like structure of the Alps is correctly represented in the sections of Studer, Lory, and Favre, can, we think, no longer admit of doubt. Another explanation was, however, possible; the dipping of the beds on either side toward the centre of the mass might indicate synclinal mountains, lying between two eroded anticlinals. Such a mountain-structure appears not to be uncommon in regions where the undulations are moderate; and is, according to Lesley, frequent in the anthracite region of Pennsylvania. Snowdon in Wales, according to Sedgwick, and Ben Nevis and Ben Lawers in the Scottish Highlands, according to Murchison, are also examples of this structure, the summits of all of these being composed of newer strata, beneath which, on either side, dip the older formations. When, therefore, geologists of authority from Bertrand and Keferstein to Murchison and Lyell maintained that the crystalline rocks of Mont Blanc were newer than the limestones of the valleys on either side, and even declared them to be altered sediments of the tertiary period, it was difficult to regard Mont Blanc as anything else than a synclinal mountain similar in general structure and origin to those just mentioned. Hence it was that in 1860 (American Journal of Science (2), XXIX. 118) I remarked, "the weight of evidence now tends to show that the crystalline nucleus of the Alps, so far from being an extruded mass of so-called primitive rock, is really an altered sedimentary deposit more recent than many of the fossiliferous strata upon their flanks, so that the Alps, as a whole, have a general synclinal structure." This view of the recent age of the crystalline rocks of this region, supported though it has been by the authority of great names, must now, we conceive, be abandoned, and their great antiquity, as maintained by the learned professor of Geneva, admitted. It however remains true that the extrusion and laying bare of these ancient crystalline rocks is, as we have seen, an event geologically very recent.

It would greatly exceed our present limits to notice our author's learned discussion of the superficial geology, including the glacial phenomena, of the Alpine region. His views upon some of the most keenly controverted questions with regard to glacial action will be found set forth in his letter to Sir R. I. Murchison on the Origin of Alpine Lakes and Valleys, which appeared in the London, Edinburgh, and Dublin Philosophical Magazine for March, 1865.

This beautiful work of Professor Favre is accompanied by an atlas of thirty-two folio plates, embracing maps, sections, views, and figures of organic remains, which elucidate the text in a very complete manner. It is a magnificent monument to the industry, acumen, and scientific zeal of one who for a quarter of a century has devoted his time and his fortune to the pursuit of science, and has worthily completed the task which his illustrious countryman De Saussure commenced.

APPENDIX.

[THE crystalline rocks in the line of the Mont Cenis Tunnel, consisting of micaceous limestones, dolomites, gypsums, and anhydrites, with talcose schists, serpentines, and quartzite, have been, as we have seen, regarded by all observers as altered mesozoic strata. According to Elie de Beaumont and Sismonda, they are metamorphosed jurassic, and the uncrystalline anthraciferous strata in contact with them near Modane are unaltered rocks belonging to the same period. Favre, on the other hand, while maintaining the carboniferous age of the latter, followed Lory in regarding the crystalline strata as more recent than these, and, in fact, as metamorphosed triassic. These conclusions as to the age of the crystalline rocks I have ventured in the preceding pages to call in question, and have compared them with certain ancient crystalline schists of Scandinavia. A letter from Professor Favre, dated February, 1872, admits the justice of my strictures ; he now rejects the notion that they are altered fossiliferous strata, and regards them as of unknown age, citing the recently expressed opinion of Gastaldi that they are older than the carboniferous and are altered palæozoic. The existence of such rocks of palæozoic age is, however, improbable, and those to which I have compared them are eozoic.

Professor Favre writes, with reference to my ideas as expressed in the above review and also in my address at Indianapolis (*ante*, pages 286 – 312), as to the possible alteration of palæozoic and more recent strata to crystalline schists : "Je vois avec grand plaisir que vous n'y croyez guère, puisque vous ne voyez nulle part des schistes cristallins dont on puisse dire que ce sont des schistes paléozoiques altérés. Je suis arrivé à croire qu'il n'y a pas de métamorphisme pour les terrains en grand, au moins bien peu, et que tous les terrains se sont déposés a peu près dans l'état où nous les voyons."*

* "I see with great pleasure that you have little belief in it" (the alteration of palæozoic and more recent strata to crystalline schists), "since you nowhere recognize crystalline schists of which it can be said that they are altered palæozoic schists. I have come to believe that there is little or no metamorphism for the great formations, and that all these formations were deposited very nearly in the state in which we see them." With the above extract from

He then proceeds to explain his view that the crystalline schists, the dolomites, and the serpentines have been deposited as such, or have only undergone a subsequent molecular change, such as I have described on pages 300 and 305 of the present volume. It is gratifying to record such testimony to the views I have so long advocated, from the learned geologist of Geneva, who has devoted his life to the study of what is generally regarded as the classic region of rock-metamorphism.

The dip of the strata of the whole section of the Mont Cenis Tunnel is, according to Sismonda and Elie de Beaumont, to the northwest, but, according to Favre and to Pillet, the carboniferous rocks at Modane dip to the southward, suggesting (what might here be looked for), a want of conformity between the crystalline and uncrystalline series. The ancient views of Elie de Beaumont and of Sismonda, according to whom the anthraciferous rocks of this region belong to a single great series of jurassic age, which includes at the same time crystalline schists, a carboniferous flora, a jurassic fauna, and nummulitic beds, appear to be still maintained by these geologists, and are set forth by De Beaumont in a communication to the French Academy of Science, in 1871, on the rocks of the Mont Cenis Tunnel. The publication of this in the Comptes Rendus called forth an energetic protest from Pillet in behalf of the Academy of Sciences of Savoy, in December, 1871. He there complains of the persistent maintenance of views which he declares to have been set aside by the labors of Favre and others, as shown in the work reviewed above, and adds: "The opening of the Mont Cenis Tunnel might have been expected to put an end to the discussion, since we see at St. André, near Modane, the primitive granitic rock overlaid by the coal formation with anthracite, by the trias, and by the liassic schists with belemnites, all placed in their normal order and succession."]

Favre's letter to me, written in February, 1872, may be compared Gümbel's conclusions, cited in a note to page 305, from his letter to me, also written early in 1872.

XV.

HISTORY OF THE NAMES CAMBRIAN AND SILURIAN IN GEOLOGY.

The present essay appeared in the Canadian Naturalist for April and July, 1872, and the first two parts of it were reprinted in Nature for May of the same year, and subsequently in the Geological Magazine in 1873, while a French translation of the entire paper by Dewalque, with notes and additions, is announced as about to appear in Belgium.

Having been desired, in 1872, to prepare for publication a notice of the scientific labors of Murchison, it became necessary for me to examine critically the whole ground of the Cambrian and Silurian controversy, a task which proved much more serious than I had supposed, and brought to light facts which both surprised and pained me. In the interest of truth I determined to write the history as I have here given it, and I had the great pleasure of laying this statement, in its completed form, before the venerable Sedgwick, who, in several letters written to me during the last months of his life, testified his gratitude for the manner in which justice had at length been done to him and to his labors, and, moreover, warmly acknowledged it in the Preface to a new Catalogue of the Cambridge Fossils, dictated by him a few months before his death, which took place in his eighty-eighth year, at Trinity College, Cambridge, January 27, 1873. That Preface contains a more circumstantial and complete account of the personal history of the controversy than had previously appeared.

Such a history as this of the Cambrian and Silurian rocks of the Old World was not complete without an account of the progress of our knowledge regarding the similar rocks of North America; and I have, therefore, in the third part, endeavored to set forth in an impartial manner the share of each investigator in the working out of this important chapter in the geological history of our continent. I have, in the present reprint, made several important additions, and some changes with the view of rendering more complete, both for Great Britain and North America, the history of these older palæozoic rocks. The additions and the important changes, whether in notes or in the text, are distinguished by being enclosed in brackets.

It is proposed in the following pages to give a concise account of the progress of investigation of the lower palæozoic rocks during the last forty years. The subject may naturally be divided into three parts: 1. The history of Silurian and Upper Cambrian in Great Britain from 1831 to 1854; 2. That of the still more ancient palæozoic rocks in Scandinavia, Bohemia, and Great Britain up to the present time, including the recognition by Barrande of the so-called primordial palæozoic fauna; 3. The history of the lower palæozoic rocks of North America.

I. Silurian and Upper Cambrian in Great Britain.

Less than forty years since, the various uncrystalline sedimentary rocks beneath the coal-formation in Great Britain and in continental Europe were classed together under the common name of graywacke or grauwacké, a term adopted by geologists from German miners, and originally applied to sandstones and other coarse sedimentary deposits, but extended so as to include associated argillites and limestones. Some progress had been made in the study of this great Graywacke formation, as it was called, and organic remains had been described from various parts of it; but to two British geologists was reserved the honor of bringing order out of this hitherto confused group of strata, and establishing on stratigraphical and palæontological grounds a succession and a geological nomenclature. The work of these two investigators was begun independently and simultaneously in different parts of Great Britain. In 1831 and 1832, Sedgwick, aided in the early part of his labors by Mr. Charles Darwin, made a careful section of the rocks of North Wales from the Menai Strait across the range of Snowdon to the Berwyn hills, thus traversing in a southeastern direction Caernarvon, Denbigh, and Merionethshire. Already, he tells us, he had in 1831 made out the relations of the Bangor group (including the Llanberris slates and the overlying Harlech grits), and showed that the fossiliferous strata of Snowdon occupy a synclinal, and are stratigraphically several thousand feet above the horizon of the latter. Following up this investigation in 1832, he established the great Merioneth anticlinal, which brings up the lower rocks on the southeast side of Snowdon, and is the key to the structure of North Wales. From these, as a base, he constructed a section along the line already indicated, over Great Arenig to the Bala limestone, the whole forming an ascending series of enormous thickness. This limestone in the Berwyn hills is overlaid by many thousand feet of strata as we proceed eastward along the line of section, until at length the eastern dip of the strata is exchanged for a westward one, thus giving to the Berwyn chain, like that of Snowdon, a synclinal structure. As a consequence of this, the limestone of Bala reappears on the eastern side of the Berwyns, underlaid as before by a descending series of slates and porphyries. These results, with sections, were brought before the British Association for the Advancement of Science at its meeting at Oxford, in 1832, but only a brief and imperfect account of the communication of Sedgwick on this occasion appears in the Proceedings of the Association. He did not at this time give any distinctive name to the series of rocks in question. (L. E. & D. Philos. Mag. [1854] (4), VIII. 495.)

Meanwhile, in the same year, 1831, Murchison began the examination of the rocks on the river Wye, along the southern border of Radnorshire. In the next four years he extended his researches through this and the adjoining counties of Hereford and Salop, distinguishing in this region four separate geological formations, each characterized by peculiar fossils. These formations were, moreover, traced by him to the southwestward, across the counties of Brecon and Caermarthen; thus forming a belt of fossiliferous rocks stretching from near Shrewsbury to the mouth of the river Towey, a distance of about one hundred miles along the northwest border of the great Old Red sandstone formation, as it was then called, of the west of England.

The results of his labors among the rocks of this region for the first three years were set forth by Murchison in two papers presented by him to the Geological Society of London in January, 1834. (Proc. Geol. Soc., II. 11.) The formations were then named as follows in descending order: 1. Ludlow, 2. Wenlock; constituting together an upper group; 3. Caradoc, 4. Llandeilo (or Builth); forming a lower group. The Llandeilo formation, according to him, was underlaid by what he called the Longmynd and Gwastaden rocks. The non-fossiliferous strata of the Longmynd hills in Shropshire were described as rising up to the east from beneath the Llandeilo rocks; and as appearing again in South Wales at the same geological horizon, at Gwastaden in Breconshire, and to the west of Llandovery in Caermarthenshire; constituting an underlying series of contorted slaty rocks many thousand feet in thickness, and destitute of organic remains. The position of these rocks in South Wales was, however, to the northwest, while the strata of the Longmynd, as we have seen, appear to the east of the fossiliferous formations.

In the L. E. & D. Philosophical Magazine for July, 1835, Murchison gave to the four formations above named the designation of Silurian, in allusion, as is well known, to the ancient British tribe of the Silures. It now became desirable to find a suitable name for the great inferior series, which, according to Murchison, rose from beneath his lowest Silurian formations to the northwest, and appeared to be widely spread in Wales. Knowing that Sedgwick had long been engaged in the study of these rocks, Murchison, as he tells us, urged him to give them a British geographical name. Sedgwick accordingly proposed for this great series of Welsh rocks the appropriate designation of Cambrian, which was at once adopted by Murchison for the strata supposed by him to underlie his Silurian system. (Murchison, Anniv. Address, 1842; Proc. Geol. Soc., III. 641.) This was almost simultaneous with the giving of the name of Silurian, for in August, 1835, Sedgwick and Murchison made communications to the British Association at Dublin on Cambrian and Silurian Rocks. These, in the volume of Proceedings (pp. 59, 60), appear as a joint paper, though from the text they would seem to have been separate. Sedgwick then described the Cambrian rocks of North Wales as including three divisions: First, the Upper Cambrian, which occupies the greater part of the chain of the Berwyns, where, according to him, it was connected with the Llandeilo formation of the Silurian. To the next lower division, Sedgwick gave the name of Middle Cambrian, making up all the higher mountains of Caernarvon and Merionethshire, and including the roofing-slates and flagstones of this region. This middle group, according to him, afforded a few organic remains, as at the top of Snowdon. The inferior division, designated as

Lower Cambrian, included the crystalline rocks of the southwest coast of Caernarvon and a considerable portion of Anglesea, and consisted of chloritic and micaceous schists, with slaty quartzites and subordinate beds of serpentine and granular limestone; the whole without organic remains.

These crystalline rocks were, however, soon afterwards excluded by him from the Cambrian series, for in 1838 (Proc. Geol. Soc., II. 679) Sedgwick describes further the section from the Menai Strait to the Berwyns, and assigns to the chloritic and micaceous schists of Anglesea and Caernarvon a position inferior to the Cambrian, which he divides into two parts; namely, Lower Cambrian, comprehending the old slate series, up to the Bala limestone beds; and Upper Cambrian, including the Bala beds, and the strata above them in the Berwyn chain, to which he gave the name of the Bala group. The dividing line between the two portions was subsequently extended downwards by Sedgwick to the summit of the Arenig slates and porphyries. The lower division was afterwards subdivided by him into the Bangor group (to which the name of Lower Cambrian was henceforth to be restricted), including the Llanberris roofing-slates and the Harlech grits or Barmouth sandstones; and the Festiniog group, which included the Lingula flags and the succeeding Tremadoc slates.

In the communication of Murchison to the same Dublin meeting, in August, 1835, he repeated the description of the four formations to which he had just given the name of Silurian; which were, in descending order, Ludlow and Wenlock (Upper Silurian), and Caradoc and Llandeilo (Lower Silurian). The latter formation was then declared by Murchison to constitute the base of the Silurian system, and to offer in many places in South Wales distinct passages to the underlying slaty rocks, which latter were, according to him, the Upper Cambrian of Sedgwick.

Meanwhile, to go back to 1834, we find that after Murchison had, in his communication to the Geological Society, defined the relation of his Llandeilo formation to the underlying slaty series, but before the names of Silurian and Cambrian had been given to these respectively, Sedgwick and Murchison visited together the principal sections of these rocks from Caermarthenshire to Denbighshire. The greater part of this region was then unknown to Sedgwick, but had been already studied by Murchison, who interpreted the sections to his companion in conformity with the scheme already given; according to which the beds of the Llandeilo were underlaid by the slaty rocks which appear along their northwestern border. When, however, they entered the region which had already been examined by Sedgwick, and reached the section on the east side of the Berwyns, the fossiliferous beds of Meifod were at once pronounced by Murchison to be typical Caradoc, while others in the vicinity were regarded as Llandeilo. The beds of Meifod had, on palæontological grounds, been by Sedgwick identified with those of Glyn Ceirog, which are seen to be immediately overlaid by Wenlock rocks. These determinations of Murchison were, as Sedgwick tells us, accepted by him with great reluctance, inasmuch as they involved the upper part of his Cambrian section in most perplexing difficulties. When however, they crossed together the Berwyn chain to Bala, the limestones in this locality were found to contain fossils nearly agreeing with those of the so-called Caradoc of Meifod. The examination of the section here presented showed, however, that these limestones are overlaid by a series of several thousand feet of strata, bearing no resemblance either in fossils or in physical characters to the Wenlock formation, which overlies the Caradoc beds of Glyn Ceirog. This series, was, therefore, by Murchison supposed to be identical with the rocks which, in South Wales, he had placed beneath the Llandeilo, and he expressly declared that the Bala group could not be brought within the limits of his Silurian system. It may here be added that in 1842 Sedgwick re-examined this region, accompanied by that skilled palæontologist, Salter, confirming the accuracy of his former sections, and showing, moreover, by the evidence of fossils that the beds of Meifod, Glyn Ceirog, and Bala are very nearly on one parallel. Yet, with the evidence of the fossils before him, Murchison, in 1834, placed the first two in his Silurian system, and the last deep down in the Upper Cambrian; and consequently was aware that on palæontological grounds it was impossible to separate the lower portion of Silurian system from the Upper Cambrian of Sedgwick. (These names are here used for convenience, although we are speaking of a time when they had not been applied to designate the rocks in question.)

This fact was repeatedly insisted upon by Sedgwick, who, in the Syllabus of his Cambridge lectures, published very early in 1837, enumerated the principal genera and species of Upper Cambrian fossils, many of which were by him declared to be the same with those of the Lower Silurian rocks of Murchison. Again, in enumerating in the same Syllabus the characteristic species of the Bala limestone, it is added by Sedgwick: "all of which are common to the Lower Silurian system." This was again insisted upon by him in 1838, and 1841. (Proc. Geol. Soc., II. 679; III. 548.) It was not until 1840 that Bowman announced the same conclusion, which was reiterated by Sharpe in 1842. (Ramsay, Mem. Geological Survey, III. Part II. p. 6.)

In 1839, Murchison published his Silurian System, dedicated to Sedgwick, a magnificent work in two volumes quarto, with a separate map, numerous sections, and figures of fossils. The succession of the Silurian rocks, as there given, was precisely that already set forth by the author in 1834, and again in 1835; being, in descending order, Ludlow and Wenlock, constituting the Upper Silurian, and Caradoc and Llandeilo (including the Lower Llandeilo beds or Stiper-stones), the Lower Silurian. These are underlaid by the Cambrian rocks, into which the Llandeilo was said to offer a transition marked by beds of passage. Murchison, in fact, declared that it was impossible to draw any line of separation, either lithological, zoölogical, or stratigraphical, between the base of the Silurian beds (Llandeilo) and the upper portion of the Cambrian,—the whole forming, according to him, in Caermarthenshire, one continuous and conformable series from the Cambrian to the Ludlow. (Silurian System, pages 256, 258.) By Cambrian in this connection we are to understand only the Upper Cambrian or Bala group of Sedgwick, as appears from the express statement of Murchison, who alludes to the Cambrian of Sedgwick as including all the older slaty rocks of Wales, and as divided into three groups, but proceeds to say that in his present work (the Silurian System) he shall notice only the highest of these three.

Since January, 1834, when Murchison first announced the stratigraphical relations of the lower division of what he afterwards called the Silurian system, the aspect of the case had materially changed. This division was no longer underlaid, both to the east in Shropshire and to the west in Wales, by a great unfossiliferous series. His observations in the vicinity of the Berwyn hills with Sedgwick in 1834, and the subsequently published statements of the latter, had shown that this supposed older series was not without fossils; but on the contrary, in North Wales, at least, held a fauna identical with that characterizing the Lower Silurian. Hence the assertion of Murchison in his work on the Silurian System, in 1839, that it was not possible to draw any line of demarcation between them. The position was very embarrassing to the author of the Silurian System, and, for the moment, not less so to the discoverer of the Upper Cambrian series. Meanwhile, the latter, as we have seen, in 1842 re-examined with Salter his Upper Cambrian sections in North Wales, and satisfied himself of the correctness, both structurally and palæontologically, of his former determinations. Murchison, in his anniversary address as President of the Geological Society in 1842, after recounting, as we have already done, the history of the naming by Sedgwick, in 1835, of the Cambrian series, which Murchison supposed to underlie his Silurian system, proceeded as follows: "Nothing precise was then known of the organic contents of this lower or Cambrian system except that some of the fossils contained in its upper members in certain prominent localities were published Lower Silurian species. Meanwhile, by adopting the word Cambrian, my friend and myself were certain that whatever might prove to be its zoölogical distinctions, this great system of slaty rocks being evidently inferior to those zones which had been worked out as Silurian types, no ambiguity could hereafter arise. . . . . In regard, however, to a descending zoölogical order, it still remained to be proved whether there was any type of fossils in the mass of the Cambrian rocks different from those of the Lower Silurian series. If the appeal to nature should be answered in the negative, then it was clear that the Lower Silurian type must be considered the true base of what I had named the protozoic rocks; but if characteristic new forms were discovered, then would the Cambrian rocks, whose place was so well established in the descending series, have also their own fauna, and the palæozoic base would necessarily be removed to a lower horizon." If the first of these alternatives should be established, or in other words, if the fauna of the Cambrian rocks was found to be identical with that of the Lower Silurian, then, in the author's language, "the term Cambrian must cease to be used in zoölogical classification, it being, in that sense, synonymous with Lower Silurian." That such was the result of palæontological inquiry, Murchison proceeded to show by repeating the announcements already made by Sedgwick in 1837 and 1838, that the collections made by the latter from the great series of fossiliferous strata in the Berwyns, from Bala, from Snowdon and other Cambrian tracts, were identical with the Lower Silurian forms. These strata, it was said, contain throughout "the same forms of Orthis which typify the Lower Silurian rocks." It was further declared by Murchison in this address, that researches in Germany, Belgium, and Russia led to the conclusion that the "fossiliferous strata characterized by Lower Silurian Orthidæ are the oldest beds in which organic life has been detected." (Proc. Geol. Soc., III. 641, *et seq.*) The Orthids here referred to are, according to Salter, *Orthis calligramma*, Dalm, and its varieties. (Mem. Geol. Survey, III. Part II. 335 – 337.)

Meanwhile Sedgwick's views and position began to be misrepresented. In 1842, Mr. Sharpe, after calling attention to the fact that the fossils of the Bala limestone were, as Sedgwick had long before shown, identical with those of Murchison's Lower Silurian, declared that Sedgwick had placed the Upper Cambrian, in which the Bala beds were included, beneath the Silurian, and that this determination had been adopted by Murchison on Sedgwick's authority. (Proc. Geol. Soc., IV. 10.) This statement Murchison suffered to pass uncorrected in a complimentary review of Sharpe's paper in his next annual address (1843). Subsequently, in his Siluria, first edition, page 25 (1854), he spoke of the term Cambrian as applied (in 1835) by Sedgwick and himself "to a vast succession of *fossiliferous* strata containing undescribed fossils, the whole of which were supposed to rise up from beneath well-known Silurian rocks. The government geologists have shown that this *supposed* order of superposition was erroneous," etc. The italics are the author's. Such language, coupled with Mr. Sharpe's assertion noticed above, helped to fix upon Sedgwick the responsibility of Murchison's error. Although the historical sketch, which precedes, clearly shows the real position of Sedgwick in the matter, we may quote further his own words: "I have often spoken of the great Upper Cambrian group of North Wales as inferior to the Silurian system, ... on the sole authority of the Lower Silurian sections, and the author's many times repeated explanations of them before they were published. So great was my confidence in his work, that I received it as perfectly established truth that his order of superposition was unassailable. . . . . I asserted again and again that the Bala limestone was near the base of the so-called Upper Cambrian group. Murchison asserted and illustrated by sections the unvarying fact that his Llandeilo flag was superior to the Upper Cambrian group. There was no difference between us, until his Llandeilo sections were proved to be wrong." (Philos. Mag. (4), VIII. 506.) That there must be a great mistake either in Sedgwick's or in Murchison's sections was evident, and the government surveyors, while sustaining the correctness of those of Sedgwick, have shown the sections of Murchison to have been completely erroneous.

The first step towards an exposure of the errors of the Silurian sections is, however, due to Sedgwick and McCoy. In order better to understand the present aspect of the question, it will be necessary to state in a few words some of the results which have been arrived at by the government surveyors in their studies of the rocks in question, as set forth by Ramsay in the Memoirs of the Geological Survey. In the section of the Berwyns, the thin bed of about twenty feet of Bala limestone, which (as originally described by Sedgwick) they have found outcropping on both sides of the synclinal chain, is shown to be intercalated in a vast thickness of Caradoc rocks; being overlaid by about 3,300 and underlaid by 4,500 feet of strata belonging to this formation. Beneath these are 4,500 feet additional of beds described as Llandeilo, which rest unconformably upon the Lingula flags just to the west of Bala; thus making a thickness of over 12,000 feet of strata belonging to the Bala group of Sedgwick. A small portion of rocks referred to the Wenlock formation occupies the synclinal above mentioned. (Memoirs, III. Part III. 214, 222.) The second member, in ascending order, of the Silurian system, to which the name of Caradoc was given by him in 1839, was originally described by Murchison under the names of the Horderley and May Hill sandstone. The higher portions of the Caradoc were subsequently distinguished by the government surveyors as the Lower and Upper Llandovery rocks; the latter (constituting the May Hill sandstone, and known also as the Pentamerus beds), being by them regarded as the summit of the Caradoc formation. In 1852, however, Sedgwick and McCoy showed from its fauna that the May Hill sandstone belongs rather to the overlying Wenlock than to the Caradoc formation, and marks a distinct palæontological horizon.

This discovery led the geological surveyors to re-examine the Silurian sections, when it was found by Aveline that there exists in Shropshire a complete and visible want of conformity between the underlying formations and the May Hill sandstone; the latter in some places resting upon the nearly vertical Longmynd rocks, and in others upon the Llandeilo flags, the Caradoc proper or Bala group, and the Lower Llandovery beds. Again, in South Wales, near Builth, the May Hill sandstone or Upper Llandovery rests upon Lower Llandeilo beds ; while at Noeth Grug the overlying formation is traced transgressively from the Lower Llandovery across the Caradoc to the Llandeilo. These important results were soon confirmed by Ramsay and by Sedgwick. (Ibid., 4, 236.) The May Hill sandstone often includes, near its base, conglomerate beds made up of the ruins of the older formation. To the northeast, in the typical Silurian country, it is of great thickness and continuity, but gradually thins out towards the southwest.

There exists, moreover, another region where not less curious discoveries were made. About forty miles to the eastward of the typical region in South Wales appear some important areas of Silurian rocks. These are the Woolhope beds, appearing through the Old Red sandstone, and the deposits of Abberley, the Malverns, and May Hill, rising along its eastern border, and covered along their eastern base by the newer Mesozoic sandstone. The rocks of these localities were by Murchison in his Silurian System described as offering the complete sequence. When, however, it was found that his Caradoc included two unconformable series, examination showed that there was no representative of the older Caradoc or Bala group in these eastern regions, but that the so-called Caradoc was nothing but the Upper Llandovery or May Hill sandstone. The immediately underlying strata, which Murchison had regarded as Llandeilo, or rather as the beds of passage from Llandeilo to Cambrian, and had compared with the northwest parts of the Caermarthenshire sections (Silurian System, 416), have since been found to be much more ancient deposits, of Middle Cambrian age, which rest upon the crystalline hypozoic rocks of the Malverns, and are unconformably overlaid by the May Hill sandstone. We shall again revert to this region, which has been carefully studied and described by Professor John Phillips. (Mem. Geol. Sur., II. Part I.)

What then was the value and the significance of the Silurian sections of Murchison, when examined in the light of the results of the government surveyors? The Llandeilo rocks, having throughout the characteristic Orthis so much insisted upon by Murchison, were shown to be the base of a great conformable series, and to the eastward, in Shropshire, to rest on the upturned edges of the Longmynd rocks; while westward, near Bala, they overlie unconformably the Lingula flags, and in the island of Anglesea repose directly upon the ancient crystalline schists. According to the author of the Silurian System, there existed beneath the base of the Llandeilo formation a great conformable series of slaty rocks into which this formation passed, and from which it could not be distinguished either zoölogically, stratigraphically, or lithologically. The sequence, determined from what were considered typical sections in the valley of the Towey in Caermarthenshire, as given by Murchison, for several years both before and after the publication of his work, was as follows: 1. Cambrian; 2. Llandeilo flags; 3. Caradoc sandstone; 4. Wenlock and Ludlow beds; 5. Old Red sandstone; the order being from northwest to southeast. What, then, were these fossiliferous Cambrian beds underlying the Llandeilo and indistinguishable from it? Sedgwick, with the aid of the government surveyors, has answered the question in a manner which is well illustrated in his ideal section across the valley of the Towey. The whole of the Bala or Caradoc group rises in undulations to the northwest, while the Llandeilo flags at its base appear on an anticlinal in the valley, and are succeeded to the southeast by a portion of the Bala. The great mass of this group on the southeast side of the anticlinal is however concealed by the overlapping May Hill sandstone, — the base of the unconformable upper series which includes the Wenlock and Ludlow beds. (Philos. Mag. (4), VIII. 488.) The section to the southeast, commencing from the Llandeilo flags on the anticlinal, was made by Murchison the Silurian system, while the great mass of strata on the northwest side of the Llandeilo (which is the complete representative of the Caradoc or Bala beds, partially concealed on the southwest side) was supposed by him to lie beneath the Llandeilo, and was called Cambrian (the Upper Cambrian of Sedgwick). These rocks, with the Llandeilo at their base, were, in fact, identical with the Bala group studied by the latter in North Wales, and are now clearly traced through all the intermediate distance. This is admitted by Murchison, who says: "The first rectification of this erroneous view was made in 1842 by Professor Ramsay, who observed, that instead of being succeeded by lower rocks to the north and west, the Llandeilo flags folded over in those directions, and passed under superior strata, charged with fossils which Mr. Salter recognized as well-known types of the Caradoc or Bala beds." (Siluria, 4th ed., p. 57, foot-note.)

The true order of succession in South Wales was, in fact: 1. Llandeilo; 2. Cambrian (= Caradoc or Bala); 3. Wenlock and Ludlow; 4. Old Red sandstone; the Caradoc or Bala beds being repeated on the two sides of the anticlinal, but in great part concealed on the southeast side by the overlapping May Hill or Upper Llandovery rocks. These latter, as has been shown, form the true base of the upper series which, in the Silurian sections, was represented by the Wenlock and Ludlow. Murchison had, by a strange oversight, completely inverted the order of his lower series, and turned the inferior members upside down. In fact, the Llandeilo flags, instead of being, as he had maintained, superior to the Cambrian (Caradoc or Bala) beds, were really inferior to them, and were only made Silurian by a great mistake. The Caradoc, under different names, was thus made to do duty at two horizons in the Silurian system, both below and above the Llandeilo flags. Nor was this all, for by another error, as we have seen, the Caradoc in the latter position was made to include the Pentamerus beds of the unconformably overlying series. Thus it clearly appears that with the exception of the relations of the Wenlock and Ludlow beds to each other and to the overlying Old Red sandstone, which were correctly determined, the Silurian system of Murchison was altogether incorrect, and was moreover based upon a series of stratigraphical mistakes which are scarcely paralleled in the history of geological investigation.

It was thus that the Lower Silurian was imposed on the

scientific world; and we may well ask, with Sedgwick, whether geologists "would have accepted the Lower Silurian classification and nomenclature had they known that the physical or sectional evidence upon which it was based had been, from the first, positively misunderstood." Feeling that his own sections were, as has since been fully established, free from error, Sedgwick naturally thought his name of Upper Cambrian should prevail for the great Bala group. Hence the long and imbittered discussion that followed, in which Murchison, in many respects, occupied a position of vantage as against the Cambridge professor, and finally saw his name of Lower Silurian supplant almost entirely that of Upper Cambrian given by Sedgwick, who had first rightly defined and interpreted the geological relations of the group.

In a paper read before the Geological Society in June, 1843, (Proc. Geol. Soc., IV. 213 – 223) when the perplexity in which the relations of the Upper Cambrian and Lower Silurian rocks were involved had not been cleared up by the discovery of Murchison's errors in stratigraphy, Sedgwick proposed a compromise, according to which the strata from the Bala limestone to the base of the Wenlock were to take the name of Cambro-Silurian; while that of Silurian should be reserved for the Wenlock and Ludlow beds, and for those below the Bala the name of Cambrian should be retained. The Festiniog group (including what were subsequently named the Lingula flags and the Tremadoc slates) would thus be Upper instead of Middle Cambrian, the original Upper Cambrian being henceforth Cambro-Silurian; it being understood that, wherever the dividing line might be drawn, all the groups above it should be called Cambro-Silurian, and all those below it Cambrian. This compromise was rejected by Murchison, who in the map accompanying the first edition of his Siluria, in 1854, extended the Lower Silurian color so as to include all but the lowest division of the Cambrian, namely, the Bangor group. When, however, the relations of Upper Cambrian and Silurian were made known by the discoveries of Sedgwick and the government surveyors, this compromise was seen to be uncalled for, and was withdrawn in 1854 by Sedgwick, who reclaimed the name of Upper Cambrian for his Bala group.

In June, 1843, Sedgwick proposed that the whole of the fossiliferous rocks below the horizon of the Wenlock should be designated Protozoic, and on the 29th of November, 1843, presented to the Geological Society an elaborate paper on the Older Palæozoic (Protozoic) Rocks of North Wales, with a colored geological map. This paper, which embodied the results of the researches of Sedgwick and Salter, was not, however, published at length, but an abstract of it was prepared by Mr. Warburton, then president of the society, with a reduced copy of the map. (Proc. Geol. Soc., IV. 212 and 251 – 268; also Geol. Jour., I. 5 – 22.) In this map of Sedgwick's three divisions were established, namely, the hypozoic crystalline schists of Caernarvonshire, the "Protozoic" and the "Silurian." On the legend of the reduced map, as published by the Geological Society, these latter names were altered so as to read "Lower Silurian (Protozoic)" and "Upper Silurian." These changes, in conformity with the nomenclature of Murchison, were, it is unnecessary to say, made without the knowledge of Sedgwick, who did not inspect the reduced and altered map until it was appealed to as an evidence that he had abandoned his former ground, and had recognized the equivalency of the whole of his Cambrian with the Lower Silurian of Murchison. The reader will sympathize with the indignation with which Sedgwick declares that his map was "most unwarrantably tampered with," and will, moreover, learn with surprise that an inspection of the proof-sheets of Warburton's abstract of Sedgwick's paper was refused him, notwithstanding his repeated solicitations. The story of all this, and finally of the refusal to print in the pages of the Geological Journal the reclamations of the venerable and aggrieved author, make altogether a painful chapter, which will be found in the Philos. Magazine for 1854 ((4) VII. pp. 301–317, 359–370, and 483 – 506), and more fully in the Synopsis of British Palæozoic Rocks, which forms the Introduction to McCoy's British Palæozoic Fossils.

In connection with this history it may be mentioned that in March, 1845, Sedgwick presented to the Geological Society a paper on the Comparative Classification of the Fossiliferous Rocks of North Wales and those of Cumberland, Westmoreland, and Lancashire; which appears also in abstract in the same volume of the Geological Journal that contains the abstract of the essay and the map just referred to. (I. 442.) That this abstract also is made by another than the author is evident from such an expression as "the author's opinion seems to be grounded on the following facts," etc., (p. 448) and from the manner in which the terms Lower and Upper Silurian are applied to certain fossiliferous rocks in Cumberland. Yet the words of this abstract are quoted with emphasis in Siluria (1st. ed., 147), as if they were Sedgwick's own language recognizing Murchison's Silurian nomenclature.*

II. Middle and Lower Cambrian.

Investigations in continental Europe were, meanwhile, preparing the way for a new chapter in the history of the lower palæozoic rocks. A series of sedimentary beds in Sweden and Norway had long been known to abound in singular petrifications, some of which had been examined by Linnæus, who gave to them the name of *Entomolithi.* They were also studied and described by Wahlenberg and by Brongniart, the latter of whom, from two varieties of the *Entomolithus paradoxus,* Linn., established in 1822 two genera, *Paradoxides* and *Agnostus.* In 1826 appeared a memoir by Dalman on the Palæadæ, or so-called Trilobites; which was followed, in 1828, by his classic work on the same subject. (Über die Palaeaden oder so-genannten Trilobiten, 4to, with six plates, Leipsic.) In these works were described and figured, among many others, two genera, — *Olenus,* which included *Paradoxides,* Brongn.,

* [A letter to the author, written him by the late Professor Sedgwick after reading the above, confirms the opinion here expressed. The abstract in question was furnished by Murchison himself to the Geological Society, the secretary of which declined to receive the abstract offered by Sedgwick of his own paper.]

and *Battus*, including *Agnostus* of the same author. Meanwhile, Hisinger was carefully studying the strata in which these trilobites were found in Gothland, and in the same year (1828) published in his Anteckningar, or Notes on the Physical and Geognostical Structure of Norway and Sweden, a colored geological map and section of these rocks as they occur in the county of Skaraborg; where three small circumscribed areas of nearly horizontal fossiliferous strata are shown to rest upon a floor of old crystalline rocks, in some parts granitic and in others gneissic in character. The section and map, as given by Hisinger, show the succession in the principal area to be as follows, in ascending order: 1. Granite or gneiss; 2. Sandstone; 3. Alum-slates; 5. Orthoceratite-limestones; 4. Clay-slates. By a curious oversight the colors on the legend are wrongly arranged and wrongly numbered, as above; for in the map and section it is made clear that the succession is that just given, and that the clay-slates (4), instead of being below, are above the orthoceratite-limestones (5).

In 1837, Hisinger published his great work on the organic remains of Sweden, entitled *Lethœa Suecica* (4to, with forty-two plates). In this he gives a tabular view, in descending order, of the rock-formations, and of the various genera and species described. The rocks of the areas just noticed appear in his fourth or lowest division, under the head of *Formationes transitionis*, and are divided as follows:—

a. Strata calcarea recentiora Gottlandiæ.
b. Strata schisti argillacei.
c. Strata schisti aluminaris.
d. Strata calcarea antiquiora.
e. Strata saxi arenacei.

The succession thus given was, however, erroneous, and probably, like the mistake in the legend of the same author's map just mentioned, the result of inadvertence, the true position of the alum-slates (*c*) being between the older limestone (*d*) and the basal sandstone (*e*). This is shown both by Hisinger's map of 1828, and by the testimony of subsequent observers. In Murchison's work on the Geology of Russia in Europe, published in 1845, there is given (page 15 *et seq.*) an account of his visit to this region in company with Professor Loven, of Christiania; which, with figures of the sections, is reproduced in the different editions of Siluria. The hill of Kinnekulle, on Lake Wener, is one of the three areas of transition rocks delineated on the map of Hisinger above referred to. Resting upon a flat region of nearly vertical gneissic strata, we have, according to Murchison: 1. A fucoidal sandstone; 2. Alum-slates; 3. Red orthoceratite limestone; 4. Black graptolitic slates; the whole series being little over 1,000 feet in thickness, and capped by erupted greenstone. Above these higher slates there are found, in some parts of Gothland, other limestones with orthoceratites, trilobites, and corals, the newer limestone strata (*a*) of Hisinger; the whole overlaid by thin sandstone beds. These higher limestones and sandstones contain the fauna of the Wenlock and Ludlow of England; while the lower limestones and graptolitic slates afford *Calymene Blumenbachii, Orthis calligramma,* and many other species common to the Bala group of North Wales. The alum-slates below these, however, contained, according to Hisinger, none of the species then known in British rocks, but in their stead five species of *Olenus* and two of *Battus* (*Agnostus*).

In 1854, Angelin published his *Palæontologica Scandinavica,* Part I., *Crustacea formationis transitionis* [4to, forty-one plates], in which he divided the series of transition rocks above described by Hisinger into eight parts, designated by Roman numerals, counting from the base. Of these I. was named *Regio Fucoidarum,* no organic remains other than fucoids being known therein; while the remaining seven were named from their characteristic genera of trilobites, which were as follows, in ascending order, certain letters being also used to designate the parts: II. (A) Olenus; III. (B) Conocoryphe; IV. (BC) Ceratopyge; V. (C) Asaphus; VI. (D) Trinucleus; VII. (DE) Harpes; VIII. (E) Cryptonymus. In the *Regio Olenorum* (II.) was found also the allied genus *Paradoxides.* With regard to the characteristic genus of Regio III., the name of *Conocoryphe* was proposed for it by Corda in 1847, as synonymous with Zenker's name of *Conocephalus* (*Conocephalites*), already appropriated to a genus of insects.

Meanwhile, the similar crustaceans which abound in the transition rocks of Bohemia had been studied and described by Hawle, Corda, and Beyrich, when Barrande began his admirable investigations of this ancient fauna and of its stratigraphical relations. He soon found that beneath the horizon characterized by fossils of the Bala group (Llandeilo and Caradoc) there existed in Bohemia a series of strata distinguished by a remarkable fauna, entirely distinct from anything known in Great Britain, but closely allied to that of the alum-slates of Scandinavia, corresponding to Regiones II. and III. of Angelin. To this he gave the name of the first or primordial fauna, and to the rocks yielding it that of the Primordial Zone. Resting upon the old gneisses of Bohemia appears a series of crystalline schists designated by Barrande as *Étage A*, overlaid by a series of sandstones and conglomerates, *Étage B*, upon which repose the fossiliferous argillites of the primordial zone, or *Étage C*. The rocks of the Etages A and B were by Barrande regarded as azoic, but, in 1861, Fritsch of Prague, after a careful search, discovered in certain thin-bedded sandstones of B the traces of filled-up vertical double tubes; which, according to Salter (Mem. Geol. Sur., III. 243), are probably the marks of annelides, and are identical with those found in the rocks of the Bangor or Longmynd group in Great Britain, which will be shown to belong to the primordial zone. It is, therefore, probable that the Etage B, which apparently corresponds to the Regio Fucoidarum or basal sandstone of Scandinavia, should itself be included in the primordial zone. It may here be noticed that it is in the crystalline schists of A that Gümbel has found *Eozoon Bavaricum*. To the Etage C in Bohemia, Barrande assigns a thickness of about 1,200 feet, and to this his first fauna is confined, while in the succeeding divisions he distinguished a second and a third. The second fauna, which characterizes Etage D, corresponds to that of the Bala group; while the third fauna, belonging to the Etages E, F, G, and H, is that of the May Hill, Wenlock, and Ludlow formations of Great Britain.

This classification of the ancient Bohemian faunas was first set forth by Barrande in 1846, in his Notice Préliminaire, in which he declared that the first fauna was below the base of the Llandeilo of Murchison, unknown in Great Britain, and, moreover, "new and independent in relation to the two Silurian faunas (his second and third) already established in England." This opinion he reiterated in 1859. These three divisions form in Bohemia an apparently continuous series, and being connected with each other by some common species, Barrande was led to look upon the whole as forming a single stratigraphical system; and finally to assert that these three independent faunas "form by their union an indivisible triad, which is the Silurian system." (Bull. Soc. Geol. de Fr. (2), XVI. 529–545.) Already, in 1852, in his magnificent work on the Silurian System of Bohemia, Barrande had given to the strata characterized by his first fauna the name of Primordial Silurian. It is difficult to assign any just reason for thus annexing to the Silurian — already augmented by the whole Upper Cambrian or Bala group of Sedgwick (Llandeilo and Caradoc) — a great series of fossiliferous rocks lying below the base of the Llandeilo, and unsuspected by the author of the Silurian System, who persistently claimed the Llandeilo beds, with their characteristic second fauna, as marking the dawn of organic life.

Up to this time the primordial palæozoic fauna of Bohemia and of Scandinavia was, as we have said, unknown in Great Britain. The few organic remains mentioned by Sedgwick in 1835 as occurring in the region occupied by his Lower and Middle Cambrian, on Snowdon, were found to belong to Bala beds, which there rest upon the older rocks: nor was it until 1845 that Mr. Davis found in the Middle Cambrian remains of Lingula. In 1846, Sedgwick, in company with Mr. Davis, re-examined these rocks, and in December of the same year described the Lingula beds as overlaid by the Tremadoc slates and occupying a well-defined horizon in Caernarvon and Merionethshire, beneath the great mass of the Upper Cambrian rocks. (Geol. Jour., II. 75; III. 139.) Sedgwick, at the same time, noticed about this horizon certain graptolites and an Asaphus, which were supposed to belong to the Tremadoc slates, but have since been declared by Salter to pertain to the Arenig or Lower Llandeilo beds, the base of the Upper Cambrian. (Mem. Geol. Sur., III. 257, and Decade II.)

This discovery of the Lingula flags, as they were then named, and the fixing by Sedgwick of their geological horizon, was at once followed by a careful examination of them by the government surveyors, and in 1847, Selwyn detected in the Lingula flags, near Dolgelly, in Merionethshire, the remains of two crustacean forms, the one a phyllopod, which has received the name of *Hymenocaris vermicauda*, Salter, and the other a trilobite which was described by Salter in 1849 as *Olenus micrurus*. (Geol. Survey, Decade II.) A species of Paradoxides, apparently identical with *P. Forchhammeri* of Sweden, was also about this time recognized among specimens supposed to be from the same horizon. It has since been described as *P. Hicksii*, and found to belong to the basal beds of the Lingula flags, — the Menevian group.

Upon the flanks of the Malvern Hills there is found resting upon the ancient crystalline rocks of the region, and overlaid by the Pentamerus beds of the May Hill sandstone (originally called Caradoc by Murchison) a series of fossiliferous beds. These consist in their lowest part of about 600 feet of greenish sandstone, which have since yielded an Obolella and Serpulites, and are overlaid by 500 feet of black schists. In these, in 1842, Professor John Phillips found the remains of trilobites, which he subsequently described, in 1848, as three species of Olenus. (Mem. Geol. Survey, II. Part I. 55.) These black shales, which had not at that time furnished any organic remains, were by Murchison in his Silurian System (p. 416) in 1839 compared to the supposed passage-beds in Caermarthenshire between the Llandeilo and the Cambrian (Bala) rocks; which, as we have seen, were newer and not older strata than the Llandeilo flags. From their lithological characters, and their relations to the Pentamerus beds, these lower fossiliferous strata of Malvern were subsequently referred by the government geologists to the horizon of the Caradoc proper or Bala group; nor was it until 1851 that their true geological age and significance were made known. In that year, Barrande, fresh from the study of the older rocks of the continent, came to England for the purpose of comparing the British fossils with those of the primordial zone, which he had established in Bohemia and Scandinavia, and which he at once recognized in the Lingula flags of Sedgwick and in the black schists at Malvern; both of which were characterized by the presence of the genus Olenus, and were referred to the horizon of his Etage C. This important conclusion was announced by Salter to the British Association at Belfast in 1852. (Rep. Brit. Assoc., abstracts, p. 56, and Bull. Soc. Geol. de Fr. (2), XVI. 537.) [The black schists of Malvern, and the underlying greenish beds known as the Hollybush sandstones, are by Hicks regarded as the equivalents respectively of the Dolgelly and Festiniog divisions of the Lingula-flags. (Proc. Geologists, Association, Vol. III. No. 3.)]

The palæontological studies of Salter, while they confirmed the primordial character of the whole of the great mass of strata which make up the Middle Cambrian or Festiniog group of Sedgwick (consisting of the Lingula flags and the Tremadoc slates), led him to propose several subdivisions. Thus he distinguished on palæontological grounds between the upper and lower Tremadoc slates, and for like reasons divided the Lingula flags into a lower and an upper portion. For the discussion of these distinctions the reader is referred to the memoirs of the Geol. Survey (III. 240–257). Subsequent researches led to the division of the original Lingula flags into four parts, an upper, middle, and lower, to which the names of Dolgelly, Festiniog, and Maentwrog were given by Mr. Belt in 1867, and a fourth, consisting of the basal beds, which had been already separated in 1865 by Salter and Hicks, with the designation of Menevian, derived from the ancient Roman name of St. David's in Pembrokeshire.* It was here that, in

[* The researches of Mr. Belt on the Lingula Flags appeared in 1867. (Geological Magazine, Vol. IV. 483 and 536, and Vol. V. 5.) He included

1862, Salter found Paradoxides with Agnostus and Lingula in fine black shales at the base of the Lingula flags, resting conformably on the green and purple grits of the Lower Cambrian or Harlech beds. The locality was afterwards carefully studied by Hicks, and it was soon made apparent that the genus Paradoxides, both here and in North Wales, was confined to a horizon below the great mass of the Lingula flags; which, on the contrary, are characterized by numerous species of Olenus. These lower or Menevian beds are hence regarded by Salter as equivalent to the lowest portion of the Etage C of Barrande.

Beneath these Menevian beds there lies, in apparent conformity, the great Lower Cambrian series, frequently called the bottom or basement rocks by the government surveyors; represented in North Wales by the Harlech grits, and in South Wales, near St. David's, by a similar series of green and purple sandstones, considered by Murchison, and by others, as the equivalent of the Harlech rocks. They were still supposed to be unfossiliferous until in June, 1867, Salter and Hicks announced the discovery in the red beds of this lower series, at St. David's, of a Lingulella, very like *L. ferruginea* of the Menevian. (Geol. Jour., XXIII. 339; Siluria, 4th ed., 550.) This led to a further examination of these Lower Cambrian beds, which has resulted in the discovery in them of a fauna distinctly primordial in type, and linked by the presence of several identical fossils to the Menevian; but in many respects distinct, and marking a lower fossiliferous horizon than anything known in Bohemia or in Scandinavia.

The first announcement of these important results was made

under the name of Upper Cambrian the Tremadoc rocks with the Lingula flags proper, which he divided in descending order into three parts, Dolgelly, Festiniog, and Maentwrog; while he suggested the union of the basal beds, (previously separated under the name of Menevian,) with the underlying Harlech and Bangor rocks as Lower Cambrian. These divisions of Belt are now recognized by Hicks. It will be recollected that the whole of the Lingula flags were originally included in his Festiniog group by Sedgwick. All of these rocks are inverted in the vicinity of Dolgelly, the apparent succession in descending order being Festiniog, Dolgelly, Tremadoc, and Arenig.]

to the British Association at Norwich in 1868. Further details were, however, laid before the Geological Society in May, 1871, by Messrs. Harkness and Hicks, whose paper on The Ancient Rocks of St. David's Promontory appears in the Geological Journal for November, 1871. (XXVIII. 384.) The Cambrian sediments here rest upon an older series of crystalline stratified rocks, described by the geological surveyors as syenite and greenstone, and having a northwest strike. Lying unconformably upon these, and with a northeast strike, we have the following series, in ascending order: 1. Quartzose conglomerate, 60 feet; 2. Greenish flaggy sandstones, 460 feet; 3. Red flags or slaty beds, 50 feet, containing *Lingulella ferruginea,* besides a larger species, *Discina,* and *Leperditia Cambrensis;* 4. Purple and greenish sandstones, 1,000 feet; 5. Yellowish-gray sandstones, flags, and shales, 150 feet, with *Plutonia, Conocoryphe, Microdiscus, Agnostus, Theca,* and *Protospongia;* 6. Gray, purple, and red flaggy sandstones, with most of the above genera, 1,500 feet; 7. Gray flaggy beds, 150 feet, with *Paradoxides;* 8. True Menevian beds, richly fossiliferous, 500 feet. The latter are the probable equivalent of the base of Barrande's Etage C, and at St. David's are conformably overlaid by the Lingula flags; beneath which we have, including the Menevian, a conformable series of 3,370 feet of uncrystalline sediments, fossiliferous nearly to the base, and holding a well-marked fauna distinct from anything hitherto known in Great Britain or elsewhere.

The Menevian beds are connected with the underlying strata by the presence of *Lingulella ferruginea, Discina pileolus,* and *Obolella sagittatis,* which extend through the whole series; and also by the genus Paradoxides, four species of which occur in these lower strata; from which the genus Olenus, which characterizes the Lingula flags, seems to be absent. To a large tuberculated trilobite of a new genus found in these lowest rocks the name of *Plutonia Sedgwickii* has been given. Hicks has proposed to unite the Menevian with the Harlech beds, and to make the summit of the former the dividing line between the Lower and Middle Cambrian, a suggestion which has been adopted by Lyell. (Proc. Brit. Assoc. for 1868, p. 68, and Lyell, Student's Manual of Geology, 466 – 469.)

Both Phillips and Lyell give the name of Upper Cambrian to the Lingula flags and the Tremadoc slates, which together constitute the Middle Cambrian of Sedgwick, and concede the title of Lower Silurian to the Bala group or Upper Cambrian of Sedgwick. The same view is adopted by Linnarsson in Sweden, who places the line between Cambrian and Silurian at the base of the Llandeilo or the second fauna. It was by following these authorities that I, inadvertently, in my address to the American Association for the Advancement of Science in August, 1871, gave this horizon as the original division between Cambrian and Silurian.* The reader of the first part of this paper will see with how much justice Sedgwick claims for the Cambrian the whole of the fossiliferous rocks of Wales beneath the base of the May Hill sandstone, including both the first and the second fauna. I cannot but agree with the late Henry Darwin Rogers, who, in 1856, reserved the designation of "the true European Silurian" for the rocks above this horizon. (Keith Johnson's Physical Atlas, 2d ed.)

The Lingula flags and Tremadoc slates have been made the subject of careful stratigraphical and palæontological studies by the Geological Survey, the results of which are set forth by Ramsay and Salter in the third volume of the Memoirs of the Geological Survey, published in 1866, and also, more concisely, in the Anniversary Address by the former to the Geological Society in 1863. (Geol. Jour. (19), XVIII.) The Lingula flags (with the underlying Menevian, which resembles them lithologically) rest in apparent conformity upon the purple Harlech rocks both in Pembrokeshire and in Merionethshire, where the latter appear on the great Merioneth anticlinal, long since pointed out by Sedgwick. The Lingula flags, (including the Menevian) have in this region, according to Ramsay, a thickness of about 6,000 feet. Above these, near Tremadoc and Festiniog, lie the Tremadoc slates, which are here overlaid, in apparent conformity, by the Lower Llandeilo beds. At a

* Since corrected in the reprint of that address in the present volume.

distance of eleven miles to the northwest, however, the Tremadoc slates disappear, and the Lingula flags are represented by only 2,000 feet of strata; while in parts of Caernarvonshire, and in Anglesea, the whole of the Lingula flags and, moreover, the Lower Cambrian rocks are wanting, and the Llandeilo beds rest directly upon the ancient crystalline schists. In Scotland and in Ireland, moreover, the Lingula flags are wholly absent, and the Llandeilo rocks there repose unconformably upon grits regarded as of Lower Cambrian age. Thus, without counting the Tremadoc slates, which are a local formation, unknown out of Merionethshire,* we have (including the Bangor group and Lingula flags), beneath the Llandeilo, over 9,000 feet of fossiliferous strata, which disappear entirely in the distance of a few miles. From a careful survey of all the facts, the conclusion of Ramsay is irresistible, that there exists between the Lingula flags and the Llandeilo not merely one, but two great stratigraphical breaks in the succession; the one between the Lingula flags and the Lower Tremadoc slates, and the other between the Upper Tremadoc slates and the Lower Llandeilo, at the base of which were included the Arenig rocks.

This conclusion is confirmed by the fact that there exists at each of these horizons a nearly complete palæontological break.

* [This statement requires correction, since already, in 1866, Messrs. Salter and Hicks had mentioned the occurrence of rocks supposed to be of that age near St. David's in South Wales, and very recently, in the Quarterly Geological Journal for February, 1873, the latter has given a description of the localities of Tremadoc rocks in this region, with figures of the organic remains, a map, and sections. The beds have here a thickness of about 1,000 feet, and rest directly upon the Lingula flags. The apparent want of conformity between the two divisions noticed by Ramsay in North Wales is here not manifest. They are followed in seeming unconformity by the Arenig rocks, which are by Mr. Homfray considered equivalent to the Upper Tremadoc of North Wales, and contain in abundance the graptolites of the Levis formation of Canada. The beds between these and the Lingula flags hold a rich fauna closely allied to the Lower Tremadoc, including an *Orthoceras*, a new species of *Paleasterina*, and a *Dendrocrinus*, various brachiopods and lamellibranchs, trilobites of the genus *Niobe* and of a new genus, *Neseuretus*, closely allied to *Dikelocephalus*, to which Hicks refers the supposed species of *D.* described by Salter from the Upper Lingula and Lower Tremadoc rocks of North Wales; the only true *Dikelocephalus* in Wales, according to him, being *D. furca* from the Upper Tremadoc.]

The fauna of the Tremadoc slates is, according to Salter, almost entirely distinct from that of the Lingula flags, and not less distinct from that of the so-called Lower Llandeilo or Arenig rocks (the equivalents of the Skiddaw slates of Cumberland). Hence, says Ramsay, it is evident "that in these strata we have three perfectly distinct zones of organic remains, and therefore, in common terms, three distinct formations." The palæontological evidence is thus in complete accordance with that furnished by stratigraphy. We cannot leave this topic without citing the conclusion of Ramsay that "each of these two breaks necessarily implies a lost epoch, stratigraphically quite unrepresented in our area; the life of which is only feebly represented in some cases by the fossils common to the underlying and overlying formation." In connection with this remark, which we conceive to embody a truth of wide application, it may be said that stratigraphical breaks and discordances in a geological series may, *a priori*, be expected to occur most frequently in regions where this series is represented by a large thickness of strata. The accumulation of such masses implies great movements of subsidence, which, in their nature, are limited, and are accompanied by elevations in adjacent areas, from which may result, over these areas, either interruptions in the process of sedimentation, or the removal, by sub-aerial or sub-marine denudation, of the sediments already formed. The conditions of succession and distribution, it may be conceived, would be very different in a region where the period corresponding to this same geological series was marked by comparatively small accumulations of sediment upon an ocean-floor subjected to no great movements.

This contrast is strikingly seen between the conformable series of less than 2,000 feet of strata, which in Scandinavia are characterized by the first three palæozoic faunas (Cambrian and Silurian), and the repeatedly broken and discordant succession of more than 30,000 feet of sediments,* which in

* The Longmynd rocks in Shropshire are alone estimated at 20,000 feet; but their supposed equivalents, the Harlech rocks of Pembrokeshire, have a

Wales are their palæontological equivalents. It must, however, be considered that in regions of small accumulation where, as in Scandinavia, the formations are thin, there may be lost or unrepresented zoölogical periods whose place in the series is marked by no stratigraphical break. In such comparatively stable regions, movements of the surface sufficient to cause the exclusion, or the disappearance by removal, of the small thickness of strata corresponding to a zoölogical period, may take place without any conspicuous marks of stratigraphical discordance.

The attempt to establish geological divisions or horizons upon stratigraphical or palæontological breaks must always prove fallacious. From the nature of things, these, whether due to non-deposition or to subsequent removal of deposits, must be local; and we can say, confidently, that there exists no break in life or in sedimentation which is not somewhere filled up and represented by a continuous and conformable succession. While we may define one period as characterized by the presence of a certain fauna, which, in a succeeding age, is replaced by a different one, there will always be found, in some part of their geographical distribution, a region where the two faunas commingle, and where the gradual disappearance of the old before the new may be studied. The division of our stratified rocks into systems is therefore unphilosophical, if we assign any definite or precise boundaries or limitations to these. It was long since said by Sedgwick with regard to the whole succession of life through geologic time, that all belongs to one great *systema naturæ*. (Philos. Mag. (4), VIII. 359.)

We have already noticed that Barrande, as early as 1852, gave the name of Primordial Silurian to the rocks which, in

measured thickness of 3,300, while the Llanberris and Harlech rocks together, in North Wales, equal from 4,000 to 7,000 feet, and the Lingula flags and Tremadoc slates, united, about 7,000 feet. The Bala group in the Berwyns exceeds 12,000 feet, and the proper Silurian, from the base of the Upper Llandovery or May Hill sandstone, attains from 5,000 to 6,000 feet; so that the aggregate of 30,000 feet may be considered below the truth. (Mem. Geol. Survey, III., Part II. pages 72, 222; and Siluria, 4th ed. 185.) [The aggregate thickness since assigned to these rocks by Hicks is about 33,000 feet.]

Bohemia, were marked by the first fauna; although he, at the same time, recognized this as distinct from and older than the second fauna, discovered in the Llandeilo rocks, which Murchison had declared to represent the dawn of organic life. Into the reasons which led Barrande to include the rocks of the first, second, and third faunas in one Silurian system (a view which was at once adopted by the British Geological Survey and by Murchison himself), it is not our province to inquire, but we desire to call attention to the fact that the latter, by his own principles, was bound to reject such a classification. In his address before the Geological Society in 1842 (already quoted in the first part of this paper), he declared that the discussion as to the value of the term Cambrian involved the question "whether there was any type of fossils in the mass of the Cambrian rocks different from those of the Lower Silurian series. If the appeal to nature should be answered in the negative, then it was clear that the Lower Silurian type must be considered the true base of what I had named the protozoic rocks; but if characteristic new forms were discovered, then would the Cambrian rocks, whose place was so well established in the descending series, have also their own fauna, and the palæozoic base would necessarily be removed to a lower horizon."

In the event of no distinct fauna being found in the Cambrian series, it was declared that "the term Cambrian must cease to be used in zoölogical classification, it being, in that sense, synonymous with Lower Silurian." (Proc. Geol. Soc., III. 641, *et seq.*) That such had been the result of palæontological inquiry Murchison then proceeded to show. Inasmuch as the only portion of Sedgwick's Cambrian which was then known to be fossiliferous was really above, and not below, the Llandeilo rocks, which Murchison had taken for the base of his Lower Silurian, his reasoning with regard to the Cambrian nomenclature, based on a false datum, was itself fallacious; and it might have been expected that when the government surveyors had shown his stratigraphical error, Murchison would have rendered justice to the nomenclature of

Sedgwick. But when, still later, a further "appeal to nature" led to the discovery of "characteristic new forms," and established the existence of a "type of fossils in the mass of the Cambrian rocks, different from those of the Lower Silurian series," Murchison was bound by his own principles to recognize the name of Cambrian for the great Festiniog group, with its primordial fauna, even though Barrande and the government surveyors should unite in calling it Primordial Silurian.

He, however, chose the opposite course, and now attempted to claim for the Silurian system the whole of the Middle Cambrian or Festiniog group of Sedgwick, including the Tremadoc slates and the Lingula flags. The grounds of this assumption, as set forth in the successive editions of Siluria from 1854 to 1867, and in various memoirs, may be included under three heads: first, that the Lingula flags have been found to exist in some parts of his original Silurian region; second, that no clearly defined base had been assigned by him to his so-called system; and, third, that there are no means of drawing a line of demarcation between those Middle Cambrian formations and the overlying Llandeilo.

With regard to the first of these reasons, it is to be said that the only known representatives of the Lingula flags in the region described by Murchison in his Silurian system are the black slates of Malvern and some scanty outliers which, in Shropshire, lie between the old Longmynd rocks and the base of the Stiper-stones. The former were then (as has already been shown) supposed by him to belong to the Llandeilo, or rather to the passage-beds between the Llandeilo and the Cambrian (Bala); while with regard to the latter, Ramsay expressly tells us that they were not originally classed with the Silurian, but have since been included in it. (Mem. Geol. Sur., III. Part II. page 9; and 242, foot-note.)

The Llandeilo beds were by Murchison distinctly stated to be the base of the Silurian system (Silurian System, 222); and it was further declared by him that in Shropshire (unlike Caermarthenshire) "there is no passage from the Cambrian to the Silurian strata," but a hiatus, marked by disturbances which excluded the passage-beds, and caused the Lower Silurian to rest unconformably upon the Longmynd rocks. (Ibid., 256; and Plate 31, sections 3 and 6; Plate 32, section 4.) But in Siluria (1st ed. 47) the two are stated to be conformable; and in the subsequent sections of this region, made by Aveline, and published by the Geological Survey, the evidences of this want of conformity do not appear. Murchison at that time confounded the rocks of the Longmynd with the Cambrian (Bala) beds of Caermarthenshire and Brecon. (Silurian System, 416.) Hence it was that he gave the name of Cambrian to the former; and this mistake, moreover, led him to place the Cambrian of Caermarthenshire beneath the Llandeilo. It is clear that if he claimed no well-defined base to the Llandeilo rocks in this latter (their typical region), it was because he saw them passing into the overlying Bala beds. There was, in the error by which he placed *below* the Llandeilo, strata which were really *above* them, no ground whatever for afterwards including in his Silurian System, as a downward continuation of the Llandeilo rocks (which are the basal portion of the Bala group), the whole Festiniog group of Sedgwick; whose infra-position to the Bala had been shown by the latter long before it was known to be fossiliferous.

It was, however, claimed by Murchison that no line of separation can be drawn between these two groups. The results of Ramsay and of Salter, as set forth in the address of the former before the Geological Society of 1863, and more fully in the Memoirs of the Geological Survey (Vol. III. Part II.) published in 1866, with a preface by himself, as the director of the Survey, are completely ignored by Murchison. The reader familiar with these results, of which we have given a summary, finds with surprise that in the last edition of Siluria, that of 1867, they are noticed in part, but only to be repudiated. In the five pages of text which are there given to this great Middle Cambrian division, we are told that the distinction between the Lower Tremadoc and the Lingula flags "is difficult to be drawn," and that the Upper Tremadoc slate passes into and forms the lower part of the Llandeilo (under which name

Murchison included the Arenig rocks), "into which it graduates conformably." (*Loc. cit.*, 4th ed. p. 46.) In each of these cases, on the contrary, according to Ramsay, there is observed "a break very nearly complete both in genera and species, and probable unconformity"; the evidence of the palæontological break being furnished by the careful studies of Salter; while that of the stratigraphical break, as we have seen, leaves no reason for doubt. (Mem. Geol. Sur., III. Part II. pages 2, 161, 234.) The student of Siluria soon learns that in all cases where Murchison's pretensions were concerned, the book is only calculated to mislead.

The reader of this history will now be able to understand why, notwithstanding the support given by Barrande, by the geological survey of Great Britain, and by most American geologists to the Silurian nomenclature of Murchison, it is rejected, so far as the Lingula flags and the Tremadoc slates are concerned, by Lyell, Phillips, Davidson, Harkness, and Hicks in England, and by Linnarsson in Sweden. These geologists have, however, admitted the name of Lower Silurian for the Bala group or Upper Cambrian of Sedgwick; a concession which can hardly be defended, but which apparently found its way into use at a time when the yet unravelled perplexities of the Welsh rocks led Sedgwick himself to propose, for a time, the name of Cambro-Silurian for the Bala group. This want of agreement among geologists as to the nomenclature of the lower palæozoic rocks, causes no little confusion to the learner. We have seen that Henry Darwin Rogers followed Sedgwick in giving the name of Cambrian to the whole palæozoic series up to the base of the May Hill sandstone; and the same view is adopted by Woodward in his Manual of the Mollusca. The student of this excellent book will find that in the tables giving the geological range of the mollusca, on pages 124, 125, and 127, the name of Cambrian is used in Sedgwick's sense, as including all the fossiliferous strata beneath the May Hill sandstone. On page 123 it is, however, explained that Lower Silurian is a synonyme for Cambrian, and it is so used in the body of the work.

The distribution of the Lower and Middle Cambrian rocks in Great Britain may now be noticed. The former, or Bangor group, to which Murchison and the geological survey restrict the name of Cambrian, and which they sometimes call the Longmynd, bottom or basement rocks, occupy two adjacent areas in Caernarvon and Merionethshire; the one near Bangor, including Llanberris, to the northeast, and the other, including Harlech and Barmouth, to the southeast, of Snowdon; this mountain lying in a synclinal between them, and rising 3,571 feet above the sea. The great mass of grits or sandstones appears to be at the summit of the group, but in the lower part the blue roofing-slates of Llanberris are interstratified in a series of green and purple slates, grits, and conglomerates. (Some of the Welsh roofing-slates are, however, supposed to belong to the Llandeilo. Mem. Geol. Survey, III. Part II. pages 54, 258.) The Harlech rocks in this northwestern region are conformably overlaid by the Menevian, followed by the true Lingula flags, or Olenus beds, of the Middle Cambrian. Upon these repose the Tremadoc slates.

The third area of Lower Cambrian rocks known in Great Britain is that already described at St. David's in Pembrokeshire, about one hundred miles to the southwest; and the fourth, that of the Longmynd hills, about sixty miles to the southeast of Snowdon. The rocks of the Longmynd, like those of the other Lower Cambrian areas mentioned, consist principally of green and purple sandstones with conglomerates, shales, and some clay-slates. They occasionally hold flakes of anthracite, and small portions of mineral pitch exude from them in some localities. The only evidence of animal life yet found in the rocks of the Longmynd are furnished by worm-burrows, the obscure remains of a crustacean (*Palœopyge Ramsay*), and a form like *Histioderma*. This latter organic relic, with worm-burrows, and the fossils named *Oldhamia*, is found on the coast of Ireland opposite Caernarvonshire, in the rocks of Bray Head; which resemble lithologically the Harlech beds, and are regarded as their equivalents.

Still another area of the older rocks is that of the Malvern hills, on the western flanks of which, as already mentioned, the Lingula flags are represented by about 500 feet of black shales with Olenus, underlaid by 600 feet of greenish sandstones containing traces of fucoids, with Serpulites and an Obolella. It is not improbable, as suggested by Barrande and by Murchison, that these 1,100 feet of strata represent, in this region, the great mass of the Lingula flags; and, we may add, perhaps, the whole series of Lower Cambrian strata, which in Caernarvon and Pembroke underlie them [see page 371]; since these sandstones of Malvern, like those of St. David's, rest upon crystalline schists, and are in part made up of their ruins.

These crystalline schists of Malvern, which are described by Phillips as the oldest rocks in England, and by Mr. Holl are conjectured to be Laurentian, seem from the descriptions of their lithological characters to resemble those of Caernarvon and Anglesea, with which they are, by Murchison, regarded as identical. The crystalline schists of these latter localities are, by Sedgwick, described as hypozoic strata, below the base of the Cambrian. Murchison, however, in the first edition of his Siluria, adopted the suggestion of De la Beche that they themselves were altered Cambrian strata. In fact, they directly underlie the Llandeilo rocks, and were apparently conceived by Murchison to represent that downward continuation of these upon which he had insisted. This opinion is supported by ingenious arguments on the part of Ramsay. (Mem. Geol. Survey, III. Part II., *passim.*) I am, however, disposed to regard them, with Sedgwick and Phillips, as of pre-Cambrian age, and to compare them with the Huronian series of North America, which occupies a similar geological horizon, and with which, as seen in northern Michigan, and in the Green Mountains, I have found the rocks of Anglesea to offer remarkable lithological resemblances.

It may here be noticed that the gold-bearing quartz veins in North Wales are found in the Menevian beds, and also, according to Selwyn, throughout the Lingula flags. These fossiliferous strata at the gold-mine near Dolgelly appear in direct contact with diorites and chloritic and talcose schists, which are more or less cupriferous, and themselves also contain gold-bearing quartz veins. (Mem. Geol. Survey, Part II. pages 42, 45; and Siluria, 4th ed. 450, 547.)

The Table on page 386 gives a view of the lower palæozoic rocks of Great Britain and North America, together with the various nomenclatures and classifications referred to in the preceding pages. In the second column, the horizontal black lines indicate the positions of the three important palæontological and stratigraphical breaks signalized by Ramsay in the British succession. (Mem. Geol. Survey, III. Part II. page 2.)

[Very recently, in 1873, in the Proceedings of the Geologists' Association, Vol. III. Part III., Mr. Hicks has given a similar tabular view of the lower palæozoic rocks of Great Britain. The Bangor group (to which he applies the name of Longmynd or Lower Cambrian),differs from that given in the following table only in dividing the Menevian into an upper and a lower part. The Middle Cambrian or Festiniog group of Sedgwick (which Hicks calls Upper Cambrian) presents also the same subdivisions as are here given. In the next, or Upper Cambrian of Sedgwick (called by Hicks Lower Silurian), are included in ascending order Lower Arenig and Upper Arenig or Skiddaw, followed by Llandeilo, also divided into two parts, and by the Bala group, which he divides into Lower and Upper Caradoc, to which he adds, as we have done, the Lower Llandovery.]

[In the new Catalogue of the Cambridge Fossils is an important preface written from Sedgwick's dictation late in 1872, and published since his death. In this he unites the Lower Llandovery with the Upper Cambrian, and includes it, together with the Caradoc and Llandeilo, under the name of the Bala group, which he divides into Lower, Middle, and Upper Bala; while the Arenig or Skiddaw rocks are joined with the Middle Cambrian. Both the Arenig and the Tremadoc rocks, in fact, present a certain intermingling of organic forms belonging to the first and second faunas; but according to Hicks the Tremadoc beds are to be classed with the first, and the Arenig with the second. These two groups of rocks are in fact the palaeontological equivalents of the Calciferous, Levis and Chazy, which serve in North America to connect the Middle with the Upper Cambrian. As regards the extension to the Upper Cambrian of Sedgwick of the name of Lower Silurian, which, as has been shown, was given to it only through an enormous and now universally acknowledged mistake on the part of Murchison, I am constrained, notwithstanding its adoption by so many eminent geologists, to maintain for the division the name given to it by its true discoverer, Sedgwick.]

In the third column, the subdivisions are those of the New York and Canada Geological Surveys; in connection with which the reader is referred to a table which I prepared and published in 1863, in the Geology of Canada, page 932. Opposite to the Menevian I have placed the names of its principal American localities; which are Braintree, Massachusetts, St. John, New Brunswick, and St. John's, Newfoundland. The further consideration of the American subdivisions is reserved for the third part of this paper. With regard to the classification of Angelin, it is to be remarked that although he designates II. as *Regio Olenorum*, and III. as *Regio Conocorypharum*, the position of these, according to Linnarsson, is to be reversed; the Conocoryphe beds with Paradoxides being below, and not above, those holding Olenus. The *Regio Fucoidarum* in Sweden has lately furnished a brachiopodous shell, *Lingula monilifera*, besides the curious plant-like fossil, *Eophyton Linnœanum*. (Linnarsson, Geol. Magazine, 1869, VI. 393.)

LOWER PALÆOZOIC ROCKS OF EUROPE AND NORTH AMERICA.

	British Subdivisions.	North American Subdivisions.	Nomenclatures of Sedgwick and Murchison.	Barrande's Classification.	Angelin's Divisions.
15	Ludlow.	Lower Helderberg.	Silurian, *Sedgwick ;*	Third Fauna,	VIII. VII.
14	Wenlock.	Niagara, Clinton,	Upper Silurian,	including	or Regiones
13	Upper Llandovery.	Medina, Oneida.	*Murchison.*	Etages H. G. F. E.	E. and DE.
12	Lower Llandovery.	Hudson River, Utica,	Upper Cambrian or	Second Fauna,	VI. V. IV.
11	Caradoc or Bala.	Trenton, Birdseye,	Bala group, *Sedgwick ;*	including	or Regiones
10	Llandeilo.	Black River.	Lower Silurian,	Etage D.	D. C. and BC.
9	Arenig or Skiddaw.	Chazy.	*Murchison.*		
8	Upper Tremadoc.	Levis.	Middle Cambrian or	First Fauna	III. II. I.
7	Lower Tremadoc.	Calciferous.	Festiniog group,	or	or
6	Dolgelly.		*Sedgwick ;*	Primordial Fauna,	Regiones
5	Festiniog.	Potsdam.	Primordial Silurian,	including	B. and A.
4	Maentwrog.		*Murchison.*	Etage C,	and
3	Menevian.	Braintree and St. John.	Lower Cambrian or	and probably also	Regio
2	Harlech.	——— ?	Bangor group, *Sedgwick ;*	Etage B.	Fucoidarum.
1	Llanberris.	——— ?	Cambrian, *Murchison.*		

III. Cambrian and Silurian Rocks in North America.

In accordance with our plan we now proceed to sketch the history of the lower palæozoic rocks of North America. While European geologists were carrying out the researches which have been described in the first and second parts of this paper, American investigators were not idle. The geological studies of Eaton led the way to a systematic survey of the State of New York, the results of which have been the basis of most of the subsequent geological work in eastern North America, and which was begun by legislative enactment in 1836. The State was divided into four districts, the work of examining and finally reporting upon which was committed to as many geologists. The first or southeastern district was undertaken by Mather, the second or northeastern by Emmons, the third or central by Vanuxem, and the fourth or western by James Hall; the palæontology of the whole being left to Conrad, and the mineralogy to Beck. After various annual reports the final results of the survey appeared in 1842. The whole series of fossiliferous rocks known, from the basal or Potsdam sandstone to the coal-formation, was then described as the New York system.

At that time the published researches of British geologists furnished the means of comparison between the organic remains found in the rocks of New York, and those then known to exist in the palæozoic strata of Great Britain. Professor Hall was thus enabled, in his Geology of the Fourth District of New York, to declare, from the study of its fossils, that the New York system included the Devonian of Phillips, the Silurian of Murchison, and the Cambrian of Sedgwick; meaning by the latter the Upper Cambrian, or Bala group, which alone was then known to be fossiliferous. From the evidence then before him, he concluded that the Upper Cambrian was represented in the New York system by the whole of the rocks from the base of the Utica slate downward, with the probable exception of the Potsdam sandstone; while he conceived, partly on lithological grounds, that the Utica and Hudson-River groups represented the Llandeilo and Caradoc, or the Lower Silurian of Murchison. (*Loc. cit.*, pages 20, 29, 31.) The origin of the Cambrian and Silurian controversy, and the errors by which the Llandeilo and a part of the Caradoc had by Murchison been classed as a series distinct from the Bala group, were not then known; but in a note to this report (page 20) Hall informs us of the declaration of Murchison, already quoted from his address of 1842, that the Cambrian, so far as then known, could not, on palæontological grounds, be distinguished from his Lower Silurian.

Emmons meanwhile had examined in eastern New York and western New England a series of fossiliferous rocks which, on lithological and stratigraphical grounds, he regarded as older than any in the New York system; a view which had been previously maintained by Eaton. Holding, with Hall, that the lower members of the New York system were the equivalents of the Upper Cambrian of Sedgwick, he looked upon the fossiliferous rocks which he placed beneath them as the representatives of the Lower Cambrian. By this name, as we have seen, Sedgwick, in 1838, designated all those uncrystalline rocks of North Wales which he subsequently divided into Lower and Middle Cambrian, and which lie beneath the base of the Bala group. When Murchison, in 1842, in his so often quoted declaration, asserted that "the term Cambrian must cease to be used in a zoölogical classification, it being in that sense synonymous with Lower Silurian," he was speaking only on palæontological grounds, and, disregarding the great Lower and Middle Cambrian divisions of Sedgwick, had reference only to the Upper Cambrian. This, however, was overlooked by Emmons, who, feeling satisfied that the sedimentary rocks which he had examined in eastern New York were distinct from those which he, with Hall, regarded as corresponding to the Bala group or Upper Cambrian (the Lower Silurian of Murchison), and probably equivalent to the inferior portions of Sedgwick's Cambrian; and, supposing that the latter term was henceforth to be effaced from geology (as indeed was attempted shortly after, in the copy of Sedgwick's map published in 1844 by the Geological Society), devised for these rocks the name of the Taconic system, as synonymous with the Lower (and Middle) Cambrian of Sedgwick. These conclusions were set forth by him in 1842, in his report on the Geology of the Northern District of New York (page 162). See farther his Agriculture of New York (I. 49), the fifth chapter of which, "On the Taconic System," was also published separately in 1844; when the presence of distinctive organic remains in the rocks of this series was first announced.

Meanwhile to Professor Hall, after the completion of the survey, had been committed the task of studying and describing the organic remains of the State, and in 1847 appeared the first volume of his great work on the Palæontology of New York. Since 1842 he had been enabled to examine more fully the organic remains of the lower rocks of the New York system, and to compare them with those of the Old World; and in the Introduction to the volume just mentioned (page xix) he announced the important conclusion that the New York system itself contained an older fauna than the Upper Cambrian of Sedgwick. According to Hall, the organic forms of the Calciferous and Chazy formations had not yet been found in Europe, and our comparison with European fossiliferous rocks must commence with the Trenton group. He however excepted the Potsdam sandstone, which already, in 1842, he had conceived to be below the Upper Cambrian of Sedgwick, and now regarded as the probable equivalent of the Obolus or Ungulite grit of St. Petersburg. Thus Emmons, in 1842, asserted, on lithological and stratigraphical grounds, the existence, beneath the base of the New York system, of a lower and unconformable series of rocks, in which, in 1844, he announced the discovery of a distinctive fauna. Hall, on his part, asserted in 1842, and more fully in 1847, that the New York system itself held an older fauna than that hitherto known in the British rocks.

It is not necessary to recall in this place the details of the long and unfortunate Taconic controversy, which I have recently discussed in my address before the American Association for the Advancement of Science in August, 1871. (*Ante,* page 251.) It is, however, to be remarked that Hall, in common with all other American geologists, followed Henry D. Rogers in opposing the views of Emmons, whose Taconic system was supposed to represent either the whole or a part of the Champlain division of the New York system; which division included, as is well known, all of the fossiliferous rocks up to the base of the Oneida conglomerate (and also this latter, according to Emmons); thus comprehending both the first and the second palæozoic faunas; as shown in the preceding table on page 386.

Emmons, misled by stratigraphical and lithological considerations, complicated the question in a singular manner, which scarcely finds a parallel except in the history of Murchison's Silurian sections. Completely inverting, as I have elsewhere shown, the order of succession in his Taconic system, estimated by him at 30,000 feet, he placed near the base of the lower division of the system the Stockbridge or Eolian limestone, including the white marbles of Vermont; which, by their organic remains, have since been by Billings found to belong to the Levis formation. A large portion of the related rocks in western Vermont and elsewhere, which afford a fauna now known to be far more ancient than that of the Lower Taconic just referred to, and as low if not lower than anything in the New York system, were, by Emmons, then placed partly near the summit of the Upper Taconic, and partly, not only above the whole Taconic system, but above the Champlain division of the New York system. Thus we find, in 1842, in his Report on the Geology of the Northern District of New York (where Emmons defined his views on the Taconic system), that he placed above this latter horizon both the green sandstone of Sillery near Quebec and the Red sand-rock of western Vermont (which he then regarded as the representatives of the Oneida and the Medina sandstones), and described the latter as made up from the ruins of Taconic rocks (pages 124, 282). In 1844–1846, in his Report on the Agriculture of

New York (page 119), he however adopted a different view of the Red sand-rock, assigning it to the Calciferous; and in 1855, in his American Geology (II. 128), it was regarded as in part Calciferous and in part Potsdam. In 1848, Professor C. B. Adams, then director of the Geological Survey of Vermont, argued strongly against these latter views, and maintained that the Red sand-rock directly overlaid the shales of the Hudson-River group and corresponded to the Medina and Clinton formations of the New York system. (Amer. Jour. Sci. (2), V. 108.) He had before this time discovered in this sand-rock, besides what he considered an *Atrypa*, abundant remains of a trilobite, which Hall, in 1847, referred to the genus *Conocephalus* (*Conocoryphe*), remarking at the same time that inasmuch as this genus was (at that time) only described as occurring in "graywacke in Germany and elsewhere," no conclusions could be drawn from these fossils as to the geological horizon of the rocks in question. (Ibid. (2), XXXIII. 371.) In September, 1861, however, Mr. Billings, after an examination of the rocks in question, pronounced in favor of the later opinion of Emmons, declaring the Red sand-rock near Highgate Springs, Vermont, containing *Conocephalus* and *Theca*, to belong to the base of the second fauna, "if not indeed a little lower," and to be "somewhere near the horizon of the Potsdam." (Ibid. (2), XXXII. 232.)

The dark-colored fossiliferous shales which were asserted, both by Adams and by Emmons, to underlie this Red sand-rock, were, by the former, as we have seen, regarded as belonging to the Hudson-River group, while by the latter they were described as an upper member of the Taconic system; which was here declared to be unconformably overlaid by the Red sand-rock, a member of the New York system. These slates, a few years before, had afforded some trilobites, which, after remaining in the hands of Professor Hall for two years or more, were in 1859 described by him in the twelfth Report of the Regents of the University of New York, as *Olenus Thompsoni* and *O. Vermontana.* He soon, however, found them to constitute a distinct genus, for which he proposed the name of *Barrandia*, but finding this name preoccupied, suggested in 1861, in the fourteenth Regents' Report, that of *Olenellus*, which was subsequently adopted by Billings in 1865. (Palæozoic Fossils, pages 365, 419.) In 1860, Emmons, in his Manual of Geology, described the same species, but placed them in the genus *Paradoxides*, as *P. Thompsoni* and *P. Vermontana*. Hall had already, in 1847, in the first volume of his Palæontology of New York, referred to *Olenus* the *Elliptocephalus asaphoides* of Emmons, and also a fragment of another trilobite from Saratoga Lake; both of which were described as belonging to the Hudson-River group of the New York system, or to a still higher horizon. The reasons for this will appear in the sequel. The *Elliptocephalus*, with another trilobite named by Emmons *Atops* (referred by Hall to *Calymene*, and subsequently by Billings to *Conocoryphe*), occurs at Greenwich, New York. These were by Emmons, in his essay on the Taconic system (in 1844), described as characteristic of that system of rocks.

A copy of the Regents' Report for 1859 having been sent by Billings to Barrande, this eminent palæontologist, in a letter addressed to Professor Bronn of Heidelberg, July 16, 1860 (American Journal of Science (2), XXXI. 212), called attention to the trilobites therein figured, and declared that no palæontologist familiar with the trilobites of Scandinavia would "have hesitated to class them among the species of the primordial fauna, and to place the schists enclosing them in one of the formations containing this fauna. Such is my profound conviction," etc. The letter containing this statement had already appeared in the American Journal of Science for March, 1861, but Mr. Billings in his note just referred to, on the fossils of Highgate, in the same Journal for September of that year, makes no allusion to it. In March, 1862, however, he returns to the subject of the sand-rock, in a more detailed communication (Ibid. (2), XXXIII. 100), and after correcting some omissions in his former note, alludes in the following language to Mr. Barrande, and to the expressed opinion of the latter, just quoted, with regard to the fossils in question and the rocks containing them: "I must also state that Barrande first determined the age of the slates in Georgia, Vermont, holding *P. Thompsoni* and *P. Vermontana.*" He adds, "at the time I wrote the note on the Highgate fossils it was not known that these slates were conformably interstratified with the Red sand-rock. This discovery was made afterwards by the Rev. J. B. Perry and Dr. G. M. Hall of Swanton."

Mr. Billings has blamed me (Canadian Naturalist, new series, VI. 318) for having written in 1871 (*ante*, page 258), with regard to the Georgia trilobites first described as *Olenus* by Professor Hall, that Barrande "called attention to their primordial character, and thus led to a knowledge of their true stratigraphical horizon." I had always believed that the letter of Barrande and the explicit declaration of Mr. Billings, just quoted, contained the whole truth of the matter. My attention has since been called to a subsequent note by Mr. Billings in May, 1862 (Ibid. (2), XXXIII. 421), in which, while asserting that Emmons had already assigned to these rocks a greater age than the New York system, he mentions that in sending to Barrande, in the spring of 1860, the report of Professor Hall on the Georgia fossils, he alluded to their primordial character, and suggested that they might belong to what Mr. Barrande has called "a colony" in the rocks of the second fauna. This is also stated in a note by Sir William Logan in the Preface to the Geology of Canada (page viii). As the genus Olenus, to which Professor Hall had referred the fossils in question, was at that time (1860) well known to belong, both in Great Britain and in Scandinavia, to the primordial fauna, Mr. Barrande does not seem to have thought it necessary in his correspondence to refer to the very obvious remark of Mr. Billings.

Mr. Billings further showed in his paper in March, 1862, that fossils identical with those of the Georgia slates had been found by him in specimens collected by Mr. Richardson of the geological survey of Canada in the summer of 1861, on the Labrador coast, along the Strait of Belleisle; where *Olenellus* (*Paradoxides*) *Thompsoni* and *O. Vermontana* were found with *Conocoryphe* (*Conocephalus*) in strata which were by Billings referred to the Potsdam group. (See, for the further history of these fossils the Geology of Canada, pages 866, 955, and Pal. Fossils of Canada, pages 11, 419.)

The interstratification of the dark-colored fossiliferous shales holding *Olenellus* with the Red sand-rock of Vermont, announced by Mr. Billings, was further confirmed by Sir William Logan in his account of the section at Swanton, Vermont. (Geology of Canada, 281.) They were there declared to occur about 500 feet from the base of a series of 2,200 feet of strata, consisting chiefly of red sandy dolomites (the so-called sand-rock) containing *Conocephalus* throughout, while the shaly beds held, in addition, the two species of *Paradoxides* (*Olenellus*) and some brachiopods. These beds, like those of Labrador, were referred by Logan and by Billings to the Potsdam group. The conclusions here announced were of great importance for the history of the Taconic controversy. The trilobites of primordial type, from Georgia, Vermont, which by Emmons were placed in the Taconic system, lying unconformably beneath a series of rocks belonging to the lower part of the New York system, were now declared to belong to the Red sand-rock group, a member of this overlying system. Much has been said of these fossils, as if they furnished in some way a vindication of the views of Emmons, and of the Taconic system; a conclusion which can only be deduced from a misconception of the facts in the case. Emmons had, previous to 1860, on lithological and stratigraphical evidence alone, called the Georgia slates Taconic, and placed them unconformably beneath the Red sand-rock. If now both he and Billings were right in referring the Red sand-rock to the Calciferous and Potsdam formations, and if the stratigraphical determinations of Messrs. Perry and G. M. Hall, confirmed by those of Logan, were correct, namely, that the trilobites in question occur not in a system of strata lying unconformably beneath the Red sand-rock, but in beds intercalated with the sand-rock itself, it is clear that these trilobites must belong not to the Taconic, but to the New York, system. We shall return to the question of the age of these rocks.

We have seen that Professor James Hall, in 1847, and again in 1859, referred trilobites regarded by him as species of *Olenus* to the Hudson-River group, or, in other words, to the summit of the second palæozoic fauna, while it is now well known that they are characteristic of the first fauna. In this reference, in 1847, Professor Hall was justified by the singular errors which we have already pointed out in the works of Hisinger on the geology of Scandinavia. (*Ante*, page 366.) In his Anteckningar, in 1828, while the colored map and accompanying sections show the alum-slates with *Paradoxides* to lie beneath, and the clay-slates with graptolites, above the orthoceratite-limestone, the accompanying colored legend, designed to explain the map and sections, gives these two slates with the numbers 3 and 4, as if they were contiguous and *beneath* the limestone, which is numbered 5. The student who, in his perplexity, turned from this to the later work of Hisinger, his Lethæa Suecica, found the two groups of slates, as before, placed in juxtaposition, but assigned, together, to a position *above* the orthoceratite-limestone. Thus, in either case, he would be led to the conclusion that in Scandinavia the alum-slates with *Olenus*, *Paradoxides*, and *Conocephalus* (*Conocoryphe*) were closely associated with the graptolitic shales; and, upon the authority of the later work, that the position of both of these was there above the orthoceratite-limestones, and at the summit of the second fauna. The graptolitic shales of Scandinavia were already identified with those of the Utica and Hudson-River formations of the New York system. The Red sand-rock of Vermont, containing *Conocephalus*, had been, both by Emmons and Adams, alike on lithological and stratigraphical grounds, referred to the still higher Medina sandstone; a view which, as we have seen, was still maintained and strongly defended by Adams. This was in 1847, and Angelin's classification of the transition rocks of Scandinavia, fixing the position of the various trilobitic zones, did not appear until 1854.

Professor James Hall had therefore at this time the strongest reasons for assigning the rocks containing *Olenus* to the summit of the second fauna. Before we can understand his reasons for maintaining a similar view in 1859, we must notice the history of geological investigation in eastern Canada. So early as 1827, Dr. Bigsby, to whom North American geology owes so much, had given us (Proc. Geol. Soc., I. 37) a careful description of the geology of Quebec and its vicinity. He there found resting directly upon the ancient gneiss a nearly horizontal dark-colored conchiferous limestone, having sometimes at its base a calcareous conglomerate, and well displayed on the north shore of the St. Lawrence at Montmorenci and Beauport. He distinguished, moreover, a third group of rocks, described by him as a "slaty series composed of shale and graywacke, occasionally passing into a brown limestone, and alternating with a calcareous conglomerate in beds, some of them charged with fossils . . . . derived from the conchiferous limestone." (This fossiliferous conglomerate contained also fragments of clay-slate.) From all these circumstances Bigsby concluded that the flat conchiferous limestones were older than the highly inclined graywacke series; which latter was described as forming the ridge on which Quebec stands, the north shore to Cape Rouge, the island of Orleans, and the southern or Point-Levis shore of the St. Lawrence; where, besides trilobites and the fossils in the conglomerates, he noticed what he called vegetable impressions, supposed to be fucoids. These were the graptolites which, nearly thirty years later, were studied, described, and figured for the geological survey of Canada by Professor James Hall, who has shown that two of the species from this locality were described and figured under the name of fucoids by Ad. Brongniart, in 1828. (Geol. Sur. Canada, Decade II. page 60.) Bigsby, in 1827, conceived that the limestones of the north shore might belong to the carboniferous period, and noted the existence of what were called small seams of coal in the graywacke series of the south shore. This substance which I have since described is, however, entirely distinct from coal, and occurs in fissures, sometimes in the interstices of crystalline quartz. It is an insoluble hydrocarbonaceous body, brilliant, very fragile, giving a black powder, and results apparently from the alteration of a once liquid bitumen. (American Journal of Science (2), XXXV. 163.)

In 1842 the geological survey of Canada was begun by Sir William Logan, who in a Preliminary Report to the government, dated in that year but printed in 1845, says (page 19): "Of the relative age of the contorted rocks of Point Levis, opposite Quebec, I have not any good evidence, though I am inclined to the opinion that they come out from below the flat limestones of the St. Lawrence." He however subsequently adds, in a foot-note, "The accumulation of evidence points to the conclusion that the Point Levis rocks are superior to the St. Lawrence limestones." In 1845, Captain, now Admiral Bayfield maintained the same view, fortifying himself by the early observations of Bigsby, and expressing the opinion that the flat limestones of Montmorenci and Beauport passed beneath the graywacke series. These limestones, from their fossils, were declared to be low down in the Silurian, and identical with those which had been observed at intervals along the north shore of the St. Lawrence to Montreal (Geol. Journal, I. 455), the fossiliferous limestones of which were then well known to belong to the Trenton group of the New York system. The graywacke series of Quebec, which was still supposed by Bayfield to hold in its conglomerates fossils from these limestones, was therefore naturally regarded as belonging to the still higher members of that system; and, as we have seen, the green sandstone near Quebec, a member of that series, had already, in 1842, been regarded by Emmons as the representative of the Oneida or Shawangunk conglomerate, at the summit of the Hudson River group of New York.

It is to be noticed that immediately to the northeast of Quebec, rocks undoubtedly of the age of the Utica and Hudson River divisions overlie conformably the Trenton limestone, on the left bank of the St. Lawrence; while a few miles to the southwest, strata of the same age, and occupying a similar stratigraphical position, appear on both sides of the St. Lawrence, and are traced continuously from this vicinity to the valley of Lake Champlain. These, moreover, offer such lithological resemblances to what was called the graywacke series of Quebec and Point Levis (which extends thence some hundreds of miles northeastward along the right bank of the St. Lawrence), that the two series were readily confounded, and the whole of the belt of rocks along the southeast side of the St. Lawrence, from the valley of Lake Champlain to Gaspé, was naturally regarded as younger than the limestones of the Trenton group. It was in 1847 that Sir William Logan commenced his examination of the rocks of this region, and in his Report for the next year (1848, page 58) we find him speaking of the continuous outcrop "of recognized rocks of the Hudson River group from Lake Champlain along the south bank of the St. Lawrence to Cape Rosier." In his Report for 1850, these rocks were further noticed as extending from Point Levis southwest to the Richelieu, and northeast to Gaspé (pages 19, 32). They were described as consisting, in ascending sequence from the Trenton limestone and the Utica slate, of clay-slates and limestones, with graptolites and other fossils, followed by conglomerate-beds supposed to contain Trenton fossils, red and green shales and green sandstones; the details of the section being derived from the neighborhood of Quebec and Point Levis, and from the rocks first described by Bigsby. As further evidence with regard to the supposed horizon of these rocks, to which he subsequently (in 1860) gave the name of the Quebec group, we may cite a letter of Sir William Logan, dated November, 1861 (Amer. Jour. Sci. (2), XXXIII. 106), in which he says: "In 1848 and 1849, founding myself upon the apparent superposition in eastern Canada of what we now call the Quebec group, I enunciated the opinion that the whole series belonged to the Hudson River group and its immediately succeeding formation; a *Leptæna* very like *L. sericea*, and an *Orthis* very like *O. testudinaria*, and taken by me to be these species, being then the only fossils found in the Canadian rocks in question. This view supported Professor Hall in placing, as he had already done, the Olenus rocks of New York in the Hudson River group, in accordance with Hisinger's list of

Swedish rocks as given in the Lethæa Suecica in 1837, and not as he had previously given it." (*Ante*, pages 366 and 395.)

The concurrent evidence deduced from stratigraphy, from geographical distribution, from lithological and from paleontological characters, thus led Logan, from the first, to adopt the views already expressed by Bigsby, Emmons, and Bayfield, and to assign the whole of the palæozoic rocks of the southeast shore of the St. Lawrence below Montreal to a position in the New York system above the Trenton limestone. While thus, as he says, founding his opinion on the stratigraphical evidence obtained in eastern Canada, Logan was also influenced by the consideration that the rocks in question were continuous with those in western Vermont. Part of the rocks of this region had, as we have seen, originally been placed by Emmons at this horizon, while the others, referred by him to his Taconic system, were maintained by Henry D. Rogers to belong to the Hudson River group; a view which was adopted by Mather and by Hall, and strongly defended by Adams, at that time engaged in a geological survey of Vermont, with which, in 1846 and 1847, the present writer was connected.

As regards the subsequent paleontological discoveries in these rocks in Canada, it is to be said that the graptolites first noticed by Bigsby in 1827 were rediscovered by the Geological Survey at Point Levis in 1854, and having been placed in the hands of Professor James Hall (who in that year first saw the rocks in question), were partially described by him in a communication to Sir William Logan, dated April, 1855, and subsequently at length in 1858. (Report Geol. Survey for 1857, page 109, and Decade II.) They were new forms, it is true, but the horizon of the graptolites, both in New York and in Sweden, was the same as that already assigned by Logan to the Point Levis rocks. Thus these fossils appeared to sustain his view, and they were accordingly described as belonging to the Hudson River group.

Up to 1856 no other organic remains than the graptolites and the two species of brachiopods noticed by Sir William Logan were known to the geological survey as belonging to the Point Levis rocks; the trilobites long before observed by Bigsby not having been rediscovered. In 1856 the present writer, while engaged in a lithological study of the various rocks of Point Levis, found, in the vicinity of the graptolitic shales, beds of what were described by him in 1857 (Report Geol. Surv., 1853 – 1856, page 465) as "fine granular opaque limestones, weathering bluish-gray, and holding in abundance remains of orthoceratites, trilobites, and other fossils, which are replaced by a yellow-weathering dolomite." In these, which are probably what Bigsby had long before described as fossiliferous conglomerates, the dolomitic matter is so arranged as to suggest a resemblance to certain beds which are really conglomerate in character, and were at the same time described by me as interstratified with the fossiliferous limestones, and as holding pebbles of pure limestone, of dolomite, and occasionally of quartz and of argillite; the whole cemented by a yellow-weathering dolomite, and occasionally by a nearly pure carbonate of lime. (Ibid., 466.) The included fragments of argillite (previously noticed by Bigsby), which are greenish or purplish in color, with lustrous surfaces, are precisely similar to those which form great beds in the crystalline schists of the Green Mountain series of the Appalachian hills, which extend in a northeast and southwest course along the southeastern border of the rocks of the Quebec group. I conceive that these argillite fragments (like those in the Potsdam conglomerate near Lake Champlain, *ante*, page 268) are derived from the ancient schists of the Appalachians.

This rediscovery of fossiliferous limestones at Point Levis led to further exploration of the locality, and in 1857 and the following years a large collection of trilobites, brachiopods, and other organic remains was obtained from these limestones by the geological survey of Canada.

Mr. Billings, who in 1856 had been appointed paleontologist to the geological survey, at once commenced the study of these fossils from Point Levis, and at length arrived at the important conclusion that the organic remains there found belonged, not to the summit of the second fauna, but were to be assigned a position in the first or primordial fauna. This conclusion he communicated to Mr. Barrande in a letter dated July 12, 1860 (American Journal of Science (2), XXXI. 220), and gave descriptions of many of the organic forms in the Canadian Naturalist for the same year. I have already alluded, in describing the rocks of Point Levis, to the peculiarities of aspect which probably led Dr. Bigsby, in 1827, to confound these fossiliferous limestones penetrated by dolomite, with the true dolomitic conglomerates associated with them, and helped him to suppose the fossils to be derived from the limestones of the north shore, now known to be younger rocks. This mistake was a very natural one at a time when comparative paleontology was unknown.

Sir William Logan meanwhile made a careful stratigraphical examination of the rocks of Point Levis, and, notwithstanding the peculiarities of the limestones which there contain the primordial fauna, declared himself, in December, 1860, satisfied that "the fossils are of the age of the strata." In consequence of the discovery of Mr. Billings, Logan now proposed to separate from the Hudson River group the graywacke series of Bigsby and Bayfield, and ascribed to it a much greater antiquity; regarding it as "a great development of strata about the horizon of the Chazy and Calciferous, brought to the surface by an overturn anticlinal fold, with a crack and a great dislocation running along the summit," by which the rocks in question were "brought to overlap the Hudson River formation." This series, to which was assigned a thickness of from 5,000 to 7,000 feet, he named the Quebec group, which included the green sandstones of Sillery, regarded as the summit, the fossiliferous limestones and graptolitic shales at the base, which afterwards received the name of the Levis formation, and a great intermediate mass of barren shales and sandstones, called the Lauzon formation. The first account of this change in the stratigraphical views of Logan occurs in his letter to Barrande, dated December 31, 1860. (American Journal of Science (2), XXXI. 216.)

This important distinction once established, it was found

necessary to draw a line from the St. Lawrence, near Quebec, to the vicinity of Lake Champlain, separating the true Hudson River group, with its overlying Oneida or Medina rocks, on the northwest side, from the so-called Quebec group, on the south and east. This division was by Logan ascribed to a continuous dislocation, which had disturbed a great conformable palæozoic series, including the whole of the members of the New York system from the base of the Potsdam to the summit of the Hudson River group, and, throughout the whole distance of one hundred and sixty miles, had raised up the lower formations in a contorted and inclined attitude, and caused them to overlie in many cases the higher formations of the system. This dividing line was by Logan traced northeastward through the island of Orleans, the waters of the lower St. Lawrence, and along the north shore of Gaspé; and southwestward through Vermont, across the Hudson, as far at least as Virginia; separating, throughout, the rocks of the Quebec and Potsdam groups, with their primordial fauna, from those of the Trenton and Hudson River groups, with the second fauna. This is shown in the geological map of eastern America from Virginia to the St. Lawrence, which appears in the Atlas to the Geology of Canada, published in 1865. In an earlier geological map, published by Sir William Logan at Paris in 1855, before this distinction had been drawn, the region in question in eastern Canada is colored partly as the Oneida formation, and partly as the Hudson River group; while in the accompanying text the Sillery sandstone is spoken of as the equivalent of the Shawangunk grit or Oneida conglomerate of the New York system. (Esquisse Géologique du Canada. Logan and Sterry Hunt: Paris, 1855, page 51.) These rocks were by Logan traced southward across the frontier of Canada, into Vermont, where they included the Red sand-rock and its associated slates; which were thus by Logan, as well as by Adams, looked upon as occupying a position at the summit of the second fauna. When, therefore, in 1859, Professor Hall described the trilobites found in these slates in Georgia in Vermont, he referred them to the genus *Olenus*, whose primordial horizon in Europe was then well determined, but, in deference to the conclusions of Adams and of Logan, assigned them to a position at the summit of the Hudson River group; Hall himself never having examined the region stratigraphically. (American Journal of Science (2), XXXI. 221.) In justification of this position he appended to his description the following note (Ibid., pages 213, 221): "In addition to the evidence heretofore possessed regarding the position of the slates containing the trilobites, I have the testimony of Sir William Logan that the shales of this locality are in the upper part of the Hudson River group, or forming part of a series of strata which he is inclined to rank as a distinct group above the Hudson River proper. It would be quite superfluous for me to add one word in support of the opinion of the most able stratigraphical geologist of the American continent." Paleontology and stratigraphy here came into conflict, and it was not till in 1860, when Mr. Billings, in the face of the evidence adduced from the latter, asserted the primordial age of the Point Levis fauna, that Sir William Logan attempted a new explanation of the stratigraphy of the region; declaring at the same time that, "from the physical structure alone, no person would suspect the break which must exist in the neighborhood of Quebec; and without the evidence of the fossils every one would be authorized to deny it." (Ibid., page 218.)

The typical Potsdam sandstone of the New York system, as seen in the Ottawa basin in northern New York and the adjacent parts of Canada, affords but a very meagre fauna, including two species of brachiopods, one or two gasteropods, and a single crustacean, *Conocephalites* (*Conocoryphe*) *minutus*, found at Keeseville, New York. In 1852, however, David Dale Owen found and described an extensive fauna in Wisconsin, from rocks which were regarded as the equivalent of the Potsdam sandstone; while the observations of Shumard in Texas, in 1861, and the latter ones of Hayden and Meek in the Black Hills, have since still further extended our knowledge of the distribution and the organic remains of the rocks which are supposed to represent, in the west, the Potsdam and Calciferous formations of the New York system.

As early as 1842, Professor Hall, in a comparison of the lower palæozoic rocks of New York with those of Great Britain, declared the Potsdam to be lower than the base of the Upper Cambrian or Bala group of Sedgwick. In 1847, as we have seen, he extended this observation to the Calciferous and Chazy, both of which he placed below this horizon; which until a year or two previous had been looked upon as the base of the palæozoic series in Great Britain, and was subsequently made the lower limit of the second fauna of Barrande. Although from these facts it was probable that these lower members of the New York system might correspond to the primordial fauna of Barrande, we still remained, in the language of Professor Hall, without "the means of parallelizing our formations with those of Bohemia, by the fauna there known. The nearest approach to the type of the primordial trilobites was found in the Potsdam of the northwest, described by Dr. D. D. Owen; but none of these had been generically identified with Bohemian forms, and the prevailing opinion, sanctioned, as I have understood, by Mr. Barrande, was that the primordial fauna had not been discovered in this country until the rediscovery (in 1856) of *Paradoxides Harlani* at Braintree, Massachusetts. The fragmentary fossils published in Vol. I. of the Paleontology of New York, and similar forms of the so-called Taconic system, were justly regarded as insufficient to warrant any conclusions." (Amer. Jour. Sci. (2), XXXI. 225.) Such, according to Prof. Hall, was the state of the question up to 1860. The *Conocephalus*, detected by him from the Red sand-rock of Vermont, in 1847, and subsequently recognized in Europe as an exclusively primordial type, seems to have been forgotten by Hall, and overlooked by others, until it was rediscovered in the sand-rock by Billings in 1861. He had previously, in 1860, detected the same genus at Point Levis, together with Arionellus, and other purely primordial types. Associated with these, and with many other trilobites belonging to the second fauna, were found several species of Dikellocephalus and Menocephalus, genera first made known by Owen from the Potsdam of Wisconsin. It is by an error that Messrs. Harkness and Hicks, in a recent paper (Quar. Geol. Jour., XXVII. 395), have asserted that Owen, in 1852, found there, together with these genera, Conocephalus and Arionellus; the history of the first discovery of these genera in America being as above given. The limestones of Point Levis thus furnished what was hitherto wanting, — a direct connecting link between the fauna of the American Potsdam and the primordial zone of Bohemia.

The history of the *Paradoxides Harlani*, alluded to by Professor Hall, is as follows: in 1834, Dr. Jacob Green received from Dr. Richard Harlan the cast of a large trilobite occurring in a silicious slate, which was in the collection of Francis Alger of Boston, and, it was supposed, might have come from Trenton Falls, New York. Dr. Green, who at once pointed out the fact that the rock was wholly unlike any found at this locality, declared the fossil to resemble greatly the *Paradoxides Tessini*, Brongn., — the former *Entomolithus paradoxus* of Linnæus, from Westrogothia, — and named the species *P. Harlani*. (Amer. Jour. Sci. (1), XXV. 336.) In 1856, the attention of Professor William B. Rogers was called to a locality of organic remains in Braintree, on the border of Quincy, Massachusetts, where, on examination, he at once recognized the *Paradoxides Harlani* in a silicious slate similar to that of the original specimen. This was announced by him in a communication to the American Academy of Sciences (Proc., Vol. III.), as a proof of the protozoic age of some of the rocks of eastern Massachusetts. Professor Rogers then called attention to the fact that this genus of trilobites is characteristic of the primordial fauna, and noticed that Barrande had already remarked that, from the casts of *P. Harlani* in the London School of Mines and the British Museum (which had been made from the original specimen, and distributed by Dr. Green), this species appeared to be identical with *P. spinosus* from Skrey in Bohemia.

In 1858, Salter found in specimens sent to the Bristol Institution in England, by Mr. Bennett of Newfoundland, from the promontory between St. Mary's and Placentia Bays, in the southwestern part of this island, a large trilobite, described by him as *Paradoxides Bennettii* (Geol. Jour., XV. 554), which appears, according to Mr. Billings, to be identical with *P. Harlani.* On the same occasion Salter described, under the name of *Conocephalites antiquatus,* a trilobite from a collection of American fossils sent by Dr. Feuchtwanger of New York to the London Exhibition of 1851. This was said to occur in a bowlder of brown sandstone from Georgia, and, as I have been informed by Dr. Feuchtwanger, was found near the town of Columbus in that State.

The slates of St. John, New Brunswick, and its vicinity have recently yielded an abundant fauna, examined by Professor Hartt, who at once recognized its primordial character. This conclusion was first announced, on the authority of Professor Hartt, in a paper by Mr. G. F. Matthew, in May, 1865. (Geol. Jour., XXI. 426.) The rocks of this region have afforded two species of *Paradoxides* and fourteen of *Conocoryphe,* together with *Agnostus* and *Microdiscus,* all of which have been described by Professor Hartt. It may here be noticed that, in 1862, Professor Bell found in the black shales of the Dartmouth valley, in Gaspé, a single specimen of a large trilobite, which, according to Mr. Billings, closely resembles *Paradoxides Harlani,* but from its imperfectly preserved condition cannot certainly be identified with it. (Geol. Canada, page 882.)

The geological examinations of Mr. Alexander Murray in Newfoundland, since 1865, have shown that the southeastern part of that island contains a great volume of Cambrian rocks, estimated by him at about 6,000 feet in all. No traces of the Upper Cambrian or second fauna have been detected among these, but some portions contain the Paradoxides already mentioned, while others yield the fauna which Mr. Billings has called Lower Potsdam. This name was first given in an appendix (prepared by Sir William Logan) to Mr. Murray's report on Newfoundland for 1865, published in 1866 (page 46 ; see also Report of the Geol. Survey of Canada for 1866, page 236). The Lower Potsdam was there assigned a place above the Paradoxides beds of the region, which were called the St. John group, — the fossiliferous strata of St. John, New Brunswick, being referred to the same horizon; which corresponds to the Menevian of Wales, now recognized as the summit of the Lower Cambrian. The succession of the rocks containing these two faunas in southeastern Newfoundland is not yet clear; the Lower Potsdam fauna is regarded by Mr. Billings as identical with that found on the Strait of Bellisle, at Bic (on the south shore of the river St. Lawrence, below Quebec), at Georgia in Vermont, and at Troy, New York; but in none of these other localities is it as yet known to be accompanied by a Menevian fauna. The trilobites hitherto described from these rocks belong to the genera *Olenellus*, *Conocoryphe*, and *Agnostus*; neither *Paradoxides*, which characterizes the Menevian and the underlying Harlech beds in Wales, nor *Olenus*, which there abounds in the rocks immediately above this horizon, having as yet been described as occurring in the Lower Potsdam of Mr. Billings. Future discoveries may perhaps assign it a place below instead of above the Menevian horizon.

[To the above genera of trilobites occurring at Troy, Mr. Ford has since (in 1873) added *Microdiscus*, which has also been found at Bic. This genus is common to the Menevian and the underlying Harlech rocks in Wales, and is also, according to Emmons, found with graptolites in Augusta County, Virginia. The strata which contain this fauna at Troy, as described by Ford, are of considerable thickness, consisting of limestones with coarse sandstones and shales, which, as the result of a dislocation or of an overturned and eroded anticlinal, are made to overlie, in apparent conformity, the beds of the Utica or Hudson River group, the whole dipping eastward. (American Journal of Science (3), VI. 134.)]

The characteristic Menevian fauna in and near St. John, New Brunswick, is found in a band of about one hundred and fifty feet, towards the base of a series of nearly vertical sandstones and argillites, underlaid by conglomerates, and resting upon crystalline schists, in a narrow basin. The series, the total thickness of which is estimated by Messrs. Matthew and Bailey at over 2,000 feet, contains Lingula throughout, but has yielded no remains of a higher fauna. The same Menevian forms have been found in small outlying areas of similar rocks, at two or three places north of the St. John basin, but to the south of the New Brunswick coal-field. To the north of this is a broad belt of similar argillites and sandstones, which extends southwestward into the State of Maine. This belt has hitherto yielded no organic remains, but is compared by Mr. Matthew to the Cambrian rocks of the St. John basin, and to the gold-bearing series of Nova Scotia (Geol. Jour., XXI. 427), which at the same time resembles closely the Cambrian rocks of southeastern Newfoundland. This was remarked by Dr. Dawson in 1860, when he expressed the opinion that the auriferous rocks of Nova Scotia were "the continuation of the older slate series of Mr. Jukes in Newfoundland, which has afforded Paradoxides," and probably the equivalent of the Lingula flags of Wales. (Supplement to Acadian Geology (1860), page 53; also Acad. Geol., 2d ed., page 613.) Associated with these gold-bearing strata, along the Atlantic coast of Nova Scotia, occur fine-grained gneisses, and mica-schists with andalusite and staurolite; besides other crystalline schists which are chloritic and dioritic, and contain crystallized epidote, magnetite, and menaccanite. These two types of crystalline schists (which, from their stratigraphical relations, as well as from their mineral condition, appear to be more ancient than the uncrystalline gold-bearing strata) were in 1860, as now, regarded by me as the equivalents respectively of the White Mountain and Green Mountain series of the Appalachians, as will be seen by reference to Dr. Dawson's work just quoted. At that time, however, and for many years after, I held, in common with most American geologists, the opinion that these two groups of crystalline schists were altered rocks of a more recent date than that assigned to the auriferous series of Nova Scotia by Dr. Dawson, who was much perplexed by the difficulty of reconciling this view with his own. The difficulty is, however, at once removed when we admit, as I have maintained since 1870, that both of these groups are pre-Cambrian in age. (Amer. Jour. Sci. (2), L. 83; *ante*, pages 276 and 327.)

A notice by Mr. Selwyn of some of these crystalline schists in Nova Scotia will be found in the Report of the Geological Survey of Canada for 1870 (page 271). He there remarks, moreover, the close lithological resemblances of the gold-bearing strata to the Harlech grits and Lingula flags of North Wales, and announces the discovery among these strata at the Ovens gold-mine in Lunenburg, Nova Scotia, of peculiar organic markings regarded by Mr. Billings as identical with the *Eophyton Linnæanum*, which is found in the Regio Fucoidarum, at the base of the Cambrian in Sweden. In the volume just quoted (page 269) will be found some notes by Mr. Billings on this fossil, which occurs also near St. John, New Brunswick, in strata supposed to underlie the Paradoxides beds. The same form is found in Conception Bay in southeastern Newfoundland, in strata regarded by Mr. Murray as higher than those with *Paradoxides*, and containing also two new species of *Lingula*, a *Cruziana*, and several fucoids. Still more recently, *Eophyton*, accompanied by these same fucoids, has been found by Mr. Billings at St. Laurent, on the island of Orleans near Quebec, in strata hitherto referred by the geological survey, on stratigraphical grounds, to the Quebec group. The evidence adduced by Mr. Billings tends to show that this organic form, whatever its nature, belongs to a very low horizon in the Cambrian.

As regards the probable downward extension of these forms of ancient life, I cannot refrain from citing the recent language of Mr. Hicks. (Quar. Jour. Geol. Soc., May, 1872, page 174.) After a comparative study of the Lower Cambrian fauna, including that of the Harlech and Menevian rocks in Wales, and the representatives of the latter in other regions, he adds :—

"Though animal life was restricted to these few types, yet at this early period the representatives of the several orders do not show a very diminutive form, or a markedly imperfect state; nor is there an unusual number of blind species. The earliest known brachiopods are apparently as perfect as those which succeed them; and the trilobites are of the largest and best developed types. The fact also that trilobites had attained their maximum size at this period, and that forms were present representative of almost every stage in development, from the little *Agnostus* with two rings to the thorax, and *Microdiscus* with four, to *Erinnys* with twenty-four, and blind genera along with those having the largest eyes, leads to the conclusion that for these several stages to have taken place numerous previous faunas must have had an existence, and, moreover, that even at this time in the history of our globe an enormous period had elapsed since life first dawned upon it."

The facts insisted upon by Hicks do not appear to be inconsistent with the view that at this horizon the trilobites had already culminated. Such does not, however, appear to be the idea of Barrande, who in a recent learned essay upon the trilobitic fauna (1871) has drawn from its state of development at this early period conclusions strongly opposed to the theory of derivation.

The strata holding the first fauna in southeastern Newfoundland rest unconformably, according to Mr. Murray, upon what he has called the Intermediate series; which is of great thickness, consists chiefly of crystalline rocks, and is supposed by him to represent the Huronian. He has, however, included in this intermediate series several thousand feet of sandstones and argillites which, near St. John's in Newfoundland, are seen to be unconformably overlaid by the fossiliferous strata already noticed, and have yielded two species of organic forms, lately described by Mr. Billings. One of these is an *Arenicolites*, like the *A. spiralis* found in the Lower Cambrian beds of Sweden, and the other a patella-like shell, to which he has given the name of *Aspidella Terranovica*. (Amer. Jour. Science (3), III. 223.) These, from their stratigraphical position, have been regarded as Huronian; but from the lithological description of Mr. Murray, the strata containing them appear to be unlike the great mass of the Huronian rocks of the region. Their occurrence in these strata, in either case, marks a downward extension of these forms of palæozoic life.

Mr. Billings has described from the rocks of the first fauna certain forms under the name of *Archeocyathus*, one of the species of which, according to Dr. Dawson, belonged to a calcareous chambered foraminiferal organism similar in its nature to much of the *Stromatopora* of the second, and the closely related *Coenostroma* of the third fauna. All of these Dawson shows to have strong affinities to *Eozoon,* which is represented by *E. Canadense* of the Laurentian, and by similar forms in the newer crystalline schists of Hastings, Ontario, as well as by the *E. Bavaricum* of the upper crystalline schists of Bavaria. The succession of related foraminiferal organisms is further seen in the Devonian limestones of Michigan, where occur great masses like Stromatopora, which present, according to Dawson, a structure intermediate between the Eozoon of the Laurentian and the genera Parkeria and Loftusia of the Cretaceous and the Eocene. These details are taken from Dr. Dawson's presidential address to the Natural History Society of Montreal, in May, 1872, where he has announced some of the results of his studies, yet in progress, on the earlier foraminifera.

In 1856 the late Professor Emmons described (Amer. Jour. Sci. (2), XXII. 389), under the name of *Palæotrochis,* certain forms regarded by him as organic, found in North Carolina in a bed of auriferous quartzite, among rocks referred to his Taconic system. Their organic nature has also been maintained by Professor Wurtz, but from my own examinations, I agree with the opinion expressed by Professor Hall, and subsequently supported by the observations of Professor Marsh (Ibid. (2), XXIII. 278 ; XVL. 217), that the forms to which the name of Palæotrochis has been given are nothing more than silicious concretions.

As regards the geological horizon of the series of strata to which Sir William Logan has given the name of the Quebec group, the Sillery and Lauzon divisions have as yet yielded to the paleontologist only two species of *Obolella* and one of *Lingula.* Our comparisons must therefore be based upon the fauna of the Levis limestones and graptolitic shales, which have already been compared with the Middle Cambrian of Sedgwick by the combined labors of Billings and Salter.

The former has, moreover, carefully compared this fauna with that of the lower members of the New York system, in which the succession of organic life appears to have been very much interrupted. Thus, according to Mr. Billings, of the ninety species known to exist in the Chazy limestone of the Ottawa basin, only twenty-two species have been observed to pass up into the directly overlying Birdseye and Black River limestones, which form the lower part of the Trenton group. The break in the succession between the Chazy and the underlying Calciferous sand-rock in this region is still more complete; since, according to the same authority, of forty-four species in the latter only two pass up into the Chazy limestone. This latter break appears to be filled, in the region to the eastward of the Ottawa basin, by the Levis limestone; which has been studied near Quebec, and also near Phillipsburg, not far from the outlet of Lake Champlain. This formation (including the accompanying graptolitic shales) has yielded, up to the present time, two hundred and nineteen species of organic remains (comprising seventy-four of crustacea and fifty-one of graptolitidiæ), none of which, according Mr. Billings, have been found either in the Potsdam or in the Birdseye and Black River limestone. Twelve of the species of the Levis formation are, however, met with in the Calciferous, and five in the Chazy of the Ottawa basin, and the Levis is therefore regarded by Mr. Billings as the connecting link between these two formations.

With regard to the British equivalents of these rocks, the Levis limestone, according to Salter, corresponds to the Tremadoc beds; although the species of *Dikellocephalus* found in the Levis rocks are by him compared with those found in the Upper Lingula flags or Dolgelly beds. The graptolitic strata of Levis, however, clearly represent the Arenig rocks of North Wales, the Skiddaw group of Sedgwick in Cumberland, the graptolitic beds which in Esthonia, according to Schmidt, are found below the orthoceratite-limestones (Can. Naturalist (1), VI. 345) and those of Victoria in Australia (Mem. Geol. Sur., III. Part II. 255, 304). In the Arenig and Upper Tremadoc beds there appears to be, in North Wales, a mingling of forms of the first and second faunas, as in the Levis and Chazy formations. The latter was already, by Hall, in 1847, declared to be beneath the Silurian horizon then recognized in Great Britain; and it is by its fauna comparatively isolated from the strata both below and above it. According to a private communication from Professor James Hall, the Chazy limestone at Middleville, Herkimer County, New York, to the south of the Adirondacks, is wanting, and the basal beds of the Trenton group (the Birdseye limestone) there rest unconformably upon the Calciferous sand-rock.

The relations of the various members of the Quebec group to each other, and of the group, as a whole, to the succeeding Trenton and Hudson River groups, require further elucidation. If, as I am disposed to believe, the southeastward-dipping series of the older strata near Quebec exhibits the northwest side of an overturned and eroded anticlinal, in which the normal order of the strata is inverted, then the Lauzon and Sillery divisions, which there appear to overlie the Levis limestones and shales, are older rocks, occupying the position of the Potsdam or of still lower members of the Cambrian. Sir William Logan supposes the appearance of these rocks in their present attitude by the side of the strata of the Trenton and Hudson River groups, in the vicinity of Quebec, to be due to a great dislocation and uplift subsequent to the deposition of these higher rocks; but, as elsewhere suggested (*ante*, page 263), I conceive the Quebec group to have been in its present upturned and disturbed condition before the deposition of the Trenton limestones. The supposed dislocation and uplift, extending from the Gulf of St. Lawrence to Virginia, is, according to this view, but the outcrop of the rocks of the first fauna from beneath the unconformably overlying strata of the second fauna. The later movements along the borders of the Appalachian region have, however, to some extent, affected these, in their turn, and thus complicated the relations of the two series. This unconformity, which corresponds to the marked break between the Levis and Trenton faunas, is further shown by the stratigraphical break and discordance in Herkimer County, New York; and by the fact that beyond the limits of the Ottawa basin, on either side, the limestone of the Trenton group rests directly on the crystalline rocks; the older members of the New York system being altogether absent at the northern outcrop, as well as in the outliers of Trenton limestone seen at the north of Lake Ontario, and as far to the northeast as Lake St. John on the Saguenay. This distribution shows that a considerable movement, just previous to the Trenton period, took place both to the west and the east of Adirondack region, which formed the southern boundary of the Ottawa basin.

[Lesley observes, "There are certainly evidences of some obscure unconformability between the limestones of II. and the slates of III.," which immediately overlie them in the great Appalachian valley in Pennsylvania. This horizon corresponds to the base of the Trenton (see, further, page 421), and the evidence consists in the different strike of the rocks of the two divisions, that of the overlying slates being always with the valley, while the limestone-outcrops often cross it at various angles. Lesley appears to regard the discordance of no great importance; but it deserves further study in connection with the evidences of a similar want of conformity farther northward. (Proc. American Philosophical Society, December, 1864, page 469.)

[There are, as we have seen, two breaks in the succession of life in the Ottawa basin, the one at the base of the Trenton group, and the other at the base of the Chazy; and in this connection, besides the fact of the absence of the latter between the Calciferous and Trenton, observed by Hall in Herkimer County, New York, should be noticed the remarkable section near Grenville on the Ottawa, described by Logan in the Geology of Canada. Here, at what is regarded as the base of the Chazy, a conglomerate layer of seven feet, made up of limestone pebbles, rests upon beds of yellow-weathering limestone, supposed to be magnesian, and holding obscure fossils; while above it are fifty feet of sandstones, sometimes conglomerate, with layers of shales, the whole representing a period of disturbance which probably corresponds to the almost complete paleontological break between the magnesian limestones of the Calciferous and the pure limestones of the Chazy period.]

The Levis and Chazy formations, as we have seen, offer a commingling of forms of the first and second faunas, which shows them to belong to a period of transition between the two; but it is remarkable that, so far as yet observed, no representatives of the latter of these faunas are known to the east and south of the Appalachians, along the Atlantic coast; the first fauna, whether in Massachusetts, New Brunswick, or southeastern Newfoundland, being unaccompanied by any forms of the second. The third fauna, on the contrary, is represented in various localities both within and to the east of the Appalachian region, from Massachusetts to Newfoundland. In parts of Gaspé, and also in Nova Scotia, strata holding forms referred to the Clinton and Niagara divisions are met with, as well as other strata, of Lower Helderberg age, associated with species of shells and of plants which connect this fauna with that of the succeeding Lower Devonian or Erian period. To this Lower Helderberg horizon (corresponding to the Ludlow of England) appear to belong certain fossiliferous beds found along the Atlantic coast of Maine and of New Brunswick, in Nova Scotia, and probably in Newfoundland; as well as others included in the Appalachian belt in Massachusetts, New Hampshire, Vermont, and Quebec, along the Connecticut valley and its northeastern prolongation. The fossiliferous strata just noticed, both in the Connecticut valley and along the Atlantic coast, occur in small areas among the older crystalline schists, often made up of the ruins of these, and in highly inclined attitudes. The same is true in other places of the similarly situated strata of Cambrian, Devonian, and Lower Carboniferous periods. These derived strata, of different ages, have, from their lithological resemblances to the parent rocks, been looked upon as examples of a subsequent alteration of palæozoic sediments; and by a further extension of this notion, the pre-Cambrian crystalline schists themselves throughout this region have been looked upon as the result of an epigenic change of these various palæozoic strata; portions of which, here and there, were supposed to have escaped conversion, and to have retained more or less perfectly their sedimentary character, and their organic remains, elsewhere obliterated.

From the absence of the second fauna we may conclude that the great Appalachian area was, at least in New England and Canada, above the ocean during its period, and suffered a partial and gradual submergence in the time of the third fauna. This movement corresponds to the well-marked paleontological and stratigraphical break between the second and third faunas in the great continental basin to the westward, made evident by the appearance of the Oneida or Shawangunk conglomerate (apparently derived from the ruins of Lower Cambrian rocks) which, in some parts, overlies the strata of the Hudson River group. The break is elsewhere shown by the absence of this conglomerate, and of the succeeding formations up to the Lower Helderberg division. This latter, in the valley of the St. Lawrence, rests unconformably upon the strata of the second fauna, as it does upon the older crystalline rocks to the eastward.

In Ohio, according to Newberry, the base of the rocks of the third fauna (Clinton and Medina) is represented by a conglomerate which holds in its pebbles the organic remains of the underlying strata of the second fauna.

To the northeastward the island of Anticosti in the Gulf of St. Lawrence presents a succession of about 1,400 feet of calcareous strata rich in organic remains, which, according to Mr. Billings, include the species of the Medina, Clinton, and Niagara formations, and were named by him, in 1857, the Anticosti group. They rest upon nearly 1,000 feet of almost horizontal strata, consisting of limestones and shales rich in organic remains, with many included beds of limestone-conglomerate. This lower series has by the geological survey of Canada been referred to the Hudson River group; but, notwithstanding the large number of forms of the second fauna which it contains,

Professor Shaler is disposed to look upon it as younger, and belonging rather to the succeeding division. There seems not to have been any marked paleontological break between the second and third faunas in this region; and it is worthy of note, in this connection, that in the outlying basin of palæozoic rocks, found at Lake St. John, to the north of Anticosti, *Halysites catenulatus* is met with in limestones associated with many species of organic remains which are characteristic of the Trenton and referred to that group. (Geology of Canada, page 165.)

The strata to which, in 1857, Mr. Billings gave the name of the Anticosti group were at the same time designated by him Middle Silurian, in which he subsequently included the local subdivision known as the Guelph formation, which in western Ontario succeeds the Niagara; the name of Upper Silurian being thus reserved for the Lower Helderberg division and the underlying Onondaga formation. (Report Geol. Sur. Can., 1857, page 248; and Geol. Can., page 20.) Both the Guelph and the Onondaga have been omitted from the table on page 386: the Guelph, because it was not recognized in the New York system, and is by some regarded as but a subdivision of the Niagara; and the Onondaga or Salina, for the reason that it is a local deposit of magnesian limestones, with gypsums and rock-salt, destitute of organic remains.

[The name of Middle Silurian was at one time used by the geological survey of Great Britain to designate the Lower and Upper Llandovery rocks, but was not employed by Murchison either in his Silurian System or in the various editions of Siluria. It is rejected by Lyell (Students' Manual of Geology, page 452), and was referred to in 1854 by Sedgwick as a term then already abandoned. (London, Edinburgh, and Dublin Philosophical Magazine (3), VIII. 303, 367, 501.) Ramsay, moreover, though he speaks of the rocks as an intermediate series, does not make use of the term Middle Silurian. (Memoirs Geological Survey, III. Part II. page 2.) It is, however, once more applied by Hicks in 1873, and made to embrace, as before, the Lower Llandovery, included by Sedgwick in his

Upper Cambrian, and the Upper Llandovery or May Hill sandstone, the base of his Silurian. These two contiguous though discordant formations, in fact, exhibit a mingling of the forms of the second and third faunas. It is, however, to be noted that the Middle Silurian thus defined is by no means the equivalent of that of Mr. Billings, who has given the name, not to beds of passage, but to a group well defined both stratigraphically and paleontologically equivalent to the Upper Llandovery and the Wenlock of England, or, in other words, to the fossiliferous strata between the top of the Hudson River shales and the summit of the Niagara limestone (including the Guelph); thus taking the lower half of the true Silurian or the Upper Silurian of Murchison. That the group of strata comprised under this latter name (the third fauna of Barrande) really includes two very distinct faunas, was long since shown by Hall, the break between the two being marked in New York and Ontario by the interposition of the non-fossiliferous Onondaga or Salina group. This series of strata, in some parts 1,000 feet or more in thickness, consists of red and green magnesian marls with rock-salt and gypsum, overlaid by a great mass of magnesian limestones, the whole having been deposited in a vast mediterranean basin which extended from eastern New York to Ohio. The Water-lime beds, with their peculiar fossils, overlying the Salina group, consist of a magnesian limestone lithologically related to the rocks below, and represent the first invasion of life into the former dead sea, which was followed by the great deposit of non-magnesian limestones of the Lower Helderberg group. These, which attain a great thickness to the eastward, make up, with the Oriskany sandstone, a fourth palæozoic division, the equivalent of the Ludlow of England. In Gaspé a sandstone formation, without any apparent unconformity, connects the Oriskany with the great mass of Devonian sandstones; but in New York and in Ontario evidences of an interruption in the process of deposition are seen in the erosion of the Oriskany previous to the deposition of the Corniferous limestone, which there forms the base of the Devonian or the Erie division of the New York system, extending up to the base of the Carboniferous, for which Dawson has suggested the more appropriate name of Erian. (See further the author on Breaks in the American Palæozoic Series, and Hall on the Relations of the Niagara and Lower Helderberg Formations, Proc. American Association for the Advancement of Science, 1873, pages 118 and 321.)

[The name of Middle Silurian, applied by Billings to the group holding the Medina-Niagara fauna, should be rejected, for the reason that the group below it has no just title to the name of Lower Silurian, but is Upper Cambrian. The two distinct faunas included in the true Silurian rocks might with great propriety be distinguished as Lower and Upper Silurian.]

The history of the introduction of the names of Silurian and Devonian into North American geology now demands our notice. Professor James Hall, as we have seen, while recognizing in the rocks of the New York system the representatives alike of the British Cambrian, Silurian, and Devonian, wisely refrained from adopting this nomenclature, drawn from a region where wide diversities of opinion and controversies existed as to the value and significance of these divisions. Lyell, however, in the account of his first journey to the United States, published in 1845, applied the terms Lower and Upper Silurian and Devonian to our palæozoic rocks. Later, in 1846, De Verneuil, the friend and the colleague of Murchison in his Russian researches, visited the United States, and on his return to France, published, in 1847 (Bul. Soc. Geol. de Fr., II. iv, 12, 646), an elaborate comparison between the European palæozoic deposits and those of North America, as made known by Hall and others. He proposed to group the whole of the rocks of the New York system, up to the summit of the Hudson River group, in the Lower Silurian, and the succeeding members, including the Lower Helderberg and the overlying Oriskany, in the Upper Silurian; the remaining formations to the base of the Carboniferous system being called Devonian. This essay by De Verneuil was translated and abridged by Professor Hall, and published by him in the American Journal of Science (II. v, 176, 359; vii, 45, 218), with critical remarks, wherein he objected to the application of this disputed nomenclature to North American geology.

Meanwhile the geological survey of Canada was in progress under Logan, who in his preliminary Report in 1842, and in his subsequent ones for 1844 and 1846, adopted the nomenclature of the New York system, without reference to European divisions. Subsequently, however, the usage of Lyell and De Verneuil was adopted by Logan, who in his Report for 1848 (page 57) spoke of the Clinton group as the base of the "Upper Silurian series," while in that for 1850 (page 34) he declared the whole of a great series of fossiliferous rocks in eastern Canada, including the Trenton, Utica, and Hudson River divisions, and the shales and sandstones of Quebec (then supposed to be superior to these), to "belong to the Lower Silurian." In the Report for 1852 (page 64) the Lower Silurian was made by Mr. Murray to include not only the Utica and Trenton, but the Chazy limestone, the Calciferous sand-rock and the Potsdam sandstone of the New York system. From this time the Silurian nomenclature, as applied by Lyell and De Verneuil to our North American rocks, was employed by the officers of the Canadian geological survey (myself among the others), and was subsequently adopted by Professor Dana in his Manual of Geology, published in 1863.

The geological survey of Pennsylvania, under the direction of Professor Henry Darwin Rogers, was begun, like that of New York, in 1836, and the palæozoic rocks of the State were at first divided, on stratigraphical and lithological grounds, into groups, which were designated, in ascending order, by Roman numerals. Subsequently, as he informs us in the Preface to his final Report on the Geology of Pennsylvania, Professor H. D. Rogers, in concert with his brother, Professor William B. Rogers, then directing the geological survey of Virginia, considered the question of geological nomenclature. Rejecting, after mature deliberation, the classification and nomenclature both of the British and New York geological surveys, they proposed a new one for the whole palæozoic column to the top of the coal-measures, founded on the conception of a great palæozoic day, the divisions of which were designated by names taken from the sun's apparent course through the heavens. (Geology of Penn., I. vi, 105.) So far as regards the three great groups which we have recognized in the lower palæozoic rocks, the later names of Rogers, and his earlier numerical designations, with their equivalents in the New York system, were as follows: —

Primal (I.). This includes the mass of 2,500 feet or more of shales and sandstones, which in Pennsylvania and Virginia, and farther southward, form the base of the palæozoic series, and rest upon crystalline schists. The Primal division was regarded by the Messrs. Rogers as the equivalent both of the Potsdam and the still lower members of the Cambrian.

Auroral (II.). This division consists in great part of limestones, often magnesian, and corresponds to the Calciferous, Levis, and Chazy formations. Its thickness in Pennsylvania varies from 2,500 to 5,000 feet, and, with the preceding division, it includes the first fauna of Barrande. The representatives of the Primal and Auroral divisions attain a great development in southwestern Virginia and also in eastern Tennessee, where they have been studied by Safford.

Matinal (III.). In this, which represents the second fauna, were comprised the limestones of the Trenton group, together with the Utica and Hudson River shales.

Levant (IV.). This division corresponds to the Onedia and Shawangunk conglomerates and the Medina sandstone.

Surgent, Scalent, and *Pre-Meridional* (V., VI.). In these divisions were included the representatives of the Clinton, Niagara, and Lower Helderberg groups of New York, making, with division IV., the third fauna of Barrande.

The parallelism of these divisions with the British rocks was most clearly and correctly pointed out by H. D. Rogers himself, in an explanation prepared, as I am informed, with the collaboration of Professor William B. Rogers, and published in 1856, with a geological map of North America by the former, in the second edition of Keith Johnson's Physical Atlas. The palæozoic rocks of North America are there divided into several groups, of which the first, including the Primal, Auroral, and Matinal, is declared to be the near representative of "the European palæozoic deposits from the first-formed fossiliferous beds to the close of the Bala group; that is to say, the proximate representatives of the Cambrian of Sedgwick." A second group embraces the Levant, Surgent, Scalent, and Pre-Meridional. These are said to be "the very near representatives of the true European Silurian, regarding this series as commencing with the May Hill sandstone." The Levant division is further declared to be the equivalent of the sandstone just named; while the Matinal is made to correspond to the Llandeilo, Bala, or Upper Cambrian; the Auroral with the Festiniog or Middle Cambrian; and the Primal with the Lingula flags, the Obolus sandstone of Russia, and the Primordial of Bohemia.

The reader of the last few pages of this history will have seen how the Silurian nomenclature of Murchison and the British geological survey has been, through Lyell, De Verneuil, and the Canadian survey, introduced into American geology in opposition to the judgment, and against the protests of James Hall and the Messrs. Rogers, the founders of American palæozoic geology.

Three points have, I think, been made clear in the first and second parts of this sketch: first, that the series to which the name of Cambrian was applied by Sedgwick in 1835 (limited by him as to its downward extension, in 1838) was coextensive with the rocks characterized by the first and second faunas; second, that the series to which the name of Silurian was given by Murchison in 1835 included the second and third faunas, but that the rocks of the second fauna, the Upper Cambrian of Sedgwick, were only included in the Silurian system of Murchison by a series of errors and misconceptions in stratigraphy on the part of the latter, which gave him no right to claim the rocks of the second fauna as a lower member of his Silurian; third, that there was no ground whatever for subsequently annexing to the Silurian of Murchison the

Lower and Middle Cambrian divisions of Sedgwick, which the latter had separated from the Upper Cambrian on stratigraphical grounds, and which were subsequently found to contain a distinct and more ancient fauna.

The name of Silurian should therefore be restricted, as maintained by Sedgwick and by the Messrs. Rogers, to the rocks of the third fauna, the so-called Upper Silurian of Murchison; and the names of Middle Silurian, Lower Silurian, and Primordial Silurian banished from our nomenclature. The Cambrian of Sedgwick, however, includes the rocks both of the first and second faunas. To the former of these, the lower and middle divisions of the Cambrian (the Bangor and Festiniog groups of Sedgwick), Phillips, Lyell, Davidson, Harkness, Hicks, and other British geologists agree in applying the name of Cambrian. The great Bala group of Sedgwick, which constitutes his Upper Cambrian, is, however, as distinct from the last as it is from the overlying Silurian, and deserves a not less distinctive name than these two. Its original designation of Upper Cambrian, given when the zoölogical importance of Lower and Middle Cambrian was as yet unknown, is not sufficiently characteristic, and the same is to be said of the name of Lower Silurian, wrongly imposed upon it. The importance of this great Bala group in Britain, and of its North American equivalent, the Matinal of Rogers, — including the whole of the limestones of the Trenton group, with the succeeding Utica and Hudson River shales, — might justify the invention of a new and special name. That of Cambro-Silurian, at one time proposed by Sedgwick himself, and adopted by Phillips and by Jukes, was subsequently withdrawn by him, when investigations made it clear that this group had been wrongly united with the Silurian by Murchison. Deference to Sedgwick should therefore prevent us from restoring this name, which, moreover, from its composition, connects the group rather with the Silurian than the Cambrian. Neither of these objections can be urged against the similarly constructed term of Siluro-Cambrian, which, moreover, has the advantage that no other new name could possess, — of connecting the group both with the true Silurian, to which it has very generally been united, and with the Cambrian, of which, from the first, it has formed a part. I therefore venture to suggest the name of Siluro-Cambrian, as a convenient synonyme for the Upper Cambrian of Sedgwick (the Lower Silurian of Murchison), corresponding to the second fauna; reserving, at the same time, the name of Cambrian for the rocks of the first fauna, — the Lower and Middle Cambrian of Sedgwick, — and restricting, with him, the name of Silurian to the rocks of the third fauna, — the Upper Silurian of Murchison.*

The late Professor Jukes, it may be here mentioned, in his Manual of Geology, published in 1857, still retained for the Bala group the name of Cambro-Silurian (which had been withdrawn by Sedgwick in 1854), and reserved the name of the "true Silurian period" for the Upper Silurian of Murchison. In his recent and much-improved edition of this excellent Manual (1872), Professor Giekie, the director of the geological survey of Scotland, has substituted the nomenclature of Murchison; with the important exception, however, that he follows Hicks and Salter in separating the Menevian from the Lingula flags, and uniting it with the underlying Harlech rocks (as has been done in the table on page 386), giving to the two the name of Cambrian (*loc. cit.*, pages 526 – 529), and thus, on good paleontological grounds, extending this name above the horizon admitted by Murchison. Barrande, on the contrary, in his recent essay on trilobites (1871, page 250), makes the Silurian to include not only the Lingula flags proper (Maentwrog, Festiniog, and Dolgelly), but the Menevian, and even a great part of the Harlech rocks them-

* Dr. Dawson, in his address as president of the Natural History Society of Montreal, in May, 1872, has taken the occasion of the publication in the Canadian Naturalist of the first and second parts of this history, to review the subject here discussed. Recognizing the necessity of a reform in the nomenclature of the palæozoic rocks in conformity with the views of Sedgwick, he would restrict to the rocks of the third fauna the name of Silurian, making it a division equivalent to Devonian; and while reserving, with Lyell, Phillips, and others, the name of Cambrian for the first fauna only, agrees with me in the propriety of adopting the name of Siluro-Cambrian for the second fauna.

selves (the Cambrian of Murchison and the geological survey), for the reason that the primordial fauna has now been shown by Hicks to extend towards their base. This, although consistent with Barrande's previous views as to the extension of the name Silurian, is a still greater violation of historic truth. By thus making the Silurian system of Murchison to include successively the Upper Cambrian and the Middle Cambrian of Sedgwick, and finally his Lower Cambrian (the Cambrian system of Murchison himself), we seem to have arrived at a *reductio ad absurdum* of the Silurian nomenclature; and we may apply to Siluria, as Sedgwick has already done, the apt quotation once used by Conybeare with reference to the Graywacke of the older geologists, which it replaces: "*Est Jupiter quodcunque vides.*"

It would be unjust to conclude this historical sketch of the names Cambrian and Silurian in geology, without a passing tribute to the venerable Sedgwick, who to-day, at the age of eighty-seven years, still retains unimpaired his great powers of mind, and his interest in the progress of geological science.* The labors of his successors in the study of British geology, up to the present time, have only served to confirm the exactitude of his early stratigraphical determinations; and the last results of investigations on both continents unite in showing that in the Cambrian series, as defined by him more than a generation since, he laid, on a sure foundation, the bases of palæozoic geology.

* See the Preface to this paper for a notice of his death.

XVI.

THEORY OF CHEMICAL CHANGES AND EQUIVALENT VOLUMES.

(1853.)

The following paper was published under the title of Considerations on the Theory of Chemical Changes, etc., in the American Journal of Science for March, 1853. It soon after appeared in the London, Edinburgh, and Dublin Philosophical Magazine (4), V. 526, and was translated into German and appeared in the Chemisches Centralblatt of Leipsic in the same year (page 849). In the papers which follow, on The Composition and Equivalent Volume of Mineral Species, on Solution and the Chemical Process, on The Objects and Method of Mineralogy, as well as in that on The Theory of Types in Chemistry, I have attempted to develop some of the notions contained in this first essay, which, I still think, must form the basis of a rational theory of chemistry and a true mineralogical classification.

In the proposed inquiry we commence by distinguishing between the phenomena which belong to the domain of physics and those which make up the chemical history of matter. We conceive of matter as influenced by two forces, one of which produces condensation, attraction, and unity, and the other expansion, repulsion, and plurality. Weight, as the result of attraction, is a universal property of matter. Besides this, we have its various conditions of consistence, shape, and volume, with the relation of the latter to weight, constituting specific gravity, and the relations of heat, light, electricity, and magnetism. A description of these qualities and relations constitutes the physical history of matter, and the group of characters which serve to distinguish one species from another may be designated the apparent or specific form of a species, as distinguished from its essential form.

The forces above mentioned modify physically the specific characters of matter, but they have besides important relations to those higher processes which give rise to new species by a complete change in the specific phenomena of bodies. In the capacity of such complete change consists the chemical activity of matter.

It is necessary to distinguish between the production of new species differing in physical characters, and that reproduction which belongs to organic existences. The distinction arises from that individuation which marks the results of organic life, and is eminently characteristic of its higher forms. The individuality not only of the organism, but of its several parts, is more evident as we ascend the scale of organic life, while inorganic bodies have a specific existence, but no individuality; division does not destroy them. Crystallization is a commencement of individuation, and crystals like the tissues of plants and animals must be destroyed before they can become the subjects of chemical change; *corpora non agunt nisi soluta.*

That mode of generation which produces individuals like the parent can present no analogy to the phenomena under consideration; metagenesis, or alternate generation, and metamorphosis are, however, to a certain extent, prefigured in the chemical changes of bodies. Their metagenesis is effected in two ways; by condensation and union on the one hand, and by expansion and division on the other. In the first case, two or more bodies unite, and merge their specific characters in those of a new species. In the second case, this process is reversed, and a body breaks up into two or more new species. Metamorphosis is in the same manner of two kinds; in metamorphosis by condensation only one species is concerned, and in metamorphosis by expansion the result is homogeneous, and without specific difference.

The chemical history of bodies is a record of these changes; it is in fact their genealogy. The processes of union and division embrace by far the greater number of chemical changes, in which metamorphosis sustains a less important part. By union, we rise to indefinitely higher species; but in division a limit is met with in the production of species which seem incapable of further division, and these, being regarded as primary or original species, are called chemical elements. These two processes continually alternate with each other, and a species produced by the first may yield, by division, species unlike its parents. From this succession results double decomposition or equivalent substitution, which always involves a union followed by division, although under the ordinary conditions the process cannot be arrested at the intermediate stage.

The prevalence of certain modes of division in related species has given rise to the different hypotheses of copulates and radicles, which have been made the ground of systems of classification; but these hypotheses are based on the notion of dualism, which has no other foundation than the observed order of generation, and they can have no place in a theory of the science. A body may divide into two or more new species, yet it is evident that these did not pre-exist in it, from the fact that a different division may yield other species whose pre-existence is incompatible with the last; nor can the pre-existence of any species but those which we have called primary be admitted as possible. Apart from these considerations, it is to be remarked that our science has to do only with phenomena, and no hypothesis as to the noumenon or substance of a species under examination, based upon its phenomena, or those of its derived species, can ever be a subject of science, for it transcends all sensible knowledge.

For these reasons, it is conceived that the notion of pre-existing elements or groups of elements should find no place in the theory of chemistry. Of the relation which subsists between the higher species and those derived from them, we can only assert the possibility, and, under proper conditions, the certainty of producing the one from the other. Ultimate chemical analyses, and the formulas deduced from them, serve to show what changes are possible in any body, or to what new species it may give rise by its changes.

Chemical union is interpenetration, as Kant has taught, and not juxtaposition, as conceived by the atomistic chemists. When bodies unite, their bulks, like their specific characters, are lost in that of the new species. Gases and vapors unite in the proportion of one volume of each, or in some other simple ratio, and the resulting species in the gaseous state occupies one volume, so that the specific gravity of the new species is the sum of those of its factors. The converse of this is true in division, and the united volumes of the resulting species are some simple multiple of that of the parent; in metamorphosis a similar ratio is always observed.

Aside from the apparent exceptions about to be noticed, the weights of equal volumes of gases and vapors are their equivalent weights, and the doctrine of chemical equivalents is that of the equivalency of volumes. According to the atomic hypothesis, these weights represent the relative weights of the atoms, and as equal volumes contain the same number of atoms, these must have similar volumes, so that we come at last to the equivalency of volumes. As chemical combination is not a putting together of molecules, but an interpenetration of masses, the application of the atomic hypothesis to explain the law of definite proportions becomes wholly unnecessary. Chemical species are homogeneous; *tota in minimis existit natura.* Solution is chemical union, as is indicated by the attendant condensation; mechanical admixtures are not accompanied by any change of volume.

As two volumes of water-vapor yield one volume of oxygen and two of hydrogen, this has been assumed to be the equivalent of water and of hydrogen, while oxygen was represented by one volume, whose weight was 8, that of the volume of hydrogen being .5, so that the weight of the equivalent of water was 9. But two volumes of hydrogen unite without condensation with two of chlorine, and the resulting four volumes of hydrochloric gas are found to be equivalent to four volumes of chlorine, hydrogen, or water-vapor. Hence four volumes are to be taken for the equivalent of water, and it becomes H_2O_2, with an equivalent of 18, corresponding to HCl, and to volatile species generally, whose equivalents are represented by four volumes of vapor; from these, the equivalents of non-volatile species are determined by comparison.

Hydrogen, chlorine, and some other primary species offer

apparent exceptions to the general law of condensation and equivalency of volumes. When four volumes of chlorine unite with four of olefiant gas, or of naphthaline, the product is condensed into four volumes; but if the chlorine unite with the same volume of hydrogen gas, there is no condensation, and eight volumes or two equivalents of hydrochloric gas are produced. This, however, is explained when we find that four volumes of the chloro-hydrocarbon, MH,Cl_2, may break up into four of a new species MCl, and four of HCl; a change which with the chloride of olefiant gas is effected by the aid of hydrate of potash, and with the chloride of naphthaline takes place spontaneously at an elevated temperature. In the production of hydrochloric gas from chlorine and hydrogen, union takes place followed by immediate expansion without specific difference, or metamorphosis, while in the production of this acid with the hydrocarbons we observe the intermediate stage.

If an equivalent of four volumes of hydrochloric gas were to undergo a change like the chloride of naphthaline, and yield four volumes of chlorine and four of hydrogen, these species would appear with one half their observed densities; hence we conclude that they are actually condensed to one half their theoretical volumes, so that four volumes of hydrogen gas represent not H, but H_2. In the same way, if we conceive the quantity of oxygen produced from four volumes of water-vapor to represent two equivalents, it should equal eight volumes instead of two, so that it is condensed to one fourth, precisely as the vapor of sulphur is condensed to one twelfth of its theoretical volume. As there are no bodies which are known to yield for four volumes a less quantity than two volumes of oxygen, this may be taken to represent its equivalent, and the condensation of the theoretical volume is, like that of hydrogen and chlorine, one half. Water with an equivalent of four volumes is then H_2O, and its weight $2 + 16 = 18$; the same formula is deduced by those chemists who take two volumes for the equivalent, and, dividing the weight of hydrogen, write water H_2O, with an equivalent weight of 9. The condensation of these elements is that mode of metamorphosis which constitutes polymerism, and evidently offers no exception to the law of equivalent volumes.

The law of Laurent, that the number of atoms of hydrogen, or of hydrogen, chlorine, nitrogen, metals, etc., in any formula corresponding to four volumes of vapor, is always a sum divisible by two, clearly follows from the principles already laid down, and from the fact that nitrogen and the metals are subject to the same conditions as hydrogen and chlorine; the atoms have the value which has been assigned to H and to Cl in the formulas given above. The same rule of divisibility, as Laurent has already shown, necessarily holds in regard to the number of atoms of carbon, as well as to the oxygen and sulphur, if we take for their equivalent weights the numbers 6, 8, and 16 respectively.*

It is to be remarked that while the coefficients of H, Cl, or N, in formulas where these are associated, may be odd numbers, those of O, S, and C are always even. This seems a conclusive reason for doubling the equivalents of the latter, or dividing those of hydrogen, chlorine, the metals, etc., according as four or two volumes are taken for the equivalent. [*See* p. 176.]

I have elsewhere pointed out that carbon and oxygen sustain such relations that C_2H_2 may be compared with O_2H_2 and with O_2M_2, and, by the substitution of nitrogen for hydrogen, with C_2HN, prussic acid, and O_2N_2, nitrous oxide (the so-called compounds of nitrous oxide with bases are probably O_2MN, corresponding to the cyanides, C_2MN); while the peroxide of hydrogen, O_4H_2, corresponds to O_4N_2, nitric oxide, and to C_4N_2, cyanogen. This relation has important bearings on the history of the cyanic series, and the nitric derivatives of the hydrocarbons.†

The formulas of such related species as Gerhardt has designated chemical homologues differ from each other by $n\,C_2H_2$;

* Laurent, Récherches sur les combinaisons azotées, Ann. de Chimie et de Physique, November, 1846; and American Journal of Science for September, 1848, p. 174.

† See page 502 of my Introduction to Organic Chemistry, appended to Silliman's First Principles of Chemistry, Phila., 1852; and the above Journal for January, 1853, p. 151.

if now the relation between C and O be what we have supposed, it may be expected that mineral species will exhibit the same relations as those of the carbon series, and the principle of homology be greatly extended in its application. Such is really the case, and the history of mineral species affords many instances of isomorphous silicates whose formulas differ by $n\,O_2M_2$, as the tourmalines, and the silicates of alumina and magnesia, while the latter, with many zeolites, exhibit a similar difference of $n\,O_2H_2$. The relation is in fact that which exists between neutral and surbasic or hydrated salts.

Laurent has asserted that salts of the same base, with homologous acids of the type $(C_2H_2)n\,O_4$, may be isomorphous when they differ by O_2H_2, and has pointed out, besides, several instances of what he has called hemimorphism in species thus related, as well as in others differing by $n\,Cl_2$. The observations of Pasteur and Nicklès have greatly extended the application of these cases, which assume a new importance in connection with the views here brought forward, and demand further study.*

But to return: we have seen that in gases and vapors the specific gravity of a species enables us to fix its equivalent, which is often a multiple, by some whole number, of that calculated from the results of ultimate analysis. As the equivalents of non-volatile species are generally assumed to be those quantities which sustain the simplest ratio to certain volatile ones, the real equivalent weight corresponding to four volumes of vapor, and consequently the theoretical vapor-density of such species, is liable to a degree of the same uncertainty as those deduced from ultimate analysis. Having, however, determined the true equivalent of a species from the density of its vapor, the inquiry arises whether a definite and constant relation may not be discovered between its vapor-density and the specific gravity of a species in the solid state. Such a relation being established, and the value of the condensation in passing from a gaseous to a solid state being known, the equivalents of

* See Laurent, Comptes Rendus de l'Acad., Tom. XXVI. p. 353; and p. 257 of Laurent and Gerhardt's Comptes Rendus des Travaux de Chimie for 1848; also Pasteur, ibid., p. 165; and Nicklès, ibid. for 1849, p. 347.

solids, like those of vapors, might be determined from their specific gravities.

A connection between equivalent weight and density is evident in some allied and isomorphous species. H. Kopp, in dividing the assumed equivalent weights of such bodies by their specific gravities, obtained quantities which were found to be equal for some of these related species. These numbers evidently represent the volumes of equivalents, and in accordance with the atomic hypothesis are said to denote the atomic volumes. The inquiry of Kopp has been pursued by many investigators, among whom are Schroeder, Filhol, Playfair, and Joule, and, more recently, Dana. Their results show that the volumes thus calculated for related species of similar crystallization are generally identical, or sustain to each other some simple ratio; while Mr. Dana, who has compared isomorphous species of unlike chemical constitution, finds that the calculated volumes are often to each other as the number of equivalents of elements in the formulas representing the species; thus leading to the conclusion that the real equivalent weight is either a mean of that of all the elements, or some multiple of it. The reason of this appears in the fact that the formulas of those species in which this relation is apparent generally differ in the proportions of Al_2O_3, SiO_3, MgO, CaO, etc., and the quantities obtained in dividing the equivalent weights of these by the number of elements are nearly equal. If we divide by the number of elements, the equivalents calculated from the formulas of those species, it will be seen that the mean equivalents vary with the specific gravity.

These investigations have been principally confined to native and artificial mineral species, and the equivalents have been calculated from the formulas of Berzelius and Rammelsberg, which express the simplest ratios deducible from analysis. While in conformity with the dualistic notions, a mineral like calcite or magnesite was regarded as a compound of one equivalent of carbonic acid and one of lime or magnesia, dolomite was said to be composed of one equivalent of each of these carbonates, or of two to three, as the case might be, while its density was the mean of those of its constituents; thus implying that this union, unlike that observed in gases, is juxtaposition, and not interpenetration. This system of formulas has introduced such difficulties into the study of the relations before us, that we find Mr. Dana led to the conclusion that "the elemental molecules are not combined together or united with one another, in a compound, but that under their mutual influence each is changed alike, and becomes a mean result of the molecular forces in action." *

The solution of these difficulties is very simple, and will have been inferred from the plan of our inquiry. It is found in the principle that all species crystallizing in the same shape have the same equivalent volume; so that their equivalent weights, as in the case of vapors, are directly as their densities, and the equivalents of mineral species are as much more elevated than those of the carbon series, as their specific gravities are higher. The rhombohedral carbonates must be represented as salts having from twelve to eighteen equivalents of base, replaceable so as to give rise to a great number of species, and the variations in the volume of different carbonates, as observed by Kopp, indicate the existence of several homologous genera, which are isomorphous.

The researches of Playfair and Joule have led them to the conclusion that in some hydrated salts which crystallize with twenty and twenty-four equivalents of water, as the carbonate, the triphosphates and triarseniates of soda, the calculated volume coincides with that obtained by multiplying the volume of ice (9.8 for HO with an equivalent weight of 9) by the number of equivalents of water. This result is thus explained; water in these salts is in the same state of condensation as in ice, and 24 HO thus condensed would occupy the volume of $24 \times 9.8 = 235$, which is identical with that of the rhombic phosphate, as $20 \times 9.8 = 198$ is with that of the carbonate of soda, $C_2Na_2O_6$, 20 HO. Alum, crystallizing with 24 HO, has a volume which is greater than that of phosphate of soda, and, according to Playfair and Joule, equals that of the water

* American Journal of Science (2), Vol. IX. p. 245.

in the state of ice, with the addition of the bases, the acid being excluded.* In reality, the equivalent volume of alum is to that of the rhombic phosphate as 270 : 235; and 24 HO crystallizing in the monometric system would have the same volume as alum, with a specific gravity of about .8, giving for HO, 11.25 instead of 9.8.

What are called the atomic volumes of crystallized species are the comparative volumes of their crystals. In the rhombohedral system, the length of the vertical axis being constant, the volume varies with the length of the lateral axes, or, in other words, increases as the rhombohedron becomes obtuse, and diminishes as it becomes acute, the cube being the limit between the two. So in the dimetric and trimetric systems, the length of the vertical axis being unity, the volume diminishes as the base of the prism, the specific gravity increasing. Monoclinic and triclinic crystals may be calculated as if derivatives of the trimetric system, with which they will be found to correspond in volume.†

It is now necessary to determine what equivalent corresponds to a given specific gravity in any crystalline solid, or, in other words, what is the value of the condensation which takes place in the change from the gaseous to the solid state; and here a degree of uncertainty is met with, because the equivalent of a crystallized species may often be a multiple of that deduced from those chemical changes which only commence with the destruction of its crystalline individuality. The simplest formula deducible for alum is $KO\,SO_3$, $Al_2O_3\ 3\,SO_3$, $24\,H\,O$, or $S_4Kal_3O_{16}$, $12\,H_2O_2$, and, hydrogen being unity, its equivalent is at least 474.6, which, with a specific gravity of 1.75, gives a volume of about 270. Again, grape-sugar is not less than $C_{24}H_{24}O_{24}$, if we regard its combination with common salt as corresponding to one equivalent of each; and the ferrocyanides in the same way are represented by C_{12}, etc. There are reasons for believing that the equivalents of these species in the

* Chemical Society, Quarterly Journal, I. page 139.

[† The conclusions in this paragraph may be liable to correction, but I leave them as they were printed twenty-one years since.]

crystalline state correspond to some multiple of the above formulas, a question to be decided by an examination of the crystallization and specific gravity of species whose equivalents are admitted to be higher.

Favre and Silbermann, from their researches upon the heat evolved in fusion and solution, have been led to conclude: first, that crystallized salts are polymeric of these same salts in solution, that is, they are represented by formulas which are multiples of those deduced from analysis; secondly, that double salts and acid salts do not exist in solution, being produced only during crystallization; and, thirdly, that water, in crystallizing, changes from HO to nHO, n being some whole number.* These conclusions are seen to be in accordance with those deduced from a consideration of the relations of density and equivalent volume. A polymerism is evident in such salts as sulphate of potash and cyanide of potassium when their specific gravities are compared with those of alum and the ferrocyanide.

In the liquid state, the relation between specific gravity and equivalent is not so apparent as in solid species. The condensation often varies greatly, even in allied and homologous species, but still exhibits a relation of volumes. The alcohols $C_2H_4O_2$, $C_4H_6O_2$, $C_{10}H_{12}O_2$, and $C_{16}H_{18}O_2$ have very nearly the same specific gravity, so that the condensation is inversely as their vapor-equivalents. The densities of wine-alcohol, acetic acid, and aldehyde in the liquid state, vary as their equivalents, so that the calculated volumes are 57.5, 55.5, and 55. Formic and valeric acids show a similar relation in density to their respective alcohols, their calculated volumes being to these as 37.3 : 39, and 108 : 106.7. If to these we add butyric acid, which gives a volume of 90, and the density of whose alcohol has not yet been determined, the liquid volumes for the four acids, $C_2H_2O_4$, $C_4H_4O_4$, $C_8H_8O_4$, and $C_{10}H_{10}O_4$, are 37.3, 55.5, 90, and 108. These numbers approximate to multiples of the liquid volume of water H_2O_2, which is 18; or taking this as unity, are very nearly as 2, 3, 5, and 6. The interval between

* Comptes Rendus, XXII. 823 – 1140, and XXIII. 199 – 411.

3 and 5 corresponds to propionic acid $C_6H_6O_4$, of whose specific gravity I find no recorded observation. The density of many of these liquids is not accurately known, and the results of different experimenters are not precisely accordant. The specific gravity at their boiling-points should probably be chosen for the purpose of comparison, and these approximations lead us to expect that future observations will establish a simple relation between the densities of liquids and their vapors.

In a succeeding paper [XVII. of the present volume] it is proposed to apply the principles explained in the present essay in an examination of the equivalents of a number of minerals and other crystallized species.

XVII.

THE CONSTITUTION AND EQUIVALENT VOLUME OF MINERAL SPECIES.

.853 - 1863.)

A paper with the above title, of which the introduction and an analysis are given below, appeared in the American Journal of Science for September, 1853. In the Proceedings of the American Association for the Advancement of Science for 1854, the same subject is continued in an essay entitled Illustrations of Chemical Homology. From the author's abstract of this, which appeared in the American Journal of Science for September, 1854, some extracts are here given, in which will be found his views on the constitution of the feldspars, since adopted by Tschermak, and generally ascribed to him. Further illustrations are given by extracts from a later paper by the author in the Compte Rendu of the French Academy of Sciences for June 29, 1863, on saussurite and related minerals. Some general conclusions in accordance with the views here expressed will be found in Paper XIX. of the present volume.

In a recent paper [XVI. of the present volume] we endeavored to lay down some principles which may serve as the basis of a sound theory of chemistry. Having explained the nature of chemical changes, and the laws of combination, we showed that the volumes of the uniting species are always merged in that of the new one, so that the atomic theory, as applied by Dalton, which makes combination consist in juxtaposition, is untenable. It was further asserted that the simple relations of volumes which Gay Lussac pointed out in the chemical changes of gases apply to all liquid and solid species, thus leading the way to a correct understanding of the equivalent volumes of the latter. While chemists have not hesitated to assign high equivalents to bodies of the carbon series, they have been inclined to make the equivalent weights of denser mineral species correspond to formulas representing the simplest possible ratios. We endeavored, from a consideration of the theory of equivalent volumes, to point out the errors to which this method has led, and to show that we must assign to most mineral species much higher equivalent weights than have hitherto been admitted.

It was further asserted that a relation similar to that observed in the formulas of allied hydrocarbonaceous bodies, and designated as chemical homology, exists in the formulas of mineral species. We have said that these formulas, deduced from the results of analysis, are not to be looked upon as expressing any pre-existing relations in the constitution of the species, which is not to be regarded as a compound, but as an individual, in which the so-called chemical elements have no actual existence. The arrangements of these in our formulas only serve to make apparent the numerical relations which have been found to govern the transformations of the higher species.

The formulas of homologous bodies may be represented as series in arithmetical progression.* The first term may be the same as the common difference, and the series is then

$$b, 2b, 3b...nb,$$

as in the hydrocarbons C_2H_2, C_4H_4, C_6H_6, etc. If the first term is unlike the common difference, the series is

$$a, a + b, a + 2b,...a + nb,$$

of which the ammonias, NH_3, $NH_3 + C_2H_2$, $NH_3 + 2C_2H_2$, etc., are examples. Both of these cases are illustrated in the chemical history of mineral species.

In the paper already referred to it has been shown, from the relations of carbon, sulphur, and oxygen on the one hand, and of hydrogen and the metals on the other, that M_2S_2, M_2O_2, and H_2O_2 (M representing any metal) may be compared with H_2C_2. This view will be applied in extending the application of the principle of homology. The sesqui-oxides like ferric oxide, chromic oxide, and alumina, will be regarded as

[* The conception of progressive series in chemical compounds is generally ascribed to Gerhardt, who made it widely known in his Précis de Chimie Organique, but appears to have been first enunciated by Dr. James Schiel of St. Louis, in 1842, in Wöhler and Liebig's Annalen, Vol. XLIII., page 107. See farther the American Journal of Science (2), XXXII. 48, where Dr. Schiel has developed the whole question of series in a very complete manner.]

oxides of ferricum, chromicum, and aluminicum, having two thirds the equivalents ordinarily assigned to these metals, and represented by fe, cr, and al; so that Fe_2O_3 becomes 3feO, capable of replacing 3MgO, or 3FeO. In the same way arsenic and antimony, in one third their usual equivalents, may be represented by as and sb; AsO_3 then becomes 3asO. Silica, SiO_3, may also be written as 3siO, and by this means all these oxides may be reduced to the type M_2O_2.

We have further asserted that, for species crystallizing in the same form, the density varies directly as the equivalent weight, so that the quantities obtained in dividing the one by the other, and known as the atomic or equivalent volumes, will be equal. Such a relation is already recognized between species of the same genus, and we now propose, having fixed an equivalent weight for one species, to calculate, from their densities, those of the species isomorphous with it, and to show from the formulas corresponding to these equivalent weights that the different genera thus related are homologous, or exhibit other intimate relations.

[In developing the subject in the paper of which the above is the introduction, I began by considering the volume of some artificial salts the density of which has been carefully determined by Playfair and Joule, as given in their elaborate memoir on Atomic Volumes. The volume of the four prismatic arseniates and phosphates of soda, with 24HO, was found by them to be from 233.0 to 235.6; while that of four alums, with the same number of equivalents of water, varied from 271.6 to 280.5; the presumption, for obvious reasons, being in each case in favor of the greater density, and hence of the lesser volumes. With the alums were compared the equivalent volumes of the chlorides of sodium and potassium, calculated from their ordinary formulas, and the conclusion reached that the crystals of these salts possess equivalent weights which are such multiples of NaCl and KCl as would give an equivalent volume equal to that found for the alums, or more probably some multiple of the latter. With the volume of the arseniates and phosphates of soda was also compared that of ferrocyanide of potassium with $C_{12} = 230$, and lactose with $C_{24} = 234$; the equivalent weight of carbon being 6.

[An attempt was then made to fix the volume of the prismatic and rhombohedral carbon-spars, which were compared respectively with the isomorphous species bournonite and the red silver ores, proustite and pyrargyrite. The received formula of bournonite being doubled, and that of the rhombohedral sulphides made to correspond with it, we find for the prismatic species an equivalent volume of 508, and for the rhombohedral ones 546 – 564. In accordance with this the equivalent of calcite corresponds to $C_{30}Ca_{30}O_{90}$ ($C = 6$ and $O = 8$), while dolomite, chalybite, and diallogite become $C_{36}M_{36}O_{108}$, and calamine and magnesite $C_{40}M_{40}O_{120}$. For the prismatic carbonates, aragonite, like calcite, is $C_{30}M_{30}O_{90}$, while strontianite, cerusite, and bromlite are $C_{25}M_{25}O_{75}$, and witherite is $C_{22}M_{22}O_{66}$. With these were at the same time compared the homœomorphous rhombohedral and prismatic nitrates of soda and potash, from which it was suggested that the above equivalents were to be still further multiplied. That the volume above fixed for these rhombohedral species was, if not the true one, a measure of it, was soon rendered more probable by an examination of the compound of glucose and chloride of sodium, which was obtained in large rhombohedral forms isomorphous with calcite and having a density of 1.563. Doubling the empirical formula of this body,

$$C_{24}H_{24}O_{24} . NaCl . H_2O_2$$

we have for it an equivalent volume of 558.5, while that of calcite with $C_{30}M_{30}O_{90}$, and a density of $2.72 = 555.5$. (Amer. Jour. Science (2), XIX. 416.)

[From Glauber-salt and borax were deduced, in like manner, an equivalent volume of about 440, corresponding nearly with that of saccharose with $C_{48} = 430$, and with these were compared the silicates of the amphibole group, from which it was concluded that these silicates present among themselves relations similar to those of the homœomorphous carbon-spars. The attempts to deduce correct formulas for these and other silicates at that time were, however, vitiated by many incorrect analyses, and rendered uncertain by doubts as to the equivalent weight of silicon.

[An important point in the question of homology and homœomorphism was then referred to in the following language : —

"The similarity in crystallization between species whose formulas differ only in the elements of water has been pointed out by Laurent in certain salts of organic acids, and is seen in several mineral species. The chabazites, for example, give the formula $3RO,SiO_3 . 3Al_2O_3,2SiO_3$, with 15HO and 18HO, while the variety ledererite affords, according to Hayes and to Rammelsberg, but 6HO. The hydrous iolites are also cases in point, as well as aspasiolite, the serpentines, and the talcs, with their varying proportions of water. In the formulas of these species, water appears to replace magnesia, and Scheerer has shown that many different species may be referred to a common chemical type, by admitting 3HO to replace MgO, and 2HO to replace CuO, etc. These cases, to which he has given the name of polymeric isomorphism, are but instances of the partial substitution of water for other bases in homologous genera which differ by nMO."

[In the continuation of this subject, in 1854, as above referred to, the question of homologies was further illustrated by the neutral and the basic nitrates of lead, represented by a common formula $(Pb_2O_2)n . N_2O_{10}$. "These salts vary in solubility and in physical characters, but resemble each other in yielding nitric acid and oxide of lead as results of their decomposition, and are completely analogous to the homologous series of Gerhardt, which differ by $n(C_2H_2)$. From the relation between basic and hydrated salts the same view is to be extended to the latter, and species differing by $n(O_2H_2)$ and $n(O_2M_2)$ may thus be homologous. The above formulas are intended to involve no hypothesis as to the arrangement of the elements, for in the author's view, each species is an individual, in which the pre-existence of different species that may be obtained by its decomposition cannot be asserted. He regards silicates like eudialyte, sodalite, and pyrosmalite as oxychlorides, $(M_2O_2)n . MCl$, and nosean, hauyene, and lapis-lazuli as basic sulphates $(M_2O_2)n . S_2O_8$, while cancrinite, and perhaps some scapolites, are [may perhaps be] basic carbonates. All other silicates are reducible to the same type as the spinels, $n(M_2O_2)$, the formula of silica itself being written siO. . . . . Boric, titanic, tantalic, and niobic acids are reduced to the same formula as silica."

"Homœomorphous species have similar equivalent volumes, so that the density in species thus related enables us to determine their comparative equivalent weights, and to fix their positions in a homologous series. The proportion between the silica and the other oxides may vary greatly in related species, while the characters of the genus or the order are preserved. This is illustrated in hornblende, diopside, and aluminous pyroxenes like hudsonite. . . . . The triclinic feldspars, of which albite and anorthite are the representatives, furnish another example." [These, it was shown, might be reduced to a common formula $M_{64}O_{64}$, to which petalite was also referred, while orthoclase was described as belonging to a homologous genus, $M_{60}O_{60}$, represented by $(si_{45}al_{12}K_3)O_{60}$, with the remark that although this formula agrees with a large number of analyses, there are those which appear to show more alkali. Petalite was $(si_{51}al_{10}Li_3)O_{64}$, with a density of 2.45 and an equivalent volume of 401.5. The formulas then assigned to the two feldspars first named were, respectively :—]

		Density.	Eq. vol.
Anorthite . .	$(si_{32}\ al_{24}\ Ca_8)O_{64}$ . .	2.76 . .	405.0.
Albite . . .	$(si_{48}\ al_{12}\ Na_4)O_{64}$ . .	2.62 . .	402.4.

"Between anorthite and albite may be placed vosgite, labradorite, andesine, and oligoclase, whose composition and densities are such that they all enter into the same general formula with them, and have the same equivalent volume. The results of their analysis are by no means constant, and it is probable that many, if not all of them may be variable mixtures of albite and anorthite. Such crystalline mixtures are very common ; thus in the alums, aluminium, iron, and chromium, and potassium and ammonium may replace one another in indefinite proportions. . . . . Heintz has shown by fractional precipitation that there are mixtures of homologous fatty acids which cannot be separated by crystallization, and have hitherto been regarded as distinct acids. The author insists that the possibility of such mixtures of related species should be constantly kept in view in the study of mineral chemistry. The small portions of lime and potash in many albites, and of soda in anorthite, petalite, and orthoclase, are to be ascribed to mixtures of other feldspar-species."

[The above extracts are from the author's abstract of his paper in the American Journal of Science for September, 1854. There might be found reason to-day for modifying the formulas above given for petalite and orthoclase, but I leave them as they were written twenty years since.

[These views of mine with regard to the triclinic feldspars have since been generally accepted, but by an oversight they are attributed to Tschermak, who, so far as I am aware, first announced them ten years later, namely, in 1864 (K. K. Academie Wissenschaft, Wien). He there stated that with the exception of the baryta-feldspar, hyalophane, and the boric feldspar, danburite, the feldspars were reducible to three species, namely, adularia (orthoclase), albite, and anorthite, having a common formula, which, adopting the equivalent weights used by me above, becomes as follows for the two triclinic species :—

Anorthite	$Ca_4\ al_6\ al_6\ si_{16}\ O_{32}$
Albite	$Na_2\ al_6\ si_8\ si_{16}\ O_{32}$

This, which is but my common formula divided by two, is by Tschermak also assigned to orthoclase. He, while admitting that the potash-soda feldspars are made up of alternations of orthoclase and albite, as Gerhard had shown in the case of perthite, further concludes, as I had already done, "that oligoclase, andesine, and labradorite appear to be members of a great series, with many transitional forms, and may be regarded as isomorphous mixtures of albite with anorthite, sometimes with small admixtures of orthoclase."

[My views on the gradation into one another of the triclinic feldspars are again referred to in my Contributions to Lithology, in the American Journal of Science for March, 1864 (Vol. XXV. page 258), where the family of the feldspathides is made to include the scapolites or wernerites, and also beryl and iolite, which have a similar equivalent volume. The former is a glucinic feldspathide, subject, like the feldspars proper, leucite, and the scapolites, to kaolinization; while iolite is a magnesic feldspathide, having the oxygen ratios 5 : 3 : 1, corresponding to barsowite and bytownite, and intermediate between labradorite and anorthite.

[The relations of the feldspathides to the grenatides (in which are included the garnets, idocrase, epidote, and zoisite) furnish an important illustration of the notions put forward in the preceding pages. In the American Journal of Science for 1859 (XXVII. 336) will be found a memoir on Euphotide and Saussurite, in which I showed that the saussurite of Monte Rosa (the jade of De Saussure) does not belong, as previously supposed, to the feldspathides, but from its chemical and physical characters is to be regarded as a zoisite. This substance, which is very distinct, alike from the compact feldspars with which it had been confounded, and from the compact amphibole to which also the name of jade is sometimes given, has a specific gravity of 3.35 and a hardness of 7.0. It is only attacked by acids after intense ignition or fusion, by which it is converted into a soft glass having a specific gravity of 2.80. By analysis it is found to have the composition of zoisite or of meionite, these two species having the same centesimal composition. It has, however, the characters of the former, and differs widely from meionite, which is a scapolite having a specific gravity of 2.70 and a hardness of 5.5, and is readily attacked and decomposed by acids.

[The Comptes Rendus of the French Academy of Science for June 29, 1863, contains a communication from me, which is translated in the American Journal of Science for November, 1863 (page 427). In this, after giving in brief the history of euphotide and saussurite and the results of my examinations, I said as follows, referring to the memoir of 1859 :—

"In the memoir from which the foregoing results are cited

I have insisted upon the relation of isomerism, or rather of polymerism, which exists between meionite and zoisite, and have remarked that the augmentation of hardness, of density, and of chemical indifference which is seen in this last species, is doubtless to be ascribed to a more elevated equivalent, or, in other words, to a more condensed molecule. These different degrees of condensation, which are constantly kept in view in the study of organic chemistry, are besides, as I have already elsewhere shown, of great importance in mineralogy, and will form the basis of a new system of classification, which will be at the same time chemical and natural-historical. (Comptes Rendus, 1855, Vol XLI. page 79.) The different rhombohedral carbon-spars, kyanite and sillimanite, hornblende and pyroxene, offer in like manner examples of different degrees of condensation, and by their chemical composition belong to series the terms of which, like those of the hydrocarbons nC_2H_2, are both homologues and multiples of the first term. At the same time each one of these carbonates and silicates belongs to another possible series, the terms of which differ by nM_2O_2, corresponding to more or less basic salts."

"Meionite, with the oxygen ratios 3 : 2 : 1, is the most basic term known of the series of the wernerites (scapolites). The proportion of silica in these minerals augments until we reach in dipyre the ratios 6 : 2 : 1, with a density which does not exceed 2.66. We might then expect to find a silicate which should be to dipyre what zoisite or saussurite is to meionite, and Mr. Damour has recently had the good fortune to meet with such a mineral in a specimen of jade from China, of which he has given us the description and the analysis. (Comptes Rendus, May 4, 1863.) This substance closely resembles in its physical and chemical characters the saussurite or jade from Monte Rosa, of which it has the density, 3.34. It is a silicate of alumina, lime, and soda, and gives the same empirical formula as dipyre. We may expect to find between saussurite and this new species, to which Damour gives the name of jadeite, other jades having formulas which will correspond with the wernerites intermediate between meionite and dipyre. . . . . By its hardness, its specific gravity, and its indifference to acids, jadeite is completely separated from the wernerite group, and takes its place alongside of zoisite or saussurite, with the garnets, idocrase, and epidotes. The following table will serve to show the relations of the new species: —

	Density (about) 2.7	3.3
Oxygen ratio 3 : 2 : 1. . . .	Meionite .	Saussurite.
Oxygen ratio 6 : 2 : 1. . . .	Dipyre . .	Jadeite."

[The hardness of jadeite, according to Damour, is between that of orthoclase and quartz. It is lamellar in structure and exhibits two axes of polarization. Unlike saussurite, it is not attacked by acids after fusion, — a fact which is to be ascribed to the large proportion of silica which it contains.]

[Professor J. P. Cooke described, in 1860 (American Journal of Science (2), XXX. 194), some curious examples in the crystallized alloys of antimony and zinc, of considerable variations in composition without change in crystalline form. These cases, as remarked by him, do not come within the limits of isomorphism as generally understood, and hence he concludes that "the composition of a mineral species may be modified by an actual variation in the proportions of its constituents." These alloys of varying composition are to be regarded in part as examples of a progressive series of isomorphous compounds of antimony and zinc, of high equivalent, differing from each other by nZn_2, and in part, doubtless, as crystalline mixtures of these isomorphous homologous species. The principle embodied in the conception advanced by Professor Cooke, and rightly regarded by him of great importance to a correct science of mineralogy, he has named allomerism. It is evidently a case of homologous and isomorphous relations between members of a progressive series, — a general principle upon which I have insisted throughout the pages of this paper, and which includes the polymeric isomorphism of Scheerer.]

XVIII.

THOUGHTS ON SOLUTION AND THE CHEMICAL PROCESS.

(1854.)

This paper appeared in the American Journal of Science for January, 1854, and also in the Chemical Gazette for 1855, page 90.

By solution, as distinguished from fusion or volatilization, we understand in chemistry the production of a homogeneous liquid by the combination of two or more bodies, one of which must itself be in a liquid state, while the others may be liquid, solid, or gaseous. The solvent action of acids and alkalies upon bodies insoluble in water is by all admitted to be chemical in its nature; but, according to Leopold Gmelin, "mixtures of liquids, and solutions of solids in liquids (as of acids, alkalies, salts, oils, etc., in water and alcohol), are, by Berzelius, Mitscherlich, Dumas, and others of the most distinguished modern chemists, regarded as not chemical unless they take place in definite proportions." "Mitscherlich attributes such unions to *adhesion*, Dumas to *a solvent power* intermediate between cohesion and (chemical) affinity, and Berzelius refers them to *a modification of affinity*, while proper chemical combinations according to him result, not from affinity, but from electrical attraction." (Gmelin's Handbook, English ed., Vol. I. p. 34.)

The learned author of the Handbook objects to these views that "they restrict the idea of a chemical compound within too narrow limits," and he elsewhere implies that the force which produces solution is a weak degree of chemical affinity. (Ibid., Vol. I. p. 70.) The judicious Turner also speaks of ordinary solutions as instances of chemical union;* and Mr. J. J. Griffin has insisted upon the same view.† As these writers have not, however, sufficiently dwelt upon the important principle, rejected by so many names of authority, that *all solution is chemical union*, we propose to offer some considerations upon aqueous solution, and endeavor to show that the process presents all the phenomena of chemical combination. First, in the fact that the resulting saturated solutions are perfectly homogeneous. Secondly, in the condensation and more or less perfect identification of volume (*ante*, page 428) observed in the process (some anhydrous salts dissolve in water without increasing its volume). Thirdly, in the change of temperature which attends the process; thus oil of vitriol, hydrate of potash, and many anhydrous salts evolve heat when dissolved in water, while sal-ammoniac, nitre, and many hydrous salts produce cold by their solution. Fourthly, in the change of color which attends the solution of some salts, as the chlorides of nickel, cobalt, and copper.

It must not be forgotten that the liquid state of these aqueous combinations is often an accident of temperature; alum and the rhombic phosphate of soda are liquids at 212° F., and bi-hydrated sulphuric acid is a crystalline solid below 46° F. The ease with which many of these compounds are destroyed by evaporation, and even by changes of temperature, is not to be urged as an objection to the chemical nature of the union. We need only compare the corresponding silver-salts with the chloride and iodide of gold, or the hydrochlorates of morphia and ammonia with those of caffeine and piperine, which lose their acid by a gentle heat, to learn how variable is the stability of admitted chemical compounds. Chemical affinity may be very feeble in degree.

According to Gay-Lussac, one part of oil of vitriol will absorb from air saturated with moisture fifteen parts of water, or more than eighty equivalents; terchloride of arsenic requires eighteen equivalents of water to dissolve it, and the

* Elements of Chemistry, 7th ed., p. 139.

† L., E. and D. Phil. Mag. (3), Vol. XXIX. p. 299.

saturated solution unites with as much more water, evolving heat and forming a stable solution.* According to the experiments of Mr. Griffin in the paper cited above, the condensation which takes place in the solution of the acid is still perceptible with 6,000 equivalents of water to one of SO_3. There appears, however, to be with many bodies a limit beyond which the affinity for water is satisfied, and the liquids being then mechanically mixed, gradually separate by reason of their difference in density, as is observed in dilute alcohol, and probably in some saline solutions † and in metallic alloys.

Solution is a result of that tendency in nature which constantly leads to unity, condensation, identification. I have elsewhere, with Kant, defined chemical union to be interpenetration, but the conception is mechanical, and therefore fails to give an adequate idea. The definition of Hegel, that *the chemical process is an identification of the different and a differentiation of the identical,*‡ is, however, completely adequate. Chemical union involves an identification, not only of the volumes (interpenetration mechanically considered), but of the specific characters of the combining bodies, which are lost in those of the new species. Such is equally the case in aqueous solution, and we may say that all chemical union is nothing else than solution; the uniting species are, as it were, dissolved in each other, for solution is mutual.

Solution being, then, identification, the discussion as to whether metallic chlorides are changed into hydrochlorates when dissolved in water, is meaningless. Such a solution is a

* Penny and Wallace, L., E. and D. Phil. Mag., November, 1852, page 363.

† See Gmelin's Handbook, Eng. ed., Vol. I. p. 111. Gmelin throws a doubt upon these experiments; but the satisfactory results obtained on a large scale, in applying this principle to the rectification of spirit of wine by a recently patented process, were communicated to the American Association for the Advancement of Science, at Washington in May, 1854, by Dr. L. D. Gale. [This is questionable.]

‡ Stallo's Philosophy of Nature, page 453; see also page 67, where Stallo insists upon the same view. To Hegel belongs the merit of having first among modern philosophers obtained a just conception of the nature of the chemical process, although in its application he was misled by the received terminology of the science.

unity, in which we can no more assert the existence of the chloride or of water, than of chlorine, hydrochloric acid, or a metallic oxide, although these and many others are conceivable results of its differentiation. If the solution be one of chloride of potassium, evaporation resolves it into water and the chloride, but if chloride of aluminum, it is decomposed by boiling into water, hydrochloric acid, and alumina, or in the case of the corresponding magnesian salt, into hydrochloric acid and an oxychloride.

The precipitation of the sulphates of cerium, lanthanum, and calcium from their solutions by heat, and of most other salts by cold, is chemical decomposition or differentiation. Dilution may also effect decomposition in solutions; we have already said that the combination of terchloride of arsenic, $AsCl_3$, with 36HO is stable at ordinary temperatures, but a further addition of water causes the solution to divide into aqueous hydrochloric acid and crystalline oxide of arsenic. The precipitation of chloride of antimony, and of many salts of bismuth and mercury by water, is an analogous process. This decomposition of the solution of chloride of arsenic is an example of what is called double elective affinity (*attractio electiva duplex*), and is generally explained by saying that the attraction of arsenic for oxygen, and that of chlorine for hydrogen, enable the chloride and water to decompose each other. But these elemental species do not exist in the solution, although they are possible results of its decomposition, and to explain the process in this manner is to ascribe it to the affinities of yet unformed species.

I have elsewhere asserted that double decomposition always involves union followed by division (*ante*, page 428), although we cannot in every case arrest the process at the first stage. Under some changed conditions of temperature and pressure the decomposition may be the counterpart of the previous union, and thus reproduce the original species, as in the case of mercuric oxide, which is decomposed into mercury and oxygen at a temperature a little above that at which it was formed. When the division takes place in a sense different from the union, giving rise to new species, we have double decomposition. In the case of chloride of arsenic, the aqueous solution exhibits the first stage of the process. A similar condition of unstable union is observed in many other instances; thus binoxide of manganese gives, with cold hydrochloric acid, a brown solution, but the combination is by a gentle heat resolved into chlorine gas, and a rose-red solution of protochloride of manganese. So a mixture of equivalent parts of chloride of benzoyl and benzoate of soda combines at a temperature of 130° C., to form a limpid solution, and it is only on raising the temperature that the precipitation of sea-salt indicates the commencement of that decomposition which yields at the same time anhydrous benzoic acid.* It is only when looked upon as a momentary combination followed by a decomposition, that the theory of double decomposition becomes intelligible, and is in accordance with known facts.

From the narrow limits of temperature which often include the two processes, and from the ease with which light, warmth, friction, and pressure excite the decomposition of such bodies as the chloride of nitrogen, the nitrite of ammonia, the oxides of chlorine, and the metallic fulminates, we may conceive that within still narrower limits, and under conditions as yet undefined, many bodies may exhibit affinities for each other which are reversed by a very slight change of condition. In this way we may explain many of those obscure phenomena hitherto ascribed to *action by presence* or *catalysis.*

* Gerhardt, Ann. de Chimie et de Physique, 3me Serie, Tom. XXXVII. page 299.

XIX.

ON THE OBJECTS AND METHOD OF MINERALOGY.

(1867.)

This paper was read before the American Academy of Sciences in Boston, January 8, 1867, and published in the American Journal of Science in May of the same year.

Mineralogy, as popularly understood, holds an anomalous position among the natural sciences, and is by many regarded as having no claims to be considered as a distinct science, but as constituting a branch of chemistry. This secondary place is disputed by some mineralogists, who have endeavored to base a natural-history classification upon such characters as the crystalline form, hardness, and specific gravity of minerals. In systems of this kind, however, like those of Möhs and his followers, only such species as occur ready formed in nature are comprehended, and the great number of artificial species, often closely related to native minerals, are excluded. It may moreover be said in objection to these naturalists, that, in its wider sense, the chemical history of bodies takes into consideration all those characters upon which the so-called natural systems of classification are based. In order to understand clearly the question before us, we must first consider what are the real objects, and what the provinces, respectively, of mineralogy and of chemistry.

Of the three great divisions, or kingdoms of nature, the classification of the vegetable gives rise to systematic botany, that of the animal to zoölogy, and that of the mineral to mineralogy, which has for its subject the natural history of all the forms of unorganized matter. The relations of these to gravity, cohesion, light, heat, electricity, and magnetism belong to the domain of physics; while chemistry treats of their relations to each other, and of their transformations under the influences of heat, light, and electricity. Chemistry is thus to mineralogy what biology is to organography; and the abstract sciences, physics and chemistry, must precede, and form the basis of the concrete science, mineralogy. Many species are chiefly distinguished by their chemical activities, and hence chemical characters must be greatly depended upon in mineralogical classification.

Chemical change implies disorganization, and all so-called chemical species are inorganic, that is to say, unorganized, and hence really belong to the mineral kingdom. In this extended sense, mineralogy takes in not only the few metals, oxides, sulphides, silicates, and other salts which are found in nature, but also all those which are the products of the chemist's skill. It embraces not only the few native resins and hydrocarbons, but all the bodies of the carbon series made known by the researches of modern chemistry.

The primary object of a natural classification, it must be remembered, is not, like that of an artificial system, to serve the purpose of determining species, or the convenience of the student, but so to arrange bodies in genera, orders, and species as to satisfy most thoroughly natural affinities. Such a classification in mineralogy will be based upon a consideration of all the physical and chemical relations of bodies, and will enable us to see that the various properties of a species are not so many arbitrary signs, but the necessary results of its constitution. It will give for the mineral kingdom what the labors of great naturalists have already nearly attained for the vegetable and animal kingdoms.

Oken saw the necessity of thus enlarging the bounds of mineralogy, and in his Physiophilosophy attempted a mineralogical classification; but it is based on fanciful and false analogies, with but little reference either to physical or chemical characters, and in the present state of our knowledge is valueless, except as an effort in the right direction, and an attempt to give to mineralogy a natural system. With similar views as to the scope of the science, and with far higher and juster conceptions of its method, Stallo, in his Philosophy of Nature, has touched the questions before us, and has attempted to show the significance of the relations of the metals to cohesion, gravity, light, and electricity, but has gone no further.

In approaching this great problem of classification, we have to examine, first, the physical condition and relations of each species, considered with relation to gravity, cohesion, light, heat, electricity, and magnetism; secondly, the chemical history of the species, in which are to be considered its nature, as elemental or compound, its chemical relations to other species, and these relations as modified by physical conditions and forces. The quantitative relation of one mineral (chemical) species to another is its equivalent weight, and the chemical species, until it attains to individuality in the crystal, is essentially quantitative.

It is from all the above data, which would include the whole physical and chemical history of inorganic bodies, that a natural system of mineralogical classification is to be built up. Their application may be illustrated by a few points drawn from the history of certain natural families.

The variable relations to space of the empirical equivalents of non-gaseous species, or, in other words, the varying equivalent volume (obtained by dividing their empirical equivalent weights by the specific gravity), shows that there exist in different species very unlike degrees of condensation. At the same time we are led to the conclusion that the molecular constitution of gems, spars, and ores is such that those bodies must be represented by formulas not less complex, and with equivalent weights far more elevated than those usually assigned to the polycyanides, the alkaloids, and the proximate principles of plants. (*Ante*, pages 434 and 441.) To similar conclusions conduce also the researches on the specific heat of compounds.

There probably exists between the true equivalent weights of non-gaseous species and their densities a relation as simple as that between the equivalent weights of gaseous species and their specific gravities. The gas or vapor of a volatile body constitutes a species distinct from the same body in its liquid or solid state, the chemical formula of the latter being some multiple of the first; and the liquid and solid species themselves often constitute two distinct species of different equivalent weights. In the case of analogous volatile compounds, as the hydrocarbons and their derivatives, the equivalent weights of the liquid or solid species approximate to a constant quantity, so that the densities of those species, in the case of homologous or related alcohols, acids, ethers, and glycerides, are subject to no great variation. These non-gaseous species are generated by the chemical union, or identification, of a number of volumes or equivalents of the gaseous species, which number varies inversely as the density of these species. It follows from this that the equivalent weights of the liquid and solid alcohols and fats must be so high as to be a common measure of the vapor-equivalents of all the bodies belonging to these series. The empirical formula $C_{114}H_{110}C_{12}$, which is the lowest one representing the tristearic glyceride (ordinary stearine), is probably far from representing the true equivalent weight of this fat in its liquid or solid state; and if it should hereafter be found that its density corresponds to six times the above formula, it would follow that liquid acetic acid, whose density differs but slightly from that of fused stearine, must have a formula and an equivalent weight about one hundred times that which we deduce from the density of acetic-acid vapor, $C_4H_4O_4$.

Starting from these high equivalent weights of liquid and solid hydrocarbonaceous species, and their correspondingly complex formulas, we become prepared to admit that other orders of mineral species, such as oxides, silicates, carbonates, and sulphides, have formulas and equivalent weights corresponding to their still higher densities; and we proceed to apply to these bodies the laws of substitution, homology, and polymerism, which have so long been recognized in the chemical study of the members of the hydrocarbon series. The formulas thus deduced for the native silicates and carbon-spars, show that these polybasic salts may contain many atoms of different bases, and their frequently complex and varying constitution is thus rendered intelligible. In the application of the principle of chemical homology, we find a ready and natural explanation of those variations within certain limits, occasionally met with in the composition of certain crystalline silicates, sulphides, etc.; from which some have conjectured the existence of a deviation from the law of definite proportions, in what is only an expression of that law in a higher form.

The principle of polymerism is exemplified in related mineral species, such as meionite and zoisite, dipyre and jadeite, hornblende and pyroxene, calcite and aragonite, opal and quartz, in the zircons of different densities, and in the various forms of titanic acid and of carbon, whose relations become at once intelligible if we adopt for these species high equivalent weights and complex molecules. The hardness of these isomeric or allotropic species, and their indifference to chemical reagents, increase with their condensation, or, in other words, vary inversely as their empirical equivalent volumes; so that we here find a direct relation between chemical and physical properties.

It is in these high chemical equivalents of the species, and in certain ingenious but arbitrary assumptions of numbers, that is to be found an explanation of the results obtained by Playfair and Joule in comparing the volumes of various solid species with that of ice; whose constitution they assume to be represented by HO, instead of a high multiple of this formula. The recent ingenious but fallacious speculations of Dr. Macvicar, who has arbitrarily assumed comparatively high equivalent weights for mineral species, and has then endeavored, by conjectures as to the architecture of crystalline molecules, to establish relations between his complex formulas and the regular solids of geometry, are curious but unsuccessful attempts to solve some of the problems whose significance I have here endeavored to set forth. I am convinced that no geometrical grouping of atoms, such as are imagined by Macvicar and by Gaudin, can ever give us an insight into the way in which Nature builds up her units, by interpenetration and identification, and not by juxtaposition of the chemical elements.

None of the above points are presented as new, though they are for the greater part, I believe, original with myself, and have been from time to time brought forward and maintained, with numerous illustrations, chiefly in the American Journal of Science, since March, 1853, when my paper on the Theory of Chemical Changes and Equivalent Volumes (*ante*, page 426) was there published. I have, however, thought it well to present these views in a connected form, as exemplifying my notion of some of the principles which must form the basis of a true mineralogical classification.

XX.

THEORY OF TYPES IN CHEMISTRY.

(1848-1861.)

In the years 1848-1854 I published in the American Journal of Science several essays on the theory of chemical types and on related questions in the science. The first, on the Anomalies in the Atomic Volumes of Sulphur and Nitrogen and on Chemical Classification, appeared in September, 1848, and was followed in May and July, 1849, by a paper on Some Points to be considered in Chemical Classification. In January, 1850, appeared a paper on the Constitution of Leucine, with Remarks on the late Researches of Wurtz; and in March, 1852, one on the Compound Ammonias and the Bodies of the Cacodyle Series. In March, 1854, I published a summary of the views embodied in the preceding papers, commenting especially on the first one, in an essay on The Theoretical Relations of Water and Hydrogen, and vindicating for myself the priority in those views of chemical theory which had since my announcement of them been adopted by Gerhardt, Williamson, Wurtz, and other chemists. The publication by Wurtz of a criticism of Kolbe, in 1860, led me to write the following paper on the Theory of Types in Chemistry, in which I have concisely traced the history of the development of my views. It appeared in the Canadian Journal for March, 1861, and in the American Journal of Science for the same month of the same year. I have, in reprinting it, added an appendix on Nitrogen and Nitrification.

In the Annalen der Chemie und Pharmacie for March, 1860 (CXIII. 293), Mr. Kolbe has published a paper on the natural relations between mineral and organic compounds, considered as a scientific basis for a new classification of the latter. He objects to the four types admitted by Gerhardt, namely, hydrogen, hydrochloric acid, water, and ammonia, that they sustain to organic compounds only artificial and external relations, while he conceives that between these and certain other bodies there are natural relations, having reference to the origin of the organic species. Starting from the fact that all the bodies of the carbon series found in the vegetable kingdom are derived from carbonic acid, with the concurrence of water, he proceeds to show how all the compounds of carbon, hydrogen, and oxygen may be derived from the type of an oxide of carbon, which is either C_2O_4, C_2O_2, or the hypothetical C_2O.

When in the former we replace one atom of oxygen by one of hydrogen we have C_2O_3H, or anhydrous formic acid; the replacement of a second equivalent would yield $C_2O_2H_2$, or the unknown formic aldehyde; a third, C_2OH_3, the oxide of methyle; and a fourth, C_2H_4, or formene. By substituting methyle for one or more atoms of hydrogen in the previous formula, we obtain those of the corresponding bodies of the vinic series, and it will be readily seen that by introducing the higher alcoholic radicles we may derive from C_2O_4 the formulas of all the alcoholic series. A grave objection to this view is, however, found in the fact that while this compound may be made the type of the aldehydes, acetones, and hydrocarbons, it becomes necessary to assume the hypothetical C_2O_2,HO, as the type of the acids and alcohols. Oxide of carbon, C_2O_2, is, according to Kolbe, to be received as the type of hydrocarbons like olefiant gas (C_2HMe), while C_2O, in which ethyle replaces oxygen, is C_6H_5, or lipyle, the supposed triatomic base of glycerine.

The monobasic organic acids are thus derived from one atom of C_2O_4, while the bibasic acids, like the succinic, are by Kolbe deduced from a double molecule, C_4O_8, and tribasic acids, like the citric, from a triple molecule, C_6O_{12}. He moreover compares sulphuric acid to carbonic acid, and derives from it by substitution the various sulphuric organic compounds. Ammonia, arseniuretted and phosphuretted hydrogen, are regarded as so many types; and by an extension of his view of the replacement of oxygen by electro-positive groups, the ethylides, ZnEt, $PbEt_2$, and $BiEt_3$, are, by Kolbe, assimilated to the oxides, ZnO, PbO_2, and BiO_3.

Ad. Wurtz, in the Repertoire de Chimie Pure for October, 1860, has given an analysis of Kolbe's memoir (to which, not having the original before me, I am indebted for the preceding sketch), and follows it by a judicious criticism. While Kolbe adopts as types a number of mineral species, including the oxides of carbon, of sulphur, and the metals, Wurtz would maintain but three, hydrogen (H_2), water (H_2O_2), and ammonia (NH_3); and these three types, as he endeavored to show in

1855, represent different degrees of condensation of matter. The molecule of hydrogen, $H_2 = (M_2)$, corresponding to four volumes, combines with two volumes of oxygen (O_2) to form four volumes of water, and may thus be regarded as condensed to one half in its union with oxygen, and derived from a double molecule, M_2M_2. In like manner four volumes of ammonia contain two volumes of nitrogen and six of hydrogen, which, being reduced to one third, correspond to a triple molecule, M_3M_3, so that these three types and their multiples are reducible to that of hydrogen more or less condensed. (Wurtz, Annales de Chimie et de Physique (3), XLIV. 304.)

As regards the rejection of water as a type of organic compounds, and the substitution of carbonic acid, founded upon the consideration that these in nature are derived from C_2O_4, Wurtz has well remarked that water, as the source of hydrogen, is equally essential to their formation, and, indeed, that the carbonic anhydride C_2O_4, like all other anhydrous acids, may be regarded as a simple derivative of the water-type. Having then adopted the notion of referring a great variety of bodies to a mineral species of simple constitution, water is to be preferred to carbonic anhydride, first, because we can compare with it many mineral compounds which can with difficulty be compared with carbonic acid; and, secondly, because the two atoms of water being replaceable singly, the mode of derivation of a great number of compounds (acids, alcohols, ethers, etc.) is much more simple and natural than from carbonic acid. As Wurtz happily remarks, Kolbe has so fully adopted the theory of types, that he wishes to multiply them, and even admits condensed types, which are, however, molecules of carbonic acid, and not of water; "he combats the types of Gerhardt and at the same time counterfeits them."

Thus far we are in accordance with Mr. Wurtz, who has shown himself one of the ablest and most intelligent expounders of this doctrine of molecular types as above defined: — now almost universally adopted by chemists. He writes, "To my mind this idea of referring to water, taken as a type, a very great number of compounds, is one of the most beautiful conceptions of modern chemistry." (Repertoire de Chimie Pure, 1860, page 359.) And again, he declares that the idea of regarding both water and ammonia as representatives of the hydrogen-type more or less condensed, to be so simple and so general in its application that it is worthy "to form the basis of a system of chemistry." (Ibid., page 356.)

We have in this theory two important conceptions: the first is that of hydrogen and water regarded as types to which both mineral and organic compounds may be referred; and the second is the notion of condensed and derived types, according to which we not only assume two or three molecules of hydrogen or water as typical forms, but even look on water as the derivative of hydrogen, which is itself the primal type.

As to the history of these ideas, Wurtz remarks that the proposition enunciated by Kolbe, that all organic bodies are derived by substitution from mineral compounds, is not new, but known in the science for about ten years. "Williamson was the first who said that alcohol, ether, and acetic acid were comparable to water, — organic waters. Hoffman and myself had already compared the compound ammonias to ammonia itself. . . . . To Gerhardt belongs the merit of generalizing these ideas, of developing them, and supporting them with his beautiful discovery of anhydrous monobasic acids. Although he did not introduce into the science the idea of types, which belongs to M. Dumas, he gave it a new form, which is expressed and essentially reproduced by the proposition of Kolbe. Gerhardt reduced all organic bodies to four types, — hydrogen, hydrochloric acid, water, and ammonia." (Ibid., page 355.)

The historical inaccuracies of the above quotation are the more surprising, since in March, 1854, I published in the American Journal of Science (XVII. 194) a concise account of the progress of these views. This paper was republished in the Chemical Gazette (1854, page 181), and copies of it were by myself placed in the hands of most of the distinguished chemists of England, France, and Germany. In this paper I have shown that the germ of the idea of mineral types is to be found in an essay of Auguste Laurent (Sur les Combinaisons Azotées, Ann. de Chimie et Physique, November, 1846), where he showed that alcohol may be looked upon as water (H_2O_2) in which ethyle replaces one atom of hydrogen, and hydric ether as the result of a complete substitution of the hydrogen by a second atom of ethyle. Hence he observed that while ether is neutral, alcohol is monobasic and the type of the monobasic vinic acids, as water is the type of bibasic acids. In extending and developing this idea of Laurent's, I insisted in March, 1848, and again in January, 1850, upon the relation between the alcohols and water as one of homology, water being the first term in the series, and H_2 being in like manner the homologue of acetene and formene; while the bases of Wurtz were said to "sustain to their corresponding alcohols the same relation that ammonia does to water." (Am. Journal of Science, V. 265; IX. 65; XIII. 206.)

In a notice of his essay, published in September, 1848 (Ibid., VI. 173), I endeavored to show that Laurent's view might be further extended, so as to include in the type of water "*all those saline combinations (acids) which contain oxygen*"; and in a paper read before the American Association for the Advancement of Science at Philadelphia, in September, 1848, I further suggested that as many neutral oxygenized compounds which do not possess a saline character are derivatives of acids which are referable to the type H_2O_2, "*we may regard all oxygenized bodies as belonging to this type*," which I further showed in the same essay to be but a derivative of the primal type H_2. To this I referred all hydrocarbons and their chlorinized derivatives, as also the volatile alkaloids, which were regarded as "amidized species" of the hydrocarbons, in which the residue amidogen, NH_2, replaces an atom of H or Cl, or, what is equivalent, the residue NH is substituted for O_2 in the corresponding alcohols. (Ibid., VIII. 92.)

In the paper published in September, 1848, I showed that while water is bibasic, the acids which like hypochlorous and nitric acids were derived from it by a simple substitution of Cl and NO_4 for H, were necessarily monobasic, and I then pointed out the possible existence of the nitric anhydride $(NO_4)_2O_2$, which was soon after discovered by Deville. Gerhardt at this time denied the existence of anhydrides of the monobasic acids, regarding anhydrides as characteristic of polybasis acids, and indeed was only led to adopt my views by the discovery of the very anhydrides whose formation I had foreseen.*

In explaining the origin of bibasic acids I described them as produced by the replacement, in a second equivalent of water, of an atom of hydrogen by a monobasic saline group; thus sulphuric acid would be $(S_2HO_6H)O_2$. Tribasic acids in like manner are to be regarded as derived from a third equivalent of water in which a bibasic residue replaces an atom of hydrogen. The idea of polymeric types was further illustrated in the same paper, where three hydrogen types were proposed (HH), (H_2H_2), and (H_3H_3), corresponding to the chlorides MCl, MCl_3, and MCl_5. It was also illustrated by sulphur in its ordinary state, which I showed is to be regarded as a triple molecule S_3 (or $S_6 = 4$ volumes), and I referred sulphurous acid SO_2 to this type, to which also probably belongs selenic oxide. (At the same time I suggested that the odorant form of oxygen or ozone was possibly O_3.) Wurtz, in his memoir, published in 1855, adopts my view, and makes sulphur vapor at 400° C., the type of the triple molecule. I further suggested (American Journal of Science, V. 408; VI. 172) that gaseous nitrogen is NN, an anhydride amide or nitryl, corresponding to nitrite of ammonia $(NO_3,NH_4O) - H_4O_4 = NN$. This view a late writer attributes to Gerhardt, who adopted it from me. (Ann. de Chimie et Phys., LX. 381.) May not nitrogen gas, as I have elsewhere suggested, regenerate under certain conditions, ammonia and a nitrite, and thus explain not only the frequent formation of ammonia in presence of air and reducing agents, but certain cases of nitrification?†

* The anhydrides of the monobasic acids correspond to two equivalents of the acid, minus one of water, as $2(C_4H_4O_4) - H_2O_2 = C_8H_6O_6$; while one equivalent of a bibasic acid (itself derived from $2H_2O_2$) loses one of water, and becomes an anhydride, as $C_2H_2O_6 - H_2O_2 = C_2O_4$. So that both classes of anhydrides are to be referred to the type of one molecule of water, H_2O_2.

† The formation of a nitrite in the experiments of Cloez appears to be

I endeavored still further to show that hydrogen is to be looked upon as the fundamental type, from which the water-type is derived by the replacement of an atom of H by the residue HO_2. (American Journal of Science, VIII. 93.) In the same way I regarded ammonia as water in which the residue NH replaced O_2.

I have always protested against the view which regards the so-called rational formulas as expressing in any way the real structure of the bodies which are thus represented. These formulas are invented to explain a certain class of reactions, and we may construct from other points of view other rational formulas which are equally admissible. As I have elsewhere said, "the various hypotheses of copulates and radicles are based upon the notion of dualism, which has no other foundation than the observed order of generation, and can have no place in a theory of the science." All chemical changes are reducible to union (identification) and division (differentiation). When in these changes only one species is concerned, we designate the process as metamorphosis, which is either by condensation or by expansion (homogeneous differentiation). In metagenesis, on the contrary, unlike species may unite, and by a subsequent heterogeneous differentiation give rise to new species, constituting what is called double decomposition, the results of which, differently interpreted, have given origin to the hypothesis of radicles and the notion of substitution by residues, to express the relations between the parent bodies and their progeny. The chemical history of bodies is then a record of their changes; it is in fact their genealogy, and in making

independent of the presence of ammonia, and to require only the elements of air and water. (Comptes Rendus, LXI. 135.) Some experiments now in progress lead me to conclude that the appearance of a nitrite in the various processes for ozone is due to the power of nascent oxygen to destroy by oxidation the ammonia generated by the action of water on nitrogen, the nitrous nitryl; so that the odor and many of the reactions assigned to ozone or nascent oxygen are really due to the nitrous acid which is set free when the former encounters nitrogen and moisture. On the other hand, nascent hydrogen, which readily reduces nitrates and nitrites to ammonia, by destroying the regenerated nitrite of the nitryl, produces ammonia in many cases from atmospheric nitrogen. [See Appendix, page 470.]

use of typical formulas to indicate the derivation of chemical species, we should endeavor to show the ordinary modes of their generation. [See the preceding papers XVI. and XVIII.]

Keeping this principle in mind, let us now examine the theory of the formation of acids. As we have just seen, I taught in 1848 that the monobasic, bibasic, and tribasic acids are derived respectively from one, two, and three molecules of water, H_2O_2. Mr. Wurtz, seven years later, put forth an analogous view. He however supposes a monatomic radicle PO'_4, a diatomic radicle PO''_3, and a triatomic radicle PO'''_2, replacing respectively one, two, and three atoms of hydrogen in H_2O_2, H_4O_4, and H_6O_6; thus $(PO'_4H)O_2$, $(PO''_3H_2)O_4$, and $(PO'''_2H_3)O_6$. These radicles evidently correspond to PO_5, which has lost one, two, and three atoms of oxygen in reacting upon the hydrogen of the water-types, and the acids may be accordingly represented as formed by the substitution of the residues $PO_5 - O$ for H, $PO_5 - O_2$ for H_2, and $PO_5 - O_3$ for H_3.

To this manner of representing the generation of polybasic acids we object that it encumbers the science with numerous hypothetical radicles, and that it moreover fails to show the actual successive generation of the series of acids in question. When phosphoric anhydride is placed in contact with water, it combines with one equivalent. The union is followed by homogeneous differentiation, and two equivalents of metaphosphoric acid result. Two equivalents of this acid with one of water at ordinary temperatures are slowly transformed into two of pyrophosphoric acid, by a reaction precisely similar to the last; while two equivalents of pyrophosphoric acid when heated with a third equivalent of water yield, in like manner, two of tribasic phosphoric acid. The generation of the three acids may be represented as follows:—

$$2(PO_5) \text{ or } (PO_4)_2O_2 + H_2O_2 = 2(PO_4H)O_2 \text{ or } 2(PHO_6)$$
$$2(PHO_6) \text{ or } (PHO_5)_2O_2 + H_2O_2 = 2(PHO_5H)O_2 \text{ or } 2(PH_2O_7)$$
$$2(PH_2O_7) \text{ or } (PH_2O_6)_2O_2 + H_2O_2 = 2(PH_2O_6H)O_2 \text{ or } 2(PH_3O_8)$$

Gerhardt long since maintained that we cannot distinguish between polybasic salts and what are called sub-salts, which are as truly neutral salts of a particular type. Thus the bibasic and tribasic phosphates are to be looked upon as sub-salts, which sustain the same relation to the monobasic phosphates that the basic nitrates bear to the neutral nitrates. He succeeded in preparing two crystalline sub-nitrates of lead and copper, having the formulas NO_5,M_2O_2,HO (tribasic), and NO_5,M_4O_4,H_3O_3 (heptabasic), both of which retain their water of composition at 392° F.

The compounds of sulphuric acid are: 1. The true monobasic sulphate, S_2O_6MO, corresponding to the Nordhausen acid and the anhydrous bisulphates; 2. The ordinary neutral sulphates, S_2O_6,M_2O_2; 3. The so-called disulphates, S_2O_6,M_4O_4 corresponding to the glacial acid of density 1.780; 4. The type S_2O_6,M_6O_6, represented by turpeth mineral; 5. The so-called quadribasic sulphates, S_2O_6,M_8O_8. The copper-salt of this octobasic type still retains, moreover, 6HO at 392° F. (Gerhardt on Salts, Jour. de Pharmacie, 1848, Vol. XII.; American Journal of Science, VI. 337.) Without counting the still more basic sulphates described by Kane and Schindler, we have the following salts, which, in accordance with Wurtz's notation, correspond to the annexed radicles: —

1.	Unibasic	$S_2HO_7 = S_2O_5$	monatomic.
2.	Bibasic	$S_2H_2O_8 = S_2O_4$	diatomic.
3.	Quadribasic	$S_2H_4O_{10} = S_2O_2$	tetratomic.
4.	Sexbasic	$S_2H_6O_{12} = S_2$	hexatomic.
5.	Octobasic	$S_2H_8O_{14} = S_2 - O_2$	octatomic.

It is easy to apply a similar *reductio ad absurdum* to the radicle theory in the case of the oxychlorides and other basic salts, and to show that the radicles of the dualists are often merely algebraic expressions. (See further my remarks in the American Journal of Science, VII. 402 – 404.)*

* Those who are familiar with chemical literature will remember an amusing *jeu d'esprit* of Laurent's, in which he invited the attention of the advocates of the radicle theory to a newly invented electro-negative radicle, eurhizene. (Comptes Rendus des Travaux de Chimie for 1850, pages 251 and 376.) A late writer in the Chemical News (Vol. I. page 326) proposes, as a new electro-negative radicle, under the name of hydrine, the peroxide of hydrogen HO_2, the eurhizene of Laurent.

The mode of the generation of acids set forth in the case of those derived from phosphoric anhydride, which we conceive to be a simple statement of the process as it takes place in nature, dispenses alike with hypothetical radicles and residues, both of which are, however, convenient for the purposes of notation. In the selection of a typical form to which a great number of species may be referred, hydrogen or water merits the preference from its simplicity, and from the important part which it plays in the generation of species. Water and carbonic anhydride are both so directly concerned in the generation of the bodies in the carbon series, that either may be assumed as the type; but we prefer to regard C_2O_4, like the other anhydrides, as only a derivative of the type of water, and ultimately of the hydrogen-type.

These views were first put forward by myself in 1848, when I expressed the opinion that they were destined to form "the basis of a true natural system of chemical classification"; and it was only after having opposed them for four years to those of Gerhardt, that this chemist, in June, 1852, renounced his views, and, without any acknowledgment, adopted my own. (Ann. de Chim. et Phys. (3), XXXVII. 285.) Already in 1851, Williamson, in a paper read before the British Association, had developed the ideas on the water-type to which Wurtz refers above, and to him the English editor of Gmelin's Handbook ascribes the theory. The notion of condensed types, and of H_2 as the primal type, was not, so far as I am aware, brought forward by either of these, and remained unnoticed until resuscitated by Wurtz in 1855, seven years after I had first announced it, and one year after my reclamation already noticed, which was published in the American Journal of Science, in March, 1854.

My claims have not, however, been overlooked by Dr. Wolcott Gibbs. In an essay on the polyacid bases, he remarks that in a previous paper he had attributed the theory of water-types to Gerhardt and Williamson, and adds: "In this I find I have not done justice to Mr. T. Sterry Hunt, to whom is exclusively due the credit of having first applied the theory to the so-called oxygen-acids and to the anhydrides, and in whose earlier papers may be found the germs of most of the ideas on classification usually attributed to Gerhardt and his disciples." (Proc. Am. Assoc. Adv. Science, 1858, page 197.) It will be seen, from what precedes, that I not only applied the theory, as Dr. Gibbs remarks, but, except so far as Laurent's suggestion goes, invented it and published it in all its details some years before it was accepted by a single chemist.

In conclusion, I have only to ask that future historians will do justice to the memory of Auguste Laurent, and will, moreover, ascribe to whom is due the credit of having given to the science a theory which has exercised such an important influence in modern chemical speculation and research; remembering that my own publications on the subject, which cover the whole ground, were some years earlier than those of Williamson, Gerhardt, Wurtz, or Kolbe.

APPENDIX.

ON THE THEORY OF NITRIFICATION.

In connection with the foot-note on page 465 the following sketch of the theory of nitrification there indicated seems called for, the more especially as it will be seen that the late Professor G. C. Schaeffer of Washington apparently anticipated me in certain points therein. It was in the Amer. Jour. Science for May, 1848 (page 408), that I referred to Gerhardt's observation that the so-called protoxide of nitrogen corresponds to biphosphamide, PNO, and is NNO, a nitryl derived from nitrate of ammonia by the removal of $2H_2O$, and capable, when heated in contact with an alkaline hydrate, of regenerating ammonia and a nitrate. I then called attention to the similar decomposition of nitrite of ammonia, which by the loss of $2H_2O$ yields nitrogen gas, and remarked that the gas thus obtained, "apparently identical with that of the atmosphere, is really *composed of two equivalents of the element sustaining to each other the same relations as in nitrous oxide*," or in other words representing respectively the nitrous and the ammoniacal conditions. This view of the constitution of gaseous nitrogen was again set forth, in September, 1848, in the paper quoted above, as a means of explaining the apparent anomaly in the equivalent volume of nitrogen. The obvious conclusion that gaseous nitrogen might (after the manner of nitrous oxide) regenerate ammonia and a nitrite by assuming the elements of water, $2H_2O$, was not insisted upon. It was, however, for years so familiar to me and so often set forth in my lectures on chemistry before the medical classes at the Université Laval at Quebec, that I spoke of it in the above paper in March, 1861, as a view which I had elsewhere suggested, though this was, I believe, the first time that it had been enunciated by me in print. In further explanation of the subject I published in the Amer. Jour. Science for July, 1861 (page 109), a note in which, after describing the generation of ozone or active oxygen by passing air through a solution of permanganic acid, and the production of a nitrite from air thus ozonized, I referred to the conversion of gaseous nitrogen, as above, into ammonia and nitrous acid, and added: "From the instability of the compound of these two bodies, however, it becomes necessary to decompose the one at the instant of its formation in order to isolate the other. Certain reducing agents which convert nitrous acid into ammonia may thus transform nitrogen (NN) into $2NH_3$. In this way I explain the action of nascent hydrogen in forming ammonia with atmospheric nitrogen in presence of oxidizing metals and alkalies. . . . . An agent which, instead of attacking the nitrous acid, should destroy the newly formed ammonia, would permit us to isolate the nitrous acid. Houzeau has shown that active oxygen is such an agent, at once oxidizing ammonia with formation of nitrate (nitrite) of ammonia ; and thus when ozone is brought in contact with moist air, both of the atoms of nitrogen in the nitryl (NN) appear in the oxidized state. From this view it follows that the odor and many of the reactions ascribed to ozone are due to nitrous acid, which is liberated by the decomposition of atmospheric nitrogen in presence of water and nascent oxygen. We have thus the key to a new theory of nitrification and to the experiments of Cloez on the slow formation of a nitrite by the action of air exempt from ammonia upon porous bodies moistened with alkaline solutions."

On September 15, 1862, I read before the French Academy of Sciences a note on The Nature of Nitrogen and the Theory of Nitrification, published in the Comptes Rendus of that date and translated in the Philosophical Magazine for January, 1863, in which I repeated the points above given, and then proceeded to consider the results announced by Schönbein in 1862. I said : " The formation of nitrite of ammonia by the combination of the nitryl NN with H_4O_2 must necessarily be limited to very minute quantities by the instability of this ammoniacal salt, which, as is well known, decomposes readily into nitrogen and water. In order, therefore, to produce any considerable quantity of a nitrite by this reaction, there is required the presence of active oxygen, or of a fixed base to separate the ammonia. The recent experiments of Schönbein have furnished new evidences of the direct formation of a nitrite at the expense of the nitrogen of the atmosphere. According to him, when sheets of paper moistened with a feeble solution of an alkali or an alkaline carbonate are exposed to the air, especially in the presence of watery vapor and at a temperature of 50° or 60° C., the alkaline base soon fixes a sufficient quantity of nitrous acid to give the characteristic reactions. Appreciable traces of nitrite are, according to Schönbein, obtained in this way, even without the intervention of an alkali. He moreover found that distilled water mixed with a little potash or sulphuric acid, and evaporated slowly at a temperature of about

50° C., in the open air, fixes in one case a small portion of ammonia and in the other a little nitrous acid. Traces of a nitrite are also formed in pure water under similar conditions. Schönbein explains all of these results by the combination of nitrogen with the elements of the water, producing at the same time ammonia and nitrous acid. As he has well remarked, this reaction serves to explain the absorption of nitrogen by vegetation, and, through the oxidation of nitrites, the formation of nitrates in nature. By these elegant experiments he has confirmed in a remarkable manner my theory of nitrification and of the double nature of free nitrogen. It is, however, evident that since the publication of my note of March, 1861, above referred to, we cannot say, with Schönbein, that the generation of nitrite of ammonia from nitrogen and water is 'a most wonderful and wholly unexpected thing.' (Letter from Schönbein to Faraday, Philos. Magazine, June, 1862, page 467.)" Referring to the claims of Schönbein, and to my notes of March and July, 1861, the late Professor Nicklès wrote as follows in 1863, in his scientific correspondence for the American Journal of Science ((2) XXXV. 263): "Schönbein has done justice tardily to those who have preceded him in this question. Of this number is T. Sterry Hunt, who, as our readers may remember, long since showed that nitrite of ammonia may be formed by means of nitrogen and water, and thus led the way to a new theory of nitrification. This is what Böttger arrived at, who first announced that nitrite of ammonia is a product of all combustion in the air." With regard to the production of nitrite of ammonia from nitrogen and water, he further adds, "this point was entirely developed by Sterry Hunt."

The publication of the above called forth a communication from Professor G. C. Schaeffer in the Amer. Jour. Science for November, 1863, page 409, in which he draws attention to the fact that the Report of the Smithsonian Institution for 1861 contains an essay on Nitrification by Dr. B. F. Craig (written in 1856), in which the latter puts forth as the suggestion of Professor Schaeffer the same theory of nitrification as that maintained by the present writer and by Schönbein ; basing it upon the view that nitrogen gas is a nitryl capable of regenerating nitrite of ammonia in presence of water. From this it is clear that Professor Schaeffer had independently attained the same conclusion as myself from the conception of the dual nature of atmospheric nitrogen, which I had taught since 1848. He at the same time, as a contribution to the literature of the subject, called attention to his paper in the Proceedings of the American Association for the Advancement of Science for 1850, on the Detection of Nitrites and Nitrates, in which he indicated a delicate test for these salts, showed the frequent presence of nitrites in rain-water, and moreover pointed out that while nitrates are readily reduced to nitrites in solution, these, by oxidation, pass as readily into nitrates.

INDEX.

ACCUMULATION of sediments, effects of, 17, 49, 58, 66.
Acid springs, 111; of New York and Ontario, 130, 131.
Acids of volcanoes, their origin, 8, 15, 111, 112.
Adams, C. B., on the geology of Vermont, 391.
Adirondack Mountains, rocks of, 32, 241, 243.
Aerolites, constitution of, 302.
Agalmatolite rocks, 67.
Albertite, composition of, 176.
Albite, in Laurentian veins, 214; formula of, 443.
Albuminoids, constitution and artificial production of, 170.
Algæ. *See* Sea-weeds.
Alkalies, relative proportions in waters, 102; of mineral waters, 135. *See* Carbonate of Soda and Potash.
Alkaliferous silicates, decomposition of, 2, 10, 40, 102, 103.
Alkaline silicates, soluble, 7, 21, 25.
Alkaline waters, 85, 123, 156.
Alleghany River, brines of, 121.
Allomerism, 447.
Alps, geology of, 328; anthracitic system of, 332; grand section of, 343.
Alteration of rocks. *See* Metamorphism.
Alum slates of Sweden, 266, 366.
Alumina, solution and deposition of, 13, 14, 98, 142; sulphate of, 98, 133; in waters, 142. *See* Bauxite.
Aluminous silicates, origin of, 28, 296, 298.
Ammonia of volcanoes, 8, 15, 113; in rocks and soils, 113; nitrite of, its production, 471.
Andalusite rocks, 28, 32, 34, 243, 272, 282, 408.
Andrews, E. B., on petroleum, 174.
Angelin, Palæontologica Scandinavica, 367.
Anglesea, crystalline schists of, 270, 353, 383.
Anhydrites of the Alps, 335.
Anhydrous monobasic acids, 464.
Anorthite, its formula, 443.
Anortholite, 31, 32.
Anthracite, its origin, 177; of the Alps, 332, 334.
Anticlinals, their relations to mountains, 53.
Anticosti, geology of, 416.
Anticosti group, 417.
Apatite, 197, 208, 211, 213, 311.
Appalachians, geology of, 50, 51, 75, 241.
Aquatic vegetation, 2, 22, 95.
Arendal, vein-stones of, 209.
Arenig rocks, 375, 376, 381, 384.
Arkesine, 330.
Arkose, 285.
Artesian wells of London and Paris, 85.
Aspidella Terranovica, 410.
Atmosphere, primeval, 1, 20, 40, 42, 47, 301.
Atmospheric waters, 94.
Atomic hypothesis, 433, 438.
Atomic volumes, 433, 435, 440, 455.
Attrition of rocks, 20.
Auroral rocks, 247, 421; their relation to matinal, 414.
Azoic gneisses of Rogers, 246.

BABBAGE on internal heat, 14, 71.
Bala rocks, 353, 359, 362.
Bailey, L. W., geology of New Brunswick, 407.
Banded structure in veins, 193, 199.
Bangor group, 353, 382, 384.
Bark, its composition, 181.
Barrande on palæozoic geology, 253, 368, 369, 378, 392, 424.
Baryta, salts of, in waters, 87, 121, 141, 145.
Basic salts, Gerhardt on, 467.
Bauxite, 14, 98, 326.
Belœil, Quebec, water of, 151.
Belt, T., on Lingula flags, 371.
Bergmann on Mont Blanc, 338.
Bertrand on Mont Blanc, 338.
Berzelius on silicate of lime, 151.
Beryl, 199, 245; kaolin of, 101; a feldspathide, 445.
Bessarabia, salt lagoons of, 86.
Bicarbonates. *See* Carbonates.
Biddeford, Maine, granitic veins of, 198.
Bigsby, J., on Huronian rocks, 18, 269; on Cambrian, 269; on the geology of Quebec, 396, 400.
Billings, E., on the geology of Vermont, 260-265, 391-393; on the Potsdam rocks, 266; on Levis fossils, 258, 400, 403, 404, 412; on Eophyton, 409; on the Anticosti group, 416; on Middle Silurian, 417.
Bischof, G., 16; on a source of sulphuretted hydrogen, 87; on decomposition of silicates, 102, 151; on formation of silicate of magnesia, 122; on anthracite, 177; on deoxidation in nature, 302; on pseudomorphism, 287, 290, 293, 294, 323, 325; his plutonic basis, 294.
Bismuth, occurrence of, 200, 217.
Bitterns related to mineral waters, 103, 105, 109, 114, 117, 121, 156, 163.
Bitumens, 8, 169, 175, 382, 397. *See* Petroleum.
Bituminous rocks. *See* Pyroschists.
Blake, W. P., on Laurentian veins, 215, 218.
Blue Ridge, gneisses of, 217, 249; their decomposition, 250; their copper veins, 217, 250.
Blum on pseudomorphism, 287, 319, 325.
Bohemia, copper slates of, 232; geology of, 368.
Borates, their origin, 16, 112, 146.
Borax-lake, water of, 146.
Bosanquet, Ontario, pyroschists of, 179.
Bothwell, Ontario, water of, 159, 162.
Böttger on nitrification, 472.
Bouë on metamorphism, 24, 321.
Brainard, J., on silicious deposits, 89.
Braintree, Mass., Paradoxides of, 405.
Bray Head, rocks of, 382.
Brazil, crystalline rocks of, 278; their decay, 10.
Breaks in palæozoic series, 263, 375-377, 412-415, 418.
Breislak on the origin of sulphur, 87.
Brines, analyses of, 119-121.
Brittany, crystalline schists of, 273.
Bromine in waters, 142.
Brooks, T. B., crystalline rocks of Michigan, 274.
Brunswick, Maine, granite veins of, 194.
Buch, Von, on dolomites, 81, 309.
Buffon on mountains, 52.
Bunsen on eruptive rocks, 3, 66, 284; on aqueous decomposition of silicates, 102.

CAERNARVONSHIRE, crystalline rocks of, 269, 353, 383.
Calciferous sand-rock, a dolomite, 117, 155, 415; gypsum in, 117, 155; its relations to Trenton and Chazy, 412, 413.
Cagniard de la Tour on vapors, 37.
Calcium, chloride of, in waters, 117-120, 158.
Calcium, salts of. *See* Carbonate of lime, Lime-salts, and Gypsum.
Caledonia, Ontario, waters of, 123, 127, 129, 147-149.
California, borax-lake of, 146.
Calumet and Hecla mine, conglomerate of, 187.
Cambrian series, 266-269; Upper, in Great Britain, 350-365; Middle and Lower, 365-385, 409; in North America, 387-425; history of, 349.

Cambro-Silurian of Sedgwick, 363, 381, 423.
Canada geological survey, reports of, 420.
Cape Ann, Mass., granite veins of, 200.
Cape Breton, water of, 121.
Caradoc rocks, 353, 359 - 362, 384.
Carbon, its primitive condition, 23, 42, 302; anthracitic of Madoc, 217.
Carbonates. *See* Carbonate of lime, Carbonate of magnesia, Carbonate of soda, and Carbonic acid.
Carbonate of lime, its origin, 2, 23, 41, 47, 81, 83, 86, 88, 90, 109; solubility and supersaturated solutions of, 139; bicarbonate, its action on sea-water, 82, 85, 90, 109, 308; hydrous carbonate of, 140.
Carbonate of lime and magnesia. *See* Dolomite.
Carbonate of magnesia, its origin, 23, 82, 85, 88, 90, 109, 110; action of, on lime-salts, 87, 90, 139; its solubility and supersaturated solutions of, 140, 148; bicarbonate of, its solubility, 91, 109, 148; hydrous carbonate of, present in some dolomites, 107; sesquicarbonate of, 138.
Carbonate of soda, its origin, 12, 21, 85, 102; amount of, in waters, 85, 124 - 126; neutral carbonate of, 148; action of, on sea-waters, 2, 11, 12, 41, 85, 88, 90, 105, 139, 148, 307.
Carbonic acid, its action on silicates, 2, 10, 102, 150; amount of, in early atmosphere, 41, 47, 308; deoxidation of, 23, 42, 302; deficiency of, in certain waters, 149; relations of, to life and climate, 42, 46 - 48; to the formation of gypsum, 42, 308; subterranean sources of, 8, 15, 112.
Carboniferous rocks, 228; of North America, 49, 50.
Carbon-spars, their constitution, 441, 446.
Carlsbad, waters of, 85.
Carnallite, 105, 107, 118.
Cassiterite, 191, 192, 195, 200, 205; pseudomorphs of, 289, 290.
Catalysis, 452.
Celestial chemistry, 35, 37.
Chabazite, 442; action of saline waters on, 96.
Chacornac on the nebular hypothesis, 38.
Chambly, waters of, 125, 149, 152.
Champlain division, 252, 258, 264, 266.
Chamonix, jurassic rocks of, 338; synclinal of, 343.
Chatham, Ontario, water of, 145.
Chazy formation, 156, 414, 415; absent in Herkimer Co., New York, 413; relations of, to Calciferous and Trenton, 412; mineral waters from, 124, 156, 157.
Chemical change defined, 428, 450, 454, 465; elements, 37, 428; activities in former ages, 27, 42, 306; dissociation, 36.
Chemistry defined, 454.
Cheshire rock-salt, 120.
Chiastolite rocks. *See* Andalusite.
Chicago, oil-bearing limestone of, 172.
Chloride of calcium in waters, 122, 158; in primeval ocean, 11, 117, 122, 137.
Chloride of magnesium, 117, 122, 137.
Chloride of sodium, its origin, 2, 11, 41.
Chlorine in silicates, 144, 242.
Chlorine and hydrogen, union of, 430.
Chlorite, its probable origin, 296.
Chloritic rocks, 32, 243, 247, 249, 269, 270, 330, 331, 408; supposed pseudomorphic origin of, 316, 320, 326.
Chloritoid rocks, 32.
Chromium, its occurrence, 31, 32, 34, 238, 243, 249, 269, 270, 272, 297, 330.
Chrysolite and serpentine, 291, 315.
Chrysoberyl, 195, 214.
Circulation, terrestrial, 22, 225, 235.
Classification of the sciences, 35, 453.
Clays, origin of, 2, 10, 13, 20, 22, 41, 101, 228; precipitated by saline waters, 10.
Climate, primeval, 42, 46 - 48; palæozoic of North America, 76, 92, 310.
Cloez on nitrification, 465, 471.
Coal, its origin, 180, 182, 229; its relation to iron-ores, 229.
Collingwood, Ontario, pyroschists of, 178.
Colloidal bodies, solution of, 223.
Condensed types in chemistry, 468.

Concentration of metals in nature, 227, 235.
Concretionary structure, 89.
Connecticut, gneisses of, 248.
Conocephalites in North America, 260, 391, 404.
Continent, a pre-palæozoic, 75, 76.
Continental elevation, 53, 76.
Conularia, a phosphatic shell, 312.
Cooke, J. P., on allomerism, 447.
Cooling globe, its chemistry, 1, 38, 40, 60, 63, 301, 306.
Coös group, 282.
Copper-ores, origin of, 232; of Blue Ridge, 217.
Coprolites, 152, 225.
Cordier on limestones and dolomites, 81.
Corundum, 247; its supposed transformations, 326.
Cotta, Von, on granitic veins, 191.
Credner, H., on Eozoic rocks of North America, 277; on comparative geognosy, 278; on the origin of silicates, 304, 305.
Crinoids, fossil, injected with silicates, 304.
Croft, H., on various mineral waters, 130, 134, 145.
Crust of the earth, 1, 40, 60-64, 223; its flexibility, 8, 15, 57, 72; corrugations of, 57, 74; its disintegration, 63.
Crystalline aggregation of matter, 305.
Crystalline rocks, two great classes, 283; evidences of their plasticity, 4; how formed, 24, 283; evidences of life in, 13, 302.
Crystalline schists, relative ages of, 19; are pre-Cambrian, 327; origin of, 283; supposed plutonic, 294; Daubrée on, 301; Gümbel on, 305; Credner on, 305; Favre on, 347.
Crystals, rounded, 212; hollow or skeleton, 201, 212.
Cumberland, England, crystalline schists of, 273.
Cyanite rocks, 28, 34, 243, 272.
Cycles in sedimentation, 155, 241.

Dalman on trilobites, 365.
Damour, A., action of water on zeolites, 102; on jadeite, 446.
Dana, J. D., on the fluidity of the earth's interior, 56; on granite veins, 199; on pseudomorphism, 287, 291, 318, 319, 320-323; on regional metamorphism, 291, 320, 322; on equivalent volumes, 433.
Danville, Maine, granite veins of, 197.
Daubeny on volcanoes, 62.
Daubrée on the action of heated waters, 6; on the attrition of rocks, 20; on the waters of Plombières, 25; on the production of silicates, 25, 297; on silicious deposits, 89; on regeneration of feldspar, 100; on granitic veins, 191, 209; on the origin of crystalline schists, 301; on the primeval atmosphere, 301.
Davy, H., on volcanoes, 62.
Dawson, J. W., on dissolving of iron-oxide from sediments, 13; on the origin of coal, 180-182; on Eozoon Canadense, 302; on palæozoic foraminifera, 411; on the geology of Nova Scotia, 408; on Erian rocks, 419; on Cambrian and Silurian, 424.
Dead Sea, water of, 83.
Decomposition, double, in chemistry, 428, 451.
Decomposition of crystalline rocks, its antiquity, 10, 100, 250.
Delabeche on crystalline rocks, 301.
Delesse, A., on envelopment of minerals, 288, 289, 314, 315; on pseudomorphism, 292, 314-318; his change of views, 316; on the origin of serpentine, 316, 317; on protogine, 330.
Deoxidation in nature, 23, 230, 302.
Deville, H. Ste.-Claire, on dissociation, 37; on river-waters, 84; on crystalline aggregation, 305.
Diabase, 31.
Diagenesis in rocks, 305, 317, 321.
Differentiation, chemical, 450.
Dikes, distinguished from veins, 193, 202.
Diorite, 23, 26, 32, 186, 243, 247, 249, 269, 270, 330, 331, 408.
Disintegration of the primitive crust, 63.
Dissociation, chemical, 37.
Dipyre, 446.

Dolerite, 3, 23, 284; stratiform structure in, 186.
Dolomieu, decay of granite, 10.
Dolomite, origin of, 81, 307; two classes of, with and without gypsum, 87, 88, 309; fresh-water, 88; metalliferous, 88, 309; is not decomposed by gypsum, 106; with hydrate and hydrocarbonate of magnesia, 107; organic remains in, 88, 92; artificial formation of, 90, 91, 307; produced by evaporation in closed basins, 76, 85-88, 92, 101, 310; relations of carbonic acid in the atmosphere to its formation, 43, 308; supposed epigenic origin of, 81, 92, 287, 307, 325; Cordier on, 81; Von Morlot and Marignac on, 308; Von Buch on, 81, 309; Haidinger on, 325.
Donegal, Ireland, crystalline rocks of, 34, 272.
Drops, Guthrie on, 10.
Dualism in chemical theory, 428.
Dublin, Ireland, granite veins of, 199.
Ducktown, Tenn., copper veins of, 217, 250.
Dumas on chemical types, 462.
Dumont on disturbed strata, 334.
Durocher on igneous rocks, 3, 189, 190.
Dynamical geology, some points in, 70.

EARTH, compared to an organism, 236; interior of, whether liquid or solid, 7, 16, 39, 44, 56, 59, 60, 70, 71; its relation to magnetism, 60. *See* Crust of the earth.
Eaton, Amos, classification of rocks, 241; on the rocks of Vermont, 241, 252.
Ebelman, decay of silicious minerals, 100.
Eichhorn, action of saline waters on soils and aluminous double silicates, 95, 96.
Elæolite in granitic veins, 200.
Elements, chemical, distribution of, 221; in other worlds, 36; possible new ones in stars, 37.
Elevation of continents, 15, 17, 53, 76.
Elie de Beaumont on water in igneous rocks, 5, 190; on silicious deposits, 89, on granitic veins, 189; on terrestrial circulation, 225; on Alpine geology, 332, 348.
Emerald veins of New Grenada, 205.
Emery, origin and occurrence of, 13, 98.
Emmons, E., on rounded crystals, 212; on eruptive limestones, 218; on the Green Mountains, 250; on serpentine, 250; on the Taconic system, 251-253, 268, 388-390; on Cambrian, 268; on hypersthene rock, 279; on recomposed rocks, 341; on the geology of New York, 368.
Endogenous rocks, 193, 196-199.
Envelopment of minerals, 288-290, 314.
Eophyton, 385, 409.
Eozoic rocks of North America, 75, 277.
Eozoon Canadense, 302, 303, 326, 342, 411; E. Bavaricum, 368.
Epidermal tissues, their relations to coal, 181.
Epidotic rocks, 32, 243, 249, 408.
Epigenesis, 286, 313, 317.
Equilibrium of pressure, 15, 76.
Equivalent volumes, 433, 436, 440.
Equivalent weights of oxygen and carbon, 176; of compound species, 432, 441; defined, 455.
Erian rocks, 419.
Erosion as related to mountains, 52, 74.
Eruptive rocks. *See* Exotic rocks.
Esmark on norites, 279.
Essex County, New York, norites of, 279.
Euphotide, 330, 334, 445.
Eurhizene of Laurent, 467.
Evans on petroleum, 174.
Exotic rocks, 4, 9, 16, 24, 44, 58, 66, 188-190, 284; banded structure in, 186; local alteration by, 298.

FAHLERZ, 217.
Fairbairn on relations of pressure to fusion, 39.
Fan-like structure of the Alps, 342, 343.
Fariolo, Italy, granites of, 201.

Faults in strata, related to mineral springs, 154, 157.
Favre, Alph., on the geology of the Alps, 328; on metamorphism in the Alps, 342, 347.
Favre and Silbermann, thermo-chemical researches, 436.
Faye, constitution of the sun, 37.
Feldspar-porphyries, 187, 243, 250, 282.
Feldspars, their formation, 6, 25, 27, 100; decay of, 101; triclinic, 31, 67, 279, 443; aqueous origin of, 298, 299; constitution and formulas of, 443.
Feldspathides, 445.
Festiniog group, 353, 371, 379, 384.
Fire-clays, 13, 22, 228.
Fissures, veins in, 202, 203, 233.
Fitzroy, water of, 124, 142, 152.
Flora, fossil of the Alps, 333.
Fluid-cavities in crystals, 65, 205.
Flysch of the Alps, 337.
Foldings in strata, 17, 51, 55 - 57, 74.
Fontainebleau sandstone, 289.
Foraminifera, palæozoic, 411.
Forchhammer on fucoids, 96; on alkaline sulphurets, 99.
Ford, geology of Troy, New York, 407.
Formulas in chemistry, 465.
Foucou on native hydrocarbon gases, 182.
Fouqué on native hydrocarbon gases, 182.
Fournet on kaolinization, 100; on granites, 190; on skeleton crystals, 201; on veins, 202.
Fucoids, geological relations of, 2, 22, 96, 144, 226.
Fusion, when affected by pressure, 65, 66.

Garnet rock, 30.
Gaspé, geology of, 406, 415, 418.
Gas springs, hydrocarbon, 8, 112, 131, 182.
Gastaldi, geology of the Alps, 347.
Gay-Lussac, law of volumes, 438.
Gelatine, formula and constitution of, 180.
Generation of chemical species, 427, 465.
Genesee slates, pyroschists, 178.
Genth, F. A., on gold deposits, 237; on corundum, 326.
Geognosy, 240; comparative, 33, 34, 278.
Geological relations of mineral waters, 154, 156.
Geology, its scope and objects, 239.
Georgia, Vermont, fossils of, 391, 394, 402.
Gerhardt on types in chemistry, 462, 468; on basic salts, 467.
Gibbs, Wolcott, on the water-type in chemistry, 468.
Giekie on the geology of Skye, 281; on Cambrian and Silurian, 424.
Glaciation of rocks, 10.
Glass softened by heated water, 6.
Glauconite, relation of to potash, 2, 13, 136; in organic forms, 303.
Glucose and sea-salt, compound of, 441.
Gneiss defined, 188; granitoid, 185, 188, 206, 243; Laurentian, 206, 243; White Mountain, 188, 244, 282; of the Appalachians, 244 - 250; of Nova Scotia, 408; primitive of Scandinavia, 469.
Goderich, salt-wells of, 204.
Goethe, 287.
Gold, in Madoc, Ontario, 217; its solution and deposition in nature, 232, 237; in sea-water, 237, 238; of North Wales, 383; of Nova Scotia, 408.
Gothland, geology of, 366.
Granite, decay of, 10; not a primitive rock, 43; substratum of, unknown, 33, 43; intervention of water in its formation, 5, 65, 189 - 191; defined, 183; an intrusive rock, 33, 183; stratiform structure in, 186; graphic, 195; its origin. *See* Exotic rocks.
Granites of New England, 186, 188; of New Brunswick and Italy, 201; of the Alps, 331.
Granitic vein-stones, 183, 189 - 209; their aqueous and concretionary origin, 33, 192, 199; banded structure of, 198; mineralogy of, 200, 210; Laurentian, 33, 208; of White Mountain series, 194; of Sherbrooke, Nova Scotia, and of Biddeford, Maine, 198.

Graphite, its probable organic origin, 13, 301; in Laurentian veins, 210, 216; in various rocks, 32, 33, 243-245, 248; in aerolites, 301.
Graptolites of the Levis formation, 258, 396, 399, 412.
Gras on Alpine geology, 332.
Graywacke defined, 350; of Quebec, 396, 397, 401.
Green Mountain rocks, 18, 29, 32, 241, 243, 249, 274.
Grenatides, 445.
Grenville, Quebec, minerals of, 215; section of Chazy at, 414.
Groton, Connecticut, granite of, 186.
Grove on dissociation, 37.
Grüner on filling of veins, 203.
Guano deposits, 225.
Guelph formation, 417.
Gümbel on Eozoon, 303, 304; on metamorphism of rocks, 305; on diagenesis, 305, 321.
Guthrie on drops, 10.
Gypsum, origin of, 43, 86, 90; two modes of formation of, 110; from bicarbonate of lime and sulphate of magnesia, 82, 85-87, 90, 109; intervention of carbonic acid in its production, 43, 308; its action on soils, 97; does not decompose dolomite, 106; is decomposed by hydrous carbonate of magnesia, 107; its solubility in water, insolubility in brines, 83, 85, 91, 107-110, 144; occurrence of, in natural waters, 105, 132; its elimination from, by reduction, 99, 145; of fresh-water origin, 87; in Calciferous sand-rock, 117, 155; in Onondaga formation, 132; in crystalline schists in Sweden, 336; in tertiary in the Alps, 345. *See* Anhydrite.

Haidinger on pseudomorphism, 324.
Hall, James, on sources of palæozoic sediment, 49; on mountains, 51, 53-55, 73; on White Mountain rocks, 271; on Potsdam rocks, 389; on New York geology, 387, 389, 404; on rocks of Georgia, Vermont, 402; on Taconic fossils, 392; on American palæozoic nomenclature, 419.
Hallowell, Ontario, water of, 116, 142.
Halysites in the Trenton limestone, 417.
Harlech rocks, 372, 373, 377, 382.
Hartt, C. F., on the geology of Brazil, 278; of New Brunswick, 406.
Hastings County, Ontario, rocks of, 216, 274.
Haughton on the norites of Skye, 281.
Heat, internal, of the earth, 7, 9, 15, 43, 57, 59-66, 71, 72, 77, 78.
Heer, O., fossil flora of the Alps, 333.
Hegel on the chemical process, 450.
Helderberg. *See* Lower Helderberg.
Hennessey on the earth's crust, 7, 16.
Herkimer County, New York, geology of, 413.
Herschel, J. F. W., on volcanic phenomena, 8, 15, 44, 62.
Hicks on Cambrian geology, 372, 373, 375, 384, 409.
Hisinger, geology of Scandinavia, 366; errors in his works, 258, 366, 395.
Hitchcock, C. H., geology of the White Mountains, 282.
Hoboken, New Jersey, serpentines of, 248.
Hoffmann on Eozoon, 303.
Homologous or progressive series in chemistry, 431, 439, 442.
Hoosic Mountain, Emmons on, 250.
Hopkins on the earth's interior, 7, 16, 44, 60, 64.
Hornblende, its decay, 100; association of, with pyroxene, 215; rocks of, 244, 246. *See* Diorites.
Houzeau on ozone, 471.
How on mineral waters, 121.
Hudson River group, 252, 256, 258, 395, 397, 398, 402, 403; mineral waters from, 116, 124, 156.
Huggins, his spectroscopic studies, 35.
Humboldt on granites, 190.
Huronian rocks, 18, 29, 243, 269, 272, 274; their identity with the Urscheifer, 269. *See* Green Mountain series.
Hutton on metamorphism, 24; on primary schists, 338.
Hydrocarbon gases, 8, 112, 131, 182.
Hydrochloric acid, its volcanic origin, 8, 15, 44; in mineral waters, 111.

Hypersthene rock or hyperite, 29, 31, 279-281. *See* Norites.
Hypozoic rocks, 245, 246.

IDENTIFICATION, chemical, 450.
Idocrase, hollow crystal of, 212.
Igneous rocks, theory of, 1, 3, 4, 5. *See* Exotic rocks.
Indigenous rocks, 33, 193.
Internal heat. *See* Heat, internal.
Interpenetration in chemistry, 428, 450.
Inverted strata in the Alps, 334, 337; at Troy, New York, 407; at Quebec, 413.
Iodate of calcium in sea-water, 237.
Iodine in mineral waters, 143; its relation to earthy sediments, 143, 226; in sea-water, 143, 226, 237; Sonstadt on, 237.
Iolite or dichroite, 28; and aspasiolite, 315; a feldspathide, 445.
Iron in mineral waters, 128, 142.
Iron ores, origin of, 10, 13, 22, 29-31, 97, 227-229, 243; are evidences of life, 13, 302; relations of, to mineral coal, 229. *See* Bauxite.
Iron pyrites, origin of, 230, 232.
Isomorphism, 432, 440; its relations to pseudomorphism, 315; polymeric, 291, 315, 318, 442.

JACKSON, CHARLES T., on the White Mountains, 241, 275.
Jade and jadeite, 445, 446.
Jollyte, 338.
Joly, water of, 126.
Jukes, J. B., on mountains, 74; on Cambrian and Silurian, 424.

KANT on chemical union, 428, 450.
Kaolin, its formation, 10, 99-101, 445.
Keferstein, C., on igneous rocks and volcanoes, 16, 62, 71, 77; on Mont Blanc, 338.
King and Rowney on pseudomorphism, 325.
Kinnekulle, Sweden, geology of, 367.
Kolbe on chemical types, 459.
Kopp, H., on equivalent volumes, 433, 434.

LA BAIE DU FEBVRE, waters of, 124.
Labrador, geology of, 261, 393.
Labradorite rocks, 29, 31, 33, 67, 278-281. *See* Norian rocks.
Lake Elton, water of, 83.
Lambertville, New Jersey, eruptive rocks of, 186.
Lanoraie, water of, 123.
Laurent, A., on divisibility of formulas, 431; on isomorphism, 422; on chemical types, 463.
Laurentian series, 29, 30, 206; evidences of life in, 302; eruptive rocks of, 33; vein-stones of, 208-218.
Laurentian, Upper. *See* Norian.
Laurentides, 243.
Lauzon formation, 259, 401, 411, 413.
LeConte, Joseph, on dynamic geology, 70-76.
Leonhard on eruptive limestones, 218.
Lersch, Hydro-Chemie, 122.
Lesley, J. P., on mountains, 52, 53; on an apparent discordance in lower palæozoic, 414.
Lethæa Suecica, 366.
Leucite, 67,101, 210.
Levis formation, 259, 401, 413; its fauna, 411, 412, 415.
Leymerie on the origin of limestones, 82.
Liassic fossils in veins, 203.
Lignites, 176, 177, 181.
Lime-salts in the modern ocean, 107, 117, 119; in ancient oceans, 2, 11, 41, 82, 108, 109, 117; in mineral waters, 138. *See* Carbonate of lime and Carbonate of magnesia.
Lime, silicates of, 31, 151, 152.
Lime-soda feldspars, their possible origin, 97. *See* Feldspars, triclinic.
Limestones, Laurentian, 206; of White Mountain series, 196, 244; supposed eruptive, 218; origin of, 82, 311; relations of, to organic life, 311. *See* Carbonate of lime.
Limonite, organic matter in, 98.
Lingula, a phosphatic shell, 312.
Lingula flags, 266, 370, 371, 374.
Liquids, equivalent volume of, 436.
Logan, W. E., on Upper Laurentian, 29, 279; on the Appalachians, 257; on the White Mountains, 276; on

lower palæozoic rocks, 262 ; on the geology of Quebec, 256-258, 397-399 ; on the Quebec group, 259, 263, 264, 401, 403; on the geology of Vermont, 260, 394; on geological nomenclature in Canada, 420.
Loire, waters of, 84.
Longmynd rocks, 266, 380, 382.
Loraine shales. *See* Hudson River group.
Lory on the geology of the Alps, 334, 336.
Lower Helderberg rocks, 415, 418.
Lower palæozoic formations, classification of, 267; tabular view of, 386.
Ludlow rocks, 353, 361, 362, 418.
Luxeuil, water of, 205.
Lycopodium, spores of, 181.
Lyell, C., on the cause of plications in strata, 55; on Mont Blanc, 338.

MacCulloch on hypersthene rocks, 279.
Macfarlane, T., on Huronian rocks, 18, 269, 274; on the plutonic origin of crystalline schists, 294.
Macvicar on the constitution of mineral species, 457.
Madoc, gold and carbon of, 217.
Magnesian marls, *see* Sepiolite ; mica, 207; silicates, formation of, 21, 122, 151, 296, 297, 300.
Magnesite, 33, 90, 243.
Magnesium salts in mineral waters, 137, 138; chloride of, 117, 118; sulphate of, 106, 108, 119, 134. *See* Carbonate of magnesia.
Magnetic iron ore, in vein-stones, 214; veins in, 215. *See* Iron ores.
Magnetism, its relation to the earth's interior, 60, 61.
Mallet, R., on internal heat, 78; on volcanic rocks, 79.
Malvern, geology of, 360, 373, 383.
Manganese, relations of, to vegetation, 98 ; in waters, 142.
Manitoulin Island, water of, 158.
Marbles of Vermont, 311.
Marcou, J., on Taconic rocks, 251.
Marignac on dolomites, 309.
Marine salts in rocks, 103.
Marls, magnesian. *See* Sepiolite.
Marsh gas, origin of, 177, 182; relation of, to radiant heat, 46.
Mather on limestones, 218; on Taconic rocks, 254.
Matinal rocks, 421.
Matter, its chemical history, 426, 465.
Matthews, G. F., geology of New Brunswick, 407.
Meionite, 445, 446.
Melting-point, relation of, to pressure, 7, 39, 60, 65.
Menevian rocks, 266, 371-373 ; in North America, 385, 407.
Metagenesis in chemistry, 427, 465.
Metalliferous deposits, origin of, 23, 220.
Metals in sea-water, 231.
Metamorphic rocks, objections to the term, 18; chemistry of, 19.
Metamorphism of rocks, 9, 18, 19, 24-28, 286, 287, 291, 298-300, 305-307, 317, 320; not to be confounded with pseudomorphism, 24, 291 ; Hutton and Boué on, 24, 321; Dana on, 291, 320; Credner and Gümbel on, 305; Favre on, 342, 347; Naumann on, 25, 293, 295, 322, 323; local, 18, 24-26, 295, 298, 299, 307.
Metamorphosis of rocks, supposed, illustrations of the doctrine, 287, 324-326.
Metamorphosis in chemistry, 427, 465.
Meteoric stones, constitution of, 302.
Micas, conditions of their formation, 28; magnesian, of Laurentian series, 207.
Mica-schists, 28, 32, 207, 244-247, 272, 282, 326, 331, 353, 408; supposed pseudomorphic origin of, 326.
Michigan, crystalline rocks of, 274.
Mineralogy, its province, 453; classification in, 454.
Mississippi, mud of, 10; valley, geology of, 50, 75.
Mixtures in mineral species, 444.
Molasse of the Alps, 345.
Montalban rocks, 194, 282. *See* White Mountain series.
Montarville, dolerite of, 186.
Mont Blanc, geology of, 329; trias of, 331; crystalline rocks of, 330.

Mont Cenis Tunnel, 334, 347.
Montlosier, De, on mountains, 52, 74.
Montreal, dolerite of, 186, 298.
Moore, Charles, on liassic fossils in veins, 204.
Morlot, Von, on dolomite, 308.
Mountains, origin of, 49, 51, 52, 73, 74; synclinal structure of, 345.
Mud-volcanoes, 8.
Murchison, R. I., on geology of Scotland, 271; on Silurian rocks, 352, 355, 378-380; errors of his Silurian sections, 358, 362, 380; on geology of the Alps, 337.
Murray, Alex., on geology of Newfoundland, 406.

NATROLITE and orthoclase associated, 5, 192, 206.
Natron-lakes, 12, 85, 146, 158.
Naumann, C. F., on metamorphism, 25, 293, 295, 322, 323; on envelopment, 292; on origin of crystalline rocks, 294; on pseudomorphism, 292, 320, 322.
Nebular hypothesis, 36, 38, 222.
Neolite, 296.
Neptunists and plutonists, 45.
Nerepis, New Brunswick, granites of, 201.
Nevada, silicious veins in, 204.
Newberry, J. S., on cycles of sedimentation, 241; on geology of Ohio, 416.
New Brunswick, geology of, 275, 407-409, 415.
Newfoundland, geology of, 261, 275, 405-410.
New Hampshire, geology of, 242, 281.
Newport, Rhode Island, geology of, 249.
New York, geological survey of, 387-389; system of rocks, 387, 389.
Niagara limestone, 417, 418; of Chicago, 172.
Nickel in rocks, 31, 32, 34, 243, 247, 269.
Ncklès, J., on nitrification, 472.
Nicol, on geology of Scotland, 271.
Nicolet, Quebec, water of, 126.
Nitrates, reduction of, 94, 113, 472.
Nitre, hollow crystals of, 212.
Nitrification, theory of, 464, 470.
Nitrite of ammonia, its formation, 471.
Nitrogen gas, a nitryl, 464, 470.
Nitrogen of volcanoes, 8; amount of, in rocks, 113.
Norian rocks, 29, 31, 33, 278-282.
Norites, 31, 33, 279; olivine in, 31, 280.
Nova Scotia, geology of, 408, 409, 415.
Nucleus of the earth, 7, 39, 44, 56, 57, 59-61, 64.

OCEAN, primitive, 2, 11, 40, 41; palæozoic, 82, 104, 108, 109, 119, 137, 163; evaporation of its waters, 76, 83, 92, 104, 107, 108, 310; metals in waters of, 231, 237; bromine in, 142; iodine in, 144, 226, 237; potash in, 135. *See* Carbonate of soda and Carbonate of lime.
Ochre, formation of, 98, 228. *See* Iron ores.
Ohio, brines of, 120; geology of, 416.
Oken, mineralogical classification of, 454.
Oleiferous limestone of Chicago, 172.
Olivine, in norites, 31, 280. *See* Chrysolite.
Oneida conglomerate, 416.
Onondaga formation, 155, 417, 418; the oldest saliferous known, 119; mineral waters from, 163.
Ontario, petroleum of, 168-171.
Ophiolite. *See* Serpentine.
Orbicula, a phosphatic shell, 312.
Ore-deposits, 23, 233.
Organic and inorganic bodies, 427, 453.
Organic life, chemical relations of, 2, 13, 22, 42, 96, 144, 225, 226, 231, 302, 311, 312; evidences of, in crystalline rocks, 13, 302; in aerolites, 302.
Organic matters in waters, 94, 125, 152, 153; chemical relations of, 13, 22, 97-99.
Orthoclase, 12, 101, 192, 206; production of, 298, 299; formula of, 443. *See* Feldspars, Granites and Granitic vein-stones.
Orthophyre, 187, 243, 250, 282.
Ottawa basin, geology of, 412.
Ottawa River, water of, 84, 126; potash in, 136; silica in, 150; silicate of lime from, 152.

Owen, D. D., geology of Wisconsin, 403.
Oxychloride minerals, 442.
Oxygen, equivalent weight of, 176, 431; active, *see* Ozone.
Ozone, relation of, to radiant heat, 46; a triple molecule, 464; production of, 470; relation of, to nitrification, 471.

PALÆOTROCHIS, 411.
Palæozoic sediments, origin of, 10, 75.
Palæozoic formations of St. Lawrence basin, 154; of North America and England, thickness of, 50, 377; tabular view of lower, 386. *See* Cambrian and Silurian.
Palæozoic climate. *See* Climate.
Palæozoic ocean. *See* Ocean.
Paradoxides Harlani, 405.
Paragonite, 244.
Paraffines of petroleum, 182.
Paris, France, magnesian sediments of 296.
Paris, Maine, granitic vein of, 195; tourmalines of, 200, 212.
Peat, 94, 181.
Pebbles in veins, 204.
Pennsylvania, geology of, 245; geological survey of, 420.
Peristerite, 214.
Perthite, 214, 444.
Petalite, 210; formula of, 443.
Petrolia, Ontario, waters of, 161.
Petroleum, 168; surface wells of in Ontario, 171; of Chicago, 172-174; Andrews on, 174; of vegetable or of animal origin, 179; hydrocarbon gases accompanying, 182.
Phillips, J. A., silicious deposits of Nevada, 204.
Phillips, John, on igneous rocks, 3, 24, 66; on rocks of Anglesea, 270; geology of Malvern, 360, 370, 383.
Phosphates in waters, 94-96, 142; concentration of, 225; relations to organisms, 312.
Phosphatic shells, 312.
Phosphoric acids, genesis of, 466.
Phosphorus, its diffusion in nature, 222.
Plants. *See* Organic life.
Plasticity of rocks, 4, 9, 44, 65, 72, 189-191.
Playfair and Joule on equivalent volumes, 434, 440, 457.
Plication of rocks, 17, 55, 57, 72.
Plombières, water of, 25, 205, 297.
Plutonic origin of stratified rocks, 186, 294.
Plutonic rocks, sedimentary origin of, 8, 14, 43, 67, 317. *See* Exotic rocks.
Plutonists, 65; and neptunists, 45.
Point Levis, Quebec, geology of, 396, 397. *See* Levis.
Polybasic acids, their genesis, 464, 466.
Polymeric types, 464, 466; isomorphism. *See* Isomorphism, polymeric.
Polymerism in mineral species, 446, 457.
Porosity of rocks, 103; determination of, 164; table of, 166.
Porphyry, quartziferous. *See* Orthophyre.
Potash, 21, 22; fixity of compounds of, 12, 22, 95; how removed from ocean, 22, 137, 144, 226; rare in ancient seas, 137; occurrence of, in natural waters, 126, 135-137.
Potsdam formation, 254, 260, 262, 266, 268, 389, 403; Upper and Lower, 266; mineral waters from, 156.
Pratt on the solidity of the earth, 44.
Precipitation of sediments, influence of salts on, 10.
Predazzite, its relation to gypsum, 107, 133.
Pressure, its relations to fusion and solution, 7, 39, 65, 66, 204.
Primal rocks of Rogers, 255, 421.
Primordial rocks of Barrande, 266, 368. 378. *See* Silurian, primordial.
Progressive series in chemistry, 439.
Protogine of Mont Blanc, 330.
Protozoic rocks, 364.
Pseudomorphism defined, 24, 286-294; Dana on, 287, 291, 319, 320, 322; Delesse on, 288, 292, 314-318; Naumann on, 292, 322; Scheerer on, 291, 323; Warrington Smyth on, 313, 324; illustrations of, 324-326.
Pumpelly, R., orthophyres of Missouri, 250.

Pyrites, iron, origin of, 230.
Pyrognomic minerals, 5.
Pyrophyllite rocks, 28.
Pyroschists, 169, 176 - 179.
Pyroxene, 25, 186, 215, 216.
Pyroxenites, 31, 207.

QUARTZ, its origin, 2; conditions of crystallization, 6, 204, 205; chalcedonic, 89; crystalline sands of, 89; veins of, 192 - 194, 196, 199; rounded crystals of, 213.
Quartzite, supposed intrusive, 242.
Quebec, geology of, 256 - 259, 390, 396 - 403; probably inverted section at, 413.
Quebec group, 155, 259, 398, 401; its relation to the Trenton, 413; mineral waters from, 155.
Quincite, 296.

RADICLES in chemistry, 428, 466, 467.
Ramsay, A. C., on dolomites, 92; on stratigraphical breaks, 376; on the geology of North Wales, 374, 380, 381, 383.
Red sandrock of Vermont, 260, 390, 391, 394.
Rivers, waters of, 84 - 86, 126.
Rocks, porosity of, 103, 164; subaerial decay of, 2, 10, 41, 101 - 103; recomposed, 251, 285, 339, 341.
Rogers, H. D., on crystalline rocks of Pennsylvania, 245; on Taconic, 254; on Cambrian, 374, 381, 422.
Rogers, H. D. and W. B., on geology of the White Mountains, 242, 276; on nomenclature of palæozoic rocks, 420 - 422.
Rogers, W. B., on geology of Virginia, 275; on protozoic rocks in Massachusetts, 405.
Rounded crystals, 212, 213.
Royal Institution, 45.
Rutland, Vermont, geology of, 265.

SAFFORD on geology of Tennessee, 255.
Saginaw, Michigan, brines of, 120.
Salina formation. *See* Onondaga.
Salter, J. W., on geology of North Wales, 354, 362, 364, 371, 372.
Salt wells of Goderich, Ontario, 204.
Sands, silicious, crystalline and chalcedonic, 89.
Salt lagoons, 86.
Saratoga, waters of, 102, 149.
Saussurite, 445.
Scandinavia, geology of, 209, 257, 263, 266 - 269, 376, 385.
Scapolite, 28, 101, 210, 446.
Schaeffer, G. C., on nitrification, 472.
Scheerer, Th., on granites, 5, 65, 189; on envelopment of minerals, 291; on polymeric isomorphism, 291, 315, 318, 442.
Schiel, James, on progressive series in chemistry, 439.
Schönbein on nitrification, 471.
Scotland, Highlands, geology of, 34, 271, 272, 338.
Scrope, Poulett, on water in igneous rocks, 5, 65, 66, 190; on volcanoes, 60.
Sea-salt, its origin, 2, 12; its deposition, 76, 83, 86, 107, 120, 310.
Sea-water. *See* Ocean.
Sea-weed. *See* Fucoids.
Sedgwick, A., on geology of Anglesea, 270, 273; of North Wales, 350 - 365; on the Cambrian series, *see* Cambrian; on recomposed rocks, 341; on systems in geological classification, 377; his views misrepresented, 357, 364, 365; his classification of lower palæozoic rocks, 384; his death, 349.
Sediments, sources of, 10, 49, 75; related to mountains, 51, 73; condensation of, by heat, 56, 71; conversion of, to crystalline rocks, 4, 7 - 9, 14 - 16, 25 - 27, 43, 56, 57, 62 - 64, 284, 285, 317.
Selwyn, A. R. C., on deposition of gold, 237; on geology of Victoria, Australia, 273; on geology of Nova Scotia, 408.
Senarmont, H. de, on artificial formation of minerals, 221.
Sepiolite, 123, 296, 300; its relations to steatite, 317, 318. *See* Magnesian silicates.
Serpentine, Laurentian, 31, 34; of Green

Mountain series, 32, 34, 243; Silurian of Syracuse, N. Y., 310; in tertiary sediments, 303; an indigenous rock, 249, 250, 285, 317; of aqueous origin, 123, 297, 300, 318; regarded as an eruptive rock, 242, 247, 249, 316, 336; its supposed pseudomorphous origin, 287, 291, 316-319, 325; its supposed conversion into carbonate of lime, 325; Dana on, 319, 320; Delesse on, 316, 317; Credner on, 304; Favre on, 348.
Serpulites, a phosphatic shell, 312.
Shaler, N. S., on volcanoes, 60; on Anticosti group, 416.
Shales, bituminous. *See* Pyroschists.
Shawangunk conglomerate, 416.
Sherbrooke, Nova Scotia, granite vein of, 198.
Silica, sources of, 2, 10, 150; how removed from waters, 22; relations to organic life, 22, 312; deposits of, 89, 204, 234. *See* Quartz.
Silica and silicates in waters, 12, 21, 25, 84, 95, 105.
Silicate of lime from waters, 149, 151-153; its action on magnesian salts, 122.
Silicate of magnesia from waters. *See* Magnesian silicates.
Silicification of fossils, 89, 286.
Sidell on the precipitation of clays, 10.
Silver in sea-water, its separation from, and concentration, 231, 235.
Sillery formation, 256, 401, 411.
Silurian system, 352, 365, 379-381, 423, 425; Primordial, 369, 378, 423; Lower, 355-357, 363, 364, 378, 418, 420; Middle, 417, 418, 423; Upper, 355, 415, 418, 424; nomenclature in America, 419, 420.
Siluro-Cambrian, 423, 424.
Sismonda, geology of the Alps, 334, 348.
Skaraborg, geology of, 366.
Skeleton crystals, 201, 211.
Skiddaw, geology of, 273, 284, 412.
Skye, Isle of, norites of, 34, 281.
Smith, J. Lawrence, on silicate of lime from waters, 151.
Smyth, Warrington, on pseudomorphism, 313, 324.
Soda. *See* Carbonate of Soda and Sea-salt.
Soils, the chemistry of, 22, 95, 226-228.
Solution chemically considered, 429, 448; its relation to pressure, *see* Pressure.
Sonstadt on sea-water, 237.
Sorby, H. C., on liquids in crystals, 65, 205; on the relations of pressure to solution, 65, 204.
Spectroscopic studies of celestial bodies, 35, 222.
Spinel, supposed alteration of, 289.
Springs, mineral. *See* Chemistry of Natural Waters, contents of the various parts, pages 93, 116, 135.
Stallo, J. B., on chemical theory, 450, 455.
Staurolite-bearing rocks, 28, 272, 282, 331, 408.
St. Albans, Vermont, geology of, 264.
St. Catherine, Ontario, water of, 116.
St. Davids, Wales, geology of, 373, 375, 382.
St. John, New Brunswick, geology of, 406, 407.
St. Lawrence basin, mineral waters of, 153, 158; river, water of, 136, 150.
St. Leon, Quebec, water of, 123.
St. Ours, Quebec, water of, 136.
Stearine, its equivalent weight, 456.
Steatite, 32, 243, 247, 249, 250, 269, 272, 297, 300, 318, 330, 331, 334, 342; its supposed eruptive origin, 249; its supposed pseudomorphic origin, 319, 320, 322, 325.
Ste. Genevieve, Quebec, water of, 143.
Stratigraphical breaks, 262, 376, 377, 413, 414.
Streng on igneous rocks, 3.
Stromatopora, Dawson on, 411.
Strontia in waters, 141; sulphate of, 87, 117.
Sulphates, their constitution, 467; decomposition of, by heat, 108, 112; reduction of, 87, 99, 145, 163, 230; absence of, from some saline waters, 117, 144, 159. *See* Gypsum, also Alumina and Magnesia, sulphates of.
Sulphur, as a triple molecule, 464; native, origin of, 23, 87, 99, 111.

Sulphurets, origin of, 23, 111, 230; soluble, in natural waters, 145, 159-163; action of, on clays, 99.
Sulphuretted hydrogen, 8, 15, 87, 99, 163, 230.
Sulphuric acid in waters, 111, 112, 130.
Sulphurous acid, origin of, 8, 15, 111.
Sun, constitution of, 36, 37.
Sweden, geology of. *See* Scandinavia.
Syenite defined, 184, 185.
Syracuse, New York, brines of, 119; serpentine of, 310.

TABLE, of porosity of rocks, 166; of lower palæozoic formations, 386.
Tachydrite, 108, 118.
Taconic system, 155, 251-254, 264, 326, 388, 389, 391, 394; fauna of, 257, 391; distinguished from primary, 251, 326; synonymous with Lower and Middle Cambrian, 389.
Talc. *See* Steatite.
Talcose schists, 244-249, 251, 330-338, 341, 343, 383; their supposed pseudomorphic origin, 316, 320, 325.
Temperature of Earth's surface, *see* Climate; of Earth's interior, *see* Earth, its interior.
Tennessee, copper veins of, 217, 250; geology of, 255.
Terranovan series, 194, 275, 276. *See* Montalban and White Mountain series.
Terrestrial circulation, 22, 225, 235.
Teton Mountains, geology of, 262.
Thenardite, its formation, 108.
Thermal waters, 157.
Thompson, Sir William, on the earth's interior, 44, 77.
Tin, 238. *See* Cassiterite.
Titanium, 31, 192, 200, 210, 238, 251, 280.
Topsham, Maine, veins of, 194.
Tourmaline, 195, 200, 212.
Trachytes of Canada, 185.
Transmutation of minerals, 313, 325.
Travertines, origin of, 89.
Trebra, Von, on altered rocks, 339.
Tremadoc rocks, 353, 369-372, 374-376, 381, 412.
Trenton formation, 256, 412-414, 417; mineral waters from, 116, 123, 124, 155, 156, 158.
Trève on magnetism, 61.
Troy, New York, geology of, 407.
Trinidad, bitumen of, 176.
Tschermak on feldspars, 444.
Tuscarora, Ontario, water of, 130.
Tyndall, J., on heat-radiation and climate, 42, 46.

URSCHIEFER of Scandinavia, age of, 18, 269; gypsum in, 336.
Utica formation, 256, 421; a pyroschist, 178; mineral waters from, 124, 156, 157.

VALORSINE, Switzerland, conglomerate of, 339.
Vapors, relations of, to solids and liquids, 456.
Varennes, Quebec, waters of, 124.
Vegetable matter. *See* Organic matter.
Vegetation. *See* Organic life.
Veins, distinguished from dikes, 193, 202; fossils in, 203; pebbles in, 204; banded structure of, 193, 211; formation of, 233; recent origin of some, 234. *See* Granitic vein-stones and analysis of Essay XI., 183.
Vermont, geology of, 256-266, 390-395, 402.
Verneuil, De, on American palæozoic rocks, 419.
Vichy, water of, 85.
Victoria, Australia, geology of, 273.
Virginia, geology of, 249, 255, 275, 407.
Vital forces, 224, 235.
Voelcker, action of water on soils, 95.
Voellknerite. 289.
Volcanoes, phenomena and causes of, 8, 15, 44, 62-64, 77, 111; intervention of water in, 5, 61, 63, 65; distribution of, and relations to the newer formations, 9, 17, 57, 67, 68, 71; historical relations of, 68; Hall on, 58; Herschel, J. F. W., on, 8, 15, 44, 56, 62, 71; Keferstein on, 16, 56, 62, 71; LeConte on, 72, 77; Mallet on, 78, 79.
Volger on the filling of veins, 202; on pseudomorphism, 287, 324, 325.

Volumes, combining, 429; equivalent, 435, 438-443.
Vose, G. L., on internal heat of the earth, 78.

WATER as a chemical type, 461, 465, 468; solvent powers of, 5, 6, 35, 94, 223; they are increased by pressure, 65, 204, 223; in the formation of granitic rocks, 6, 33, 65, 189, 190; cohesion of, diminished by salts, 10.
Waters, action of, on soils and sediments, 12, 22, 27, 95, 284; chemistry of, 21-23; mineral, geological relations of, 154-158. *See* analysis of Essay XI., 93, 116, 135.
Water-lime formation, 418.
Way, action of waters on soils, 95.
Websterite, 98.
Whitby, Ontario, water of, 116, 142.
White, C. A., on decomposition of crys talline rocks, 250.
White Mountain rocks, 32, 188, 217, 242-244, 271-276, 282, 327; granite veins of, 194, 217; supposed palæozoic age of, 276.
Whitney, J. D., orthoclase of Lake Superior, 192.
Williamson on the water-type, 462, 468.
Wind-River Mountains, geology of, 262.
Wing, Aug., on geology of Vermont, 265.
Woodward on Cambrian and Silurian, 381.
Woody tissues, their change to coal, 177, 181.
Wurtz, Ad., on chemical types, 460, 468; on radicles, 466.
Wurtz, Henry, on a source of internal heat, 78; on gold in sea-water, 238.

ZEOLITES, solubility of, 5; modern origin of some, 25, 205, 297, 298; association of, with orthoclase, 5, 192, 206; are hydrous feldspars, 298.
Zinciferous minerals of New Jersey, 215.

THE END.

Cambridge: Electrotyped and Printed by Welch, Bigelow, and Company.

www.ingramcontent.com/pod-product-compliance
Lightning Source LLC
LaVergne TN
LVHW021316110826
845150LV00003B/584

* 9 7 8 1 4 2 5 5 5 7 8 3 6 *